The ecology of freshwater phytoplankton

The ecology of freshwater phytoplankton

C. S. REYNOLDS
Freshwater Biological Association

CAMBRIDGE UNIVERSITY PRESS
Cambridge
London New York New Rochelle
Melbourne Sydney

Published by the Press Syndicate of the University of Cambridge
The Pitt Building, Trumpington Street, Cambridge CB2 1RP
32 East 57th Street, New York, NY 10022, USA
296 Beaconsfield Parade, Middle Park, Melbourne 3206, Australia

First published 1984

Printed in Great Britain by the University Press, Cambridge

Library of Congress catalogue card number: 83-7211

British Library Cataloguing in Publication Data

Reynolds, C. S.
The ecology of freshwater phytoplankton –
(Cambridge studies in ecology)

1. Freshwater phytoplankton – Ecology
I. Title
589.4′0916 QK935

ISBN 0 521 23782 3 hard covers
ISBN 0 521 28222 5 paperback

UP

This book is dedicated to
MY WIFE, to whom its writing
represented an intrusion
into domestic life,
to Dr JOHN LUND and Mr CHARLES SINKER
who taught me so much

Contents

	Preface	*Page* ix
1	**What is phytoplankton?**	1
	Terminology	1
	Planktonic organisms	4
	Planktonic 'algae'	8
	General features of planktonic 'algae'	8
	The composition of planktonic 'algae'	28
2	**Mechanisms of suspension**	40
	The problem of suspension	40
	The nature of water movements	41
	Particle behaviour in turbulent columns	48
	Phytoplankton settling velocities	50
	Adaptive mechanisms for depressing v'	52
	Vital regulation of sinking rate	77
	Conclusions	81
3	**Spatial and temporal distribution of phytoplankton**	83
	Distribution patterns: the problem of scale	83
	Vertical distribution	87
	Horizontal distribution	105
	Temporal variations in abundance and composition of phytoplankton	112
	Intra-assemblage structure and competition	120
4	**Photosynthetic activity of phytoplankton**	123
	General features of planktonic photosynthesis	123
	Photosynthetic behaviour in isolated samples of natural phytoplankton	126
	Photosynthetic behaviour in non-isolated natural communities	137

5 Nutrients 157
Nutrient requirements of phytoplankton 157
Phosphorus 162
Nitrogen 164
Silicon 168
Other nutrients 173
Nutrient interactions 178
Eutrophication 183

6 Growth and survival 192
Optimal growth rates 193
Increase in natural waters 202
Perennation 217
Growth and survival strategies 221

7 Loss processes 225
What is a loss process? 225
Hydraulic washout 228
Sedimentation losses 233
Death and decomposition 245
Grazing 253
Loss processes and phytoplankton composition 274

8 Periodicity and change in phytoplankton composition 277
Seasonal periodicity of phytoplankton 279
Longer-term floristic changes 321

Glossary of symbols 329

References 333

Index to lakes and rivers 367

Index to genera/species 369

General index 374

Preface

The importance of phytoplankton is beyond question. Planktonic primary production provides the base upon which the aquatic food chains culminating in the natural fish populations exploited by man are founded, at the same time generating some 70% of the world's atmospheric oxygen supply. Excessive algal production in lakes and reservoirs presents expensive problems in the water industries, whilst deleterious effects upon fisheries and water-based recreation are fairly attributed to overabundance of phytoplankton. There is, therefore, a powerful economic and social need for as complete as practicable an understanding of the factors which regulate the spatial and temporal variations in the distribution and productivity of phytoplankton – or, in short, its ecology.

The volume of scientific literature devoted to phytoplankton biology is daunting and often bewildering. Many more titles are added each year. Fortunately, for both the beginner and the more seasoned student, there is a number of excellent general books and review papers describing fundamental features of plankton biology but there is a constant need for updating and revision as new principles and hypotheses become established.

I shall make no further attempt to justify the addition of yet another book to those already available. This volume is concerned mainly with the factors which determine the wax and wane of specific phytoplankton populations in standing freshwaters (lakes and reservoirs), though reference is made to marine plankton which is, generally, subject to similar controls. It is primarily intended for use by students, and I have therefore tended to oversimplify some of the more complex aspects of the subject, for the sake of ready comprehension, but further reading is recommended wherever suitable or relevant texts are available. I hope that I have adequately resisted the temptation to be unnecessarily encyclopaedic in referencing literature: there are many significant contributions that have not been cited.

Ecology is a complex science. There are probably almost as many definitions of 'ecology' as there are books on the subject. In framing this text, I have adopted the phrase to which I was introduced as a would-be ecology student: 'What lives where – and why'. Specifically, I have endeavoured to develop this theme through an overview of the structure of planktonic communities in which the functional adaptations of pelagic life are emphasised, building up to the dynamic aspects of production and seasonal periodicity. Biochemical aspects of phytoplankton are not specifically covered and I have included no more physiological information than I have found necessary. Again, appropriate further reading is referenced.

The text could not have been prepared without the considerable help of a great many friends and colleagues. I am grateful for advice on the presentation and discussion of subject matter (often outside my primary interests) freely contributed by Dr G. Fryer, F.R.S., Mr T. I. Furnass, Dr D. G. George, Dr G. H. Hall, Dr S. I. Heaney, Dr J. Hilton, Mr J. E. M. Horne, Mr G. H. M. Jaworski, Dr J. G. Jones, Dr J. W. G. Lund, F.R.S., Dr J. F. Talling, F.R.S. and Dr L. G. Willoughby. The excellence of the Freshwater Biological Association's comprehensive library facilities proved to be extremely beneficial to my background reading and literature searches; it is a pleasure to express my appreciation to the Council, Director and Library staff of the Association.

I should like to thank especially: Dr Hilda Canter-Lund F.R.P.S., not only for the selection of her excellent collection of algal photomicrographs but also for the enthusiastic and time-consuming trouble she took over their presentation; Sheila Wiseman, whose collaboration and support during the preparation of the book has been invaluable, and whose illustrative talents contributed most of the text figures herein; and Elisabeth Evans, upon whom fell the almost impossible task of turning my handwritten pages into a legible typescript.

Finally, special acknowledgements are also accorded to Mr Charles Sinker, O.B.E., M.A., and to Dr John W. G. Lund, C.B.E., F.R.S., both of whom deeply influenced my appreciation of ecology in general and of phytoplankton in particular, during my formative years. The most that I could wish for is that, through the pages of this text, I may in turn pass on the benefit of their teaching to others.

The Ferry House, Ambleside, Cumbria
January 1983

C. S. REYNOLDS

1

What is phytoplankton?

'He prayeth best who loveth best all things both great and small.'
COLERIDGE, *The Ancient Mariner*

1.1 Terminology

Biology, and hydrobiology in particular, is littered with terms whose precise meanings, either through initial misconception or through later abuse, have evolved, degenerated or otherwise departed from their original definitions. 'Plankton' is no exception. It is therefore necessary at the outset to clarify its contemporary meaning and to distinguish it adequately from other, apparently overlapping terms.

The first use of the term 'plankton' is widely attributed to the German biologist, Viktor Hensen (Ruttner, 1953; Hutchinson, 1967), who, in the latter half of the nineteenth century, began a series of expeditions to gauge the distribution, abundance and composition of microscopic organisms in the open ocean.

The existence of such organisms, however, had been demonstrated some years earlier by another investigator, Johannes Müller. Using a fine-mesh silk net to concentrate the organisms, Müller opened the door on a hitherto unknown community, of quite uncontemplated richness and diversity. Müller called this community the 'Auftrieb'; but it was Hensen's name for the same community, 'plankton', which gained popular recognition.

According to Hensen's (1887) usage, 'plankton' included all organic particles '*which float freely and involuntarily in open water, independent of shores and bottom*'. The dependence of plankton upon water movements for maintenance and transport is accurately implied in this definition (πλαηκτοσ – wandering). It immediately excluded other, larger inhabitants (e.g. fish, mammals) of open water (the *pelagic* zone) having the ability to substantially regulate their own distribution, by swimming (the 'nekton'). Hensen's 'plankton' did not specifically exclude non-living particles, so it is therefore synonymous with '*seston*' in Kolkwitz's (1912) later terminology, that continues to command common acceptance. Seston thus applies to all particulate matter maintained in the pelagic zone: abioseston, or *tripton*, is the non-living fraction; bioseston, or plankton, comprises only discrete, living organisms.

There are two more serious criticisms of Hensen's definition of what we now understand to be 'plankton', but it has taken nearly a century of subsequent investigations to resolve these. One is simply that, in general, plankton does *not* float: there are few planktonic organisms which are consistently buoyant; on the contrary, most are often or always more dense than the water they inhabit. As will be pursued more exhaustively in Chapter 2, the specific adaptations of planktonic organisms for pelagic life seem largely directed towards prolonged *maintenance in suspension*. Moreover, there are occasions when it is beneficial for planktonic organisms to be able to avoid the immediate sub-surface water layers and when a positive sinking rate is actually advantageous. This leads directly to the second point, which is that many planktonic organisms are not exclusively confined to the pelagic zone but may spend part, or even most, of their life cycle on the sediments or in other (littoral) habitats. Put another way, many organisms present in open waters are only facultatively planktonic (or *meroplanktonic*).

For these reasons, it is perhaps more useful to regard plankton as 'the community of plants and animals *adapted to suspension* in the sea or in fresh waters and which is *liable to passive movement by wind and current*'. This definition does not exclude temporary inhabitants of the plankton or chance introductions to the pelagic (which is conceptually important in the context of the evolution of the planktonic habit) but it nevertheless lays stress upon morphological and behavioural adaptions to survive there.

Plankton is a potentially functional community, of similar organizational rank implicit in the terms 'forest-' or 'grassland-communities'. The plants and animals are conveniently segregated in the terms *phytoplankton* and *zooplankton* respectively, notwithstanding differences in opinion about where the dividing line is drawn. Botanists and zoologists still dispute their respective claims over organisms like dinoflagellates, which are simultaneously autotrophic (i.e. capable of elaborating their foods from inorganic substances dissolved in the water), phagotrophic (i.e. capable of ingesting other organisms, or parts thereof, and assimilating food therefrom) and motile. Here, autotrophy is considered sufficient qualification to be included in this treatment of 'phytoplankton'.

Before considering the taxonomic range of organisms represented in the phytoplankton of fresh waters and their specific adaptive features, it is convenient to mention some other methods of subdividing the phytoplankton involving the use of prefixes which may be encountered in the literature.

The first of these attempts to distinguish between the plankton of lakes (*limnoplankton*), ponds (*heleoplankton*) and rivers (*potamoplankton*). Whilst it is true that, qualitatively, the assemblages of planktonic organisms represented in these broad categories often comprise distinct elements (presumably in response to quite different sets of dominant environmental characteristics), there are sufficient species which are common to two or all three types of habitat to prevent the classification from having more than a very generalized application. Besides, the distinction between the two kinds of standing water (*lentic* habitats: ponds and lakes) has no clearly defined dividing line. Moreover, both habitats are, to a greater or lesser extent, simply extensions of flowing (*lotic*) habitats characterized by relatively protracted hydraulic residence times. Seasonal and spatial differences in the composition of their plankton are, in any case, of greater significance, and broadly similar cycles are reproduced in all three habitats (see Chapters 3 and 8).

Various authors have introduced prefixes to categorize the phytoplankton according to the individual sizes of the organisms: the terms *nannoplankton* (Rodhe, Vollenweider & Nauwerck, 1958), *nanoplankton* (Yentsch & Ryther, 1959), *ultraplankton* (Wetzel, 1964) and *μ-algae* (Fogg & Belcher, 1961) have been used to separate the lower size ranges of individuals from the larger ('*netplankton*') forms. The latter have a longer history of study, for it was not until the end of the nineteenth century that it was realized that the meshes of the first phytoplankton nets (apertures *ca.* 70 μm) allowed many smaller organisms to pass through. Lohmann (1911) introduced a centrifugation technique, which showed that the number of species in seawater that passed through nets often exceeded the number of species retained and that, sometimes, they accounted for the greater fraction of the total biomass present. Modifications to Lohmann's technique, devised by Kolkwitz (1911), Ütermohl (1931) and others, involving the fixation of intact water samples and the precipitation of planktonic organisms therefrom, have enabled routine direct estimations of the concentrations of planktonic organisms to be made and they still form the basis of many quantitative studies at the present time.

Few workers, however, continue to physically separate the larger algae from quantitative water samples with the result that there is no longer any agreed differentiation between 'netplankton' and 'nanoplankton'. Pavoni's (1963) nanoplankton generally measured < 30 μm in any plane; Nauwerck (1963) placed the limit at a nominal 80 μm, Kalff (1972) at 64 μm, and Gliwicz & Hillbricht-Ilkowska (1972) at 50 μm. Manny (1972) and Gelin (1975) adopted a critical limit of only 10 μm!

Nevertheless, the larger and smaller planktonic organisms do show differential morphological adaptations to planktonic life (see below) and are dissimilar in their susceptibility to loss processes (see Chapter 7). These differences are conveniently encapsulated within the terms 'netplankton' and 'nanoplankton'. Their continued use is desirable, provided it is qualified by the size ranges actually adopted.

Another set of terms relating to size is sometimes encountered. '*Microplankton*' corresponds roughly with nanoplankton, '*mesoplankton*' with the netplankton, whilst the meaning of *macroplankton* (occasionally '*megaloplankton*', or '*megaplankton*') embraces those aquatic angiosperms and pteridophytes that float freely on water surfaces. Such plants include the duckweeds (*Wolffia*, *Lemna* spp.), frogbit (*Hydrocharis*), the nororious Water Hyacinth (*Eichhornia*) and the water-ferns (*Azolla*). For at least part of the year, they do genuinely float on water and their distribution is subject to the effects of winds and currents. They are scarcely in suspension, however, and this fact excludes them from the definition of plankton adopted above. They are obviously adapted for a very different existence from that shared by the true plankton. Indeed, these terms hardly enhance anyone's understanding of plankton, and are best avoided.

Lastly, an important distinction must be drawn between plankton and the community of plants and animals which inhabits the surface film of quiet backwaters, the *neuston*. Though generally small, neustic organisms are specifically adapted to exploit the surface tension of water for maintenance. Few plant species are common to both communities, although some neustic crustaceans (e.g. the cladoceran *Scapholeberis* and the ostracod *Notodroma*) are able to move freely between the surface and the underlying water. Neither, however, is regarded as being truly planktonic.

1.2 Planktonic organisms

The planktonic community comprises both plants and animals. Unlike the case of marine zooplankton, few animal phyla are represented in the planktonic fauna. The most conspicuous of these are the Crustacea and Rotatoria, together with some rhizopods and ciliophorans. One family of Insecta (Arthropoda, class Hexapoda), the Chaoboridae, is represented, and there are a few freshwater planktonic genera drawn from elsewhere in the animal kingdom. These are set out in Table 1, with named examples; some of the more familiar genera, reference to which is made in later chapters, are also included. Some of the larger animals named (e.g. *Chaoborus*, *Mysis*) which swim sufficiently strongly to be able to alternate between open water and the bottom deposits could arguably be described as 'nektonic'; but the argument is academic and they are included here for completeness.

The phytoplankton of freshwaters includes representatives of several groups of algae and bacteria, as well as the infective stages of certain actinomycetes and fungi. Of these, the most conspicuous are undoubtedly the algae, but neither the biomass of bacteria nor the importance of their contribution to the functioning of aquatic ecosystems should be underestimated. Bacterial activities mediate many of the chemical processes which characterize aquatic habitats (e.g. the cycles involving carbon, nitrogen, iron and sulphur, and the consumption of dissolved oxygen), currently referred to as '*geochemical* cycling' (more correctly, 'biogeochemical cycling'). The subject matter of this book is directed largely towards planktonic 'algae', to the exclusion of the microbiological and mycological aspects of phytoplankton. For recent general account of the biology of aquatic bacteria, the reader is referred to Brock (1979).

One discrete and distinctive group of micro-organisms, however, is featured prominently in this work: the *Cyanobacteria*. Until relatively recently this group was variously referred to as the *Myxophyta* (*Cyanophyta*, *Schizophyta*), or, simply, 'blue-green algae', but they have been long recognized to be remote from any other algal group. (In any case, 'Algae' has ceased to have any real taxonomic meaning it should be regarded as a loose blanket term for 'primitive' cryptogamic photoautotrophs.) Specifically, 'blue-green algae' lack both the structural organization of chromosomes within a separate nucleus typical of cells and the discrete pigment-containing organelles (*plastids* or *chromatophores*) characteristic of many plants. In these features, the 'blue-green algae' share obvious affinities with the *prokaryotic* organization of bacterial cells. A protracted academic wrangle among botanists as to whether blue-greens should be regarded as bacteria or as algae was rejoined by the proposal to include an order Cyanobacteriales within the Procaryotae, class Photobacteria (Gibbons & Murray, 1978). It was immediately followed by the proposal (Stanier *et al.*, 1978) that cyanobacterial nomenclature, hitherto subject to the Botanical Code based on Linnaeus' original system of binomials, be brought under the rules of the Bacteriological Code (Lapage *et al.*, 1975). In essence, this change means that material must be related to cultured type-strains rather than to herbarium specimens, or to iconotypes, backed by latin diagnoses, as previously. There are considerable difficulties, however, in matching wild cyanobacteria to cultured material, whose morphology may alter so radically under laboratory conditions from that of the original isolate as to be scarcely recognizable. Thus, the ecologist is left with little practical alternative to the older, botanical names. Incorrect though they may be, these specific names and, indeed, the term 'blue-green algae', continue to be widely accepted; they at least have the virtue of being understood by bacteriologists, phycologists and other

Table 1. *Freshwater planktonic organisms*

Phytoplankton		
Bacteria[a]	Chemotrophs, including budding/appendaged forms which show morphological adaptations to pelagic life	e.g. *Planktomyces*, *Metallogenium*, *Ochrobium*
	Aerobic phototrophs	*Thiopedia*, *Chromatium*
	Oxidizing chemolithotrophs and organisms utilizing C_1 compounds	*Nitrosomonas*, *Methylomonas*, *Thiobacillus*
	Anaerobic chemo-organotrophs	*Desulfovibrio*
	Uncertain status	*Pelonema*, *Peloploca*
Cyanobacteria	See Table 2	
Actinomycetes	Presumably dispersed through open waters[b]	*Actinoplanes*
Fungi	Parasitic members of the Chytridales, Saprolegniales, dispersed through open water	*Rhizophydium*, *Blastocladiella*, *Zygorhizidium*, *Podochytrium*
Viruses[c]		
'Algae'	See Table 2	
Zooplankton		
Rhizopoda	A few planktonic amoeboids exist	*Pelomyxa*, *Asterocaelum*
Ciliophora	Ciliates, including holotrichs, oligotrichs and peritrichs	*Coleps*, *Ophryoglena*, *Tintinnidium*, *Vorticella*, *Epistylis*, *Metopus*, *Frontonia*
	Suctorians	*Acineta*
Mastigophora	The holozoic 'euglenoids'	*Peranema*
Coelenterata	One or two genera of trachyline hydrozoan polyps have planktonic medusae	*Craspedacusta*, *Limnocnida*
Rotatoria	About two dozen genera, almost wholly drawn from the Order Monogononta, are truly planktonic	*Brachionus*, *Keratella*, *Kellicottia*, *Notholca*, *Trichocerca*, *Ascomorpha*

	Branchiopoda, Cladocera: includes several common and cosmopolitan genera	*Bosmina*, *Daphnia*, *Ceriodaphnia*, *Moina*, *Holopedium*, *Bythotrephes*, *Leptodora*
	Ostracoda: at least one tropical genus is planktonic	*Cypria*[(d)]
	Copepoda, Cyclopoidea: some free-living forms are entirely planktonic; certain parasites have planktonic larvae	*Mesocyclops*, *Ergasilus*
	Copepoda, Calanoidea	*Eudiaptomus*, *Limnocalanus* *Eurytemora*
	Branchiura. Ectoparasites, some of which are transmitted largely within pelagic zone.	*Argulus*
	Malacostraca, Mysidacea: mostly marine but a few species (believed relicts) enter pelagic zones of inland waters	*Mysis*
	Malacostraca, Amphipoda	*Macrohectopus*[(e)]
Arthropoda	Insecta, Diptera: one family has a larva adapted to pelagic life	*Chaoborus*
Mollusca	Lamellibranchiata: some representatives of predominantly marine families have a planktonic veliger larva	*Dreissensia*

(*a*) 'Functional' classification of the bacteria follows Buchanan & Gibbons (1974).
(*b*) See Willoughby (1976).
(*c*) Aquatic viruses are, as yet, almost unstudied, but several pathogens of phytoplankton 'algae' have been recognized (e.g. Safferman & Morris, 1963; Luftig & Haselkorn, 1967).
(*d*) *Cypria petenensis* and *C. javensis* are apparently planktonic in Laguma de Petén, Guatemala, and in several lakes in S.E. Asia respectively.
(*e*) *Macrohectopus branickii* is the endemic pelagic amphipod in L. Baykal.

biologists alike. Here, I intend to use 'algae' as meaning true, eukaryotic algae plus cyanobacteria.

1.3 Planktonic 'algae'

The 'algal' groups with planktonic representatives in freshwaters are listed separately in Table 2, together with notes on their diagnostic features and some of the generic names: all the freshwater genera mentioned in the subsequent chapters are noted at the appropriate point. There is no universally accepted phycological classification and alternative lists and group headings will be found elsewhere in the literature. None of these is necessarily correct but the variations are mainly ones of detail. The arrangement here mainly follows Christensen's (1962) scheme, although I have used alternative names for certain groupings (after Fott, 1959; Round, 1965; Bourrelly, 1966, 1968, 1970) where these have won more general acceptance. The finer points of taxonomy are not the concern of this book; the works cited should be consulted for more authoritative comments than mine!

It should be stressed that Table 2 merely conveys the diversity of planktonic representations. It should not be taken to imply that all genera occur simultaneously in all lakes or in equal quantities. The species structure of 'algal' assemblages is discussed in later chapters.

A final cautionary note might be added for the benefit of non-phycologists: in the same way that phytoplankton comprises organisms other than algae, so not all algae (or even very many) are planktonic or, for that matter, occur in freshwaters. A moment's thought is sufficient to exclude the large brown and red seaweeds, the waterside stoneworts or the green encrustations abhored by aquarists from anyone's understanding of plankton.

1.4 General features of planktonic 'algae'

Planktonic 'algae' are drawn from a diverse range of, at best, distantly related phylogenetic groups (Table 2). From an evolutionary standpoint, it may be postulated that the planktonic habit has arisen on several occasions, suggesting that adaptive radiation into pelagic environments has been backed by powerful selective pressures. That the 'algae' should have been relatively successful in exploiting the potential advantages offered by pelagic life has presumably depended upon a certain degree of pre-adaptation to a dispersed existence. Here, we may cite the generally low level of structural organization and morphological plasticity of 'algal' cells, the concomitant intracellular 'division of labour', which makes for considerable physiological and biochemical independence of individual

Table 2. *Groups of 'algae' represented in the freshwater phytoplankton and some selected genera*

KINGDOM: PROKARYOTA

CLASS: Photobacteria
CYANOBACTERIALES (blue-green bacteria/'algae')
(synonyms: CYANOPHYTA, MYXOPHYTA, SCHIZOPHYTA)
Prokaryotic 'algae' lacking typical membrane-bound nuclei and plastids

Order: *Chroococcales* Solitary or colonial coccoid 'blue-greens'

Includes: *Aphanocapsa, Aphanothece, Coelosphaerium, Gloeocapsa, Gloeothece, Gomphosphaeria, Microcystis, Synechococcus*

Order: *Nostocales* (= *Hormogonales, Oscillatoriales*) Filamentous blue greens, mostly capable of heterocyst- and akinete-formation

Includes: *Anabaena, Anabaenopsis, Aphanizomenon, Gloeotrichia, Lyngbya, Oscillatoria, Pseudanabaena, Spirulina, Trichodesmium*

KINGDOM: EUKARYOTA

Eucaryotic algae with typical nucleus and pigments localized within plastids (or chromatophores). Eight phyla (according to this treatment): two (**RHODOPHYTA** and **PHAEOPHYTA**) are without representatives in the freshwater phytoplankton

CRYPTOPHYTA
Naked, biflagellate algae, with one or two large plastids; division longitudinal; sexual reproduction unknown; assimilatory product of photosynthesis, starch. One class and one order

Order: *Cryptomonadales*

Includes: *Chilomonas, Chroomonas, Cryptomonas, Rhodomonas*

PYRRHOPHYTA (dinoflagellates)
Unicellular flagellates, rarely colonial; two flagella of different length and orientation; naked, or with cellulose cell wall, sometimes sculptured into plates. Numerous discoid plastids or colourless; assimilation products, starch or oil. Mostly marine

CLASS: Dinophyceae
Biflagellate cells, flagella located in transverse and longitudinal furrows. Planktonic representatives included within one order

Order: *Peridiniales*

Includes: *Ceratium, Glenodinium, Gonyaulax, Gymnodinium, Peridinium, Woloszynskia*

CLASS: Adinophyceae (= Desmophyceae)
Naked or cellulose-walled cells composed of two watchglass-shaped halves. One order represented in freshwater phytoplankton

Order: *Prorocentrales*

Includes: *Exuviella, Pyrocystis*

Table 2. (*cont.*)

RAPHIDOPHYTA (chloromonads)
Uniflagellate, cellulose-walled cells; numerous plastids, assimilatory product lipid. One class and one order

Order: *Raphidomonadales* (= *Chloromonadales*)
Includes: *Gonyostomum*

CHRYSOPHYTA
Unicellular, colonial, filamentous or siphonaceous algae, with a preponderance of carotenoid pigments; cell walls pectinaceous, often in two pieces, sometimes impregnated with silica (especially in the Bacillariophyceae); assimilatory products, chrysose, chrysolaminarin, leucosin, lipids but never starch. Five classes, all with planktonic representatives

CLASS: Chryosphyceae
Mainly unicellular or colonial; plastids brown, usually two; cell wall sometimes silicified or calcified; isogamous sexual reproduction; mainly freshwater. Three (out of eleven) orders represented in freshwater phytoplankton

Order: *Ochromonadales* (=*Chrysomonadales*) Unicellular and colonial Chrysophyceae, without a rigid cell wall, but often bearing siliceous scales
Includes: *Dinobryon*, *Mallomonas*, *Synura*, *Uroglena*

Order: *Chromulinales* Generally unicellular biflagellates without a rigid cell wall
Includes: *Chromulina*, *Chrysococcus*, *Kephyrion*, *Pseudopedinella*, *Stenocalyx*

Order: *Stichogloeales* Palmelloid colonial chrysophytes
Includes: *Stichogloea*

CLASS: Haptophyceae
Mainly unicellular flagellates most of which possess a rigid, flagellum-like haptonema. Mostly marine; class includes the (marine) coccolithophorids (chalk-forming algae). One (of at least two) orders represented in the plankton of fresh- (and brackish-) water forms
Includes: *Chrysochromulina*, *Prymnesium*

CLASS: Craspedophyceae (choanoflagellates)
Mostly epiphytic or epizoic chrysophytes with single flagellum, and distinctive periflagellar collar. One order: some planktonic representatives

Order: *Monosigales*
Includes: *Stylochromonas*

CLASS: Bacillariophyceae (diatoms)
Unicellular and colonial algae usually with numerous discoid plastids; pectinaceous cell wall, impregnated with silica, in two distinct halves (valves); assimilatory products, chrysose, oils. Never flagellate. Two large orders, both with planktonic representatives

Table 2. (*cont.*)

Order: *Biddulphiales* Centric diatoms, sometimes forming filaments by adhesion of valve surfaces

Includes: *Attheya, Cyclotella, Melosira, Rhizosolenia, Stephanodiscus*

Order: *Bacillariales* Pennate diatoms, sometimes forming filaments or coenobia

Includes: *Asterionella, Diatoma, Fragilaria, Nitzschia, Surirella, Synedra, Tabellaria*

CLASS: Xanthophyceae (yellow-green algae)
Unicellular, colonial, filamentous or siphonaceous algae. Motile cells usually unequally biflagellate; cell wall in two pieces but not obviously so; disc-like plastids; assimilation product, lipids (oils). Five (or six) orders mainly freshwater of which two have planktonic representatives

Order: *Mischococcales* (= *Heterococcales*) Rigid-walled unicellular or colonial xanthophytes

Includes: *Goniochloris, Monodus, Nephrodiella, Ophiocytium*

Order: *Tribonematales* (= *Heterotrichales*) Filamentous xanthophytes

Includes: *Tribonema*

EUGLENOPHYTA
Unicellular flagellates with one long and one very short flagellum; plastids numerous and irregular; reproduction by longitudinal fission; assimilatory products, paramylum, oil. One class (Euglenophyceae) and one order (a second order, Peranemales, are now regarded as protozoa)

Order: *Euglenales*

Includes: *Euglena, Lepocinclis, Phacus, Trachelomonas*

CHLOROPHYTA (green algae)
Green pigmented unicellular, colonial, filamentous, siphonaceous and thalloid algae. One or more plastids, with pyrenoids; assimilation product, starch. Authorities differ on number and extent of classes; these different classifications are avoided here by proceeding to the level of orders (though various ascriptions to classes are noted)

Order: *Pedinomonadales* A small group of unicellular biflagellates with distinctive plastids: Christensen (1962) placed these in a separate class (Loxophyceae), but Bourrelly (1966) included them in the Order Volvocales

Includes: *Pedinomonas*

Order: *Pyramimonadales* Mostly motile unicells with two or four flagella arising from a depression; non-motile forms have zoospores of similar structure. Bourrelly also includes these in the Order Volvocales, but their inclusion together with the marine *Halosphaerales* within a separate class, Prasinophyceae (cf. Round, 1965), receives general acceptance

Includes: *Pyramimonas*

Table 2. (*cont.*)

Order: *Volvocales* Unicellular or colonial biflagellates. Their inclusion within the class Chlorophyceae (Euchlorophyceae of Bourrelly, 1966) is undisputed

Includes: *Carteria, Chlamydomonas, Eudorina, Gonium, Pandorina, Phacotus, Volvox*

Order: *Tetrasporales* Non-flagellate cells embedded in mucilaginous colonies, but with motile flagellate propagules. Christensen (1962) included these in the Volvocales

Includes: *Gloeocystis, Gemellicystis* (= *Pseudosphaerocystis*), *Paulschulzia*

Order: *Chlorococcales* Non-flagellate, free-living or colonial (sometimes mucilaginous) Chlorophyceae

Includes: *Ankistrodesmus, Ankyra, Botryococcus, Chlorella, Coelastrum, Coenococcus, Crucigenia, Dictyosphaerium, Elakatothrix, Kirchneriella, Lagerheimia, Monoraphidium, Oocystis, Pediastrum, Quadrigula, Radiococcus, Scenedesmus, Selenastrum, Sphaerocystis, Tetrastrum, Tetraedron*

Order: *Ulotrichales* Mostly unbranched filamentous algae, plastids band-shaped; a few unicellular genera. (Bourrelly grouped this with 'related' orders in the class Ulotrichophyceae)

Includes: *Geminella, Raphidonema, Stichococcus*

Order: *Zygnematales* Unicellular or filamentous green algae, reproducing isogamously by conjugation. Christensen retains these in the Chlorophyceae, but separate class status has been proposed called Zygophyceae (by Bourrelly, 1966) or Conjugatophyceae (by Fott, 1959). Planktonic genera are all members of the family Desmidaceae, mostly unicellular or (rarely) filamentous algae in which the cells are constricted into two semi-cells separated by an interconnecting isthmus

Includes: *Arthrodesmus, Closterium, Cosmarium, Euastrum, Spondylosium, Staurastrum, Staurodesmus, Xanthidium*

cells, and the fact that the potential step into open water from adjacent aquatic habitats has existed for a long time.

This is not to say that planktonic 'algae' are necessarily simple organisms; indeed much of this book deals with the problems of planktonic life and the often sophisticated means by which 'algae' overcome them. Moreover, a cursory examination of the range of planktonic 'algae' (see Figures 1–5) reveals the existence of a diversity of form, function and adaptive strategies which is entirely consistent with the variety of ancestries.

What features *are* characteristic of these 'algae'? The overriding requirement of any organism must be to increase and multiply. Since algae

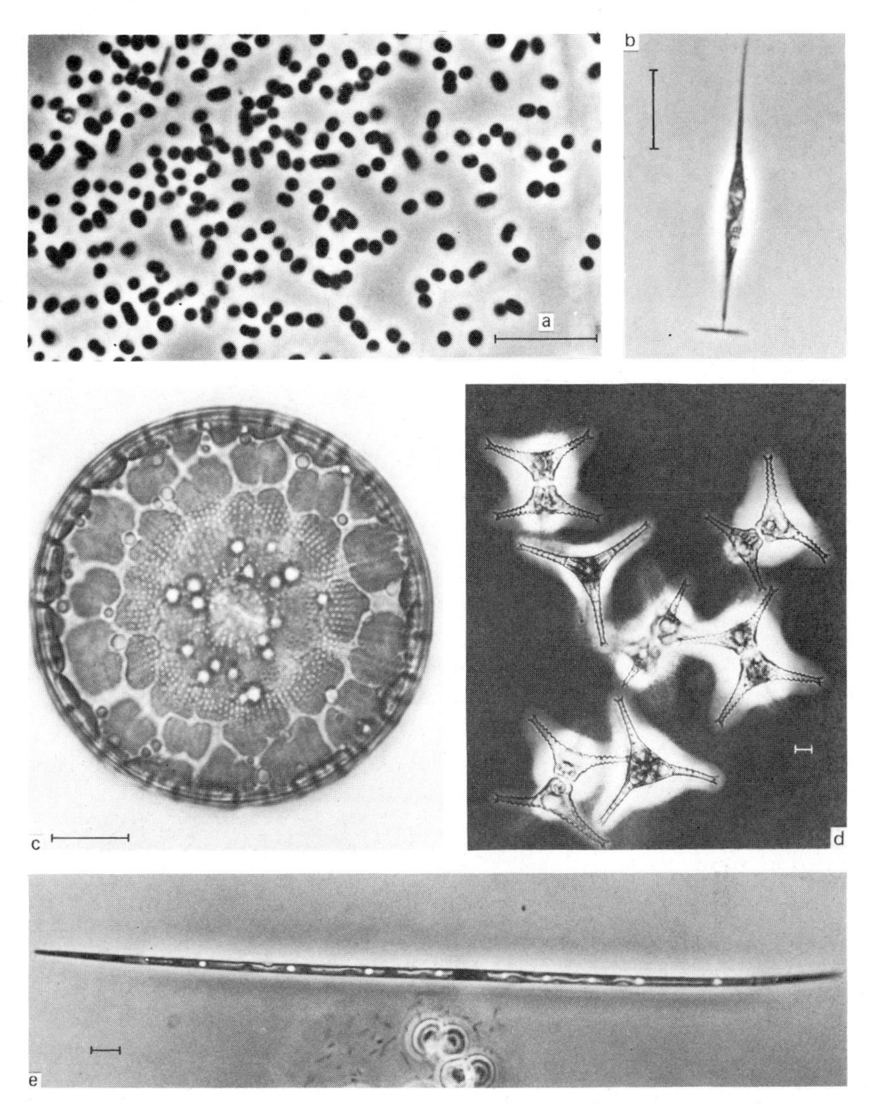

Figure 1. Non-motile unicellular planktonic 'algae': (a) *Synechococcus* sp.; (b) *Ankyra judayi*; (c) *Stephanodiscus astraea* (= *S. rotula*); (d) *Staurastrum pingue*; (e) *Closterium acutum*-type. Scale bar, 10 μm. (Original photographs by Dr H. M. Canter-Lund.)

are largely, if not exclusively, photo-autotrophic, their growth inevitably depends upon intercepting sufficient light energy to sustain photosynthetic carbon fixation in excess of immediate respiratory needs. However, radiant energy of suitable wavelengths (*photosynthetically-active radiation*, or *PhAR*) is neither universally nor uniformly available in water but it is attenuated hyperbolically with depth, through absorbance by the water

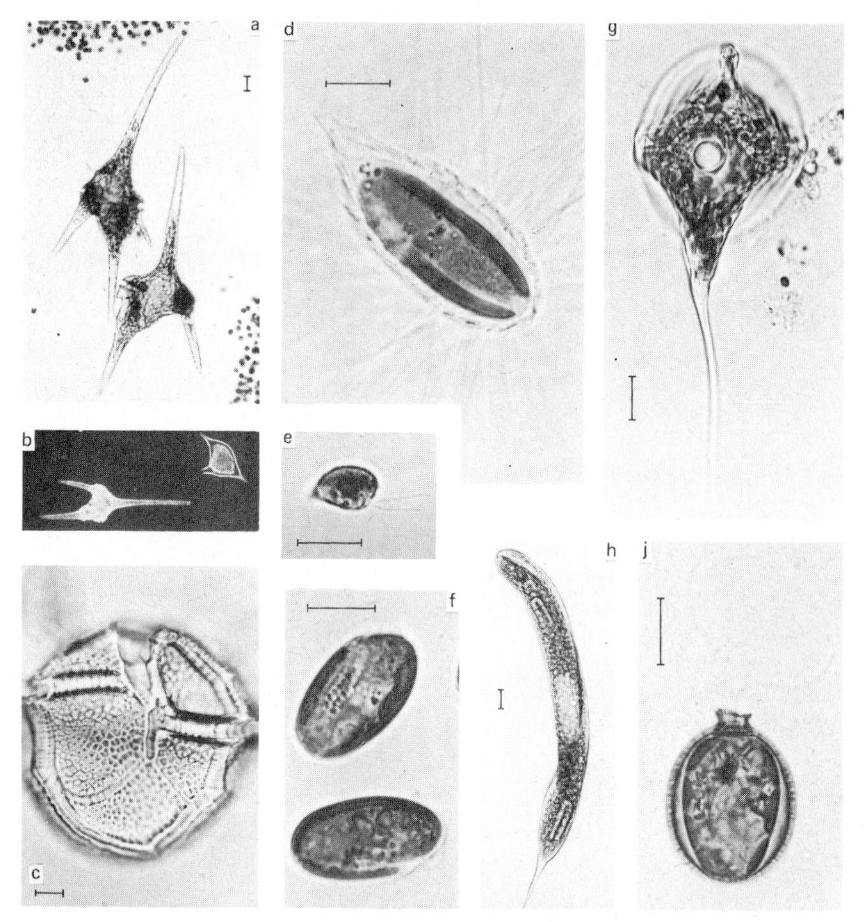

Figure 2. Planktonic unicellular flagellates: (a) Two variants of *Ceratium hirundinella*; (b) overwintering cyst of *C. hirundinella*, with vegetative cell for comparison; (c) empty case of *Peridinium* cf. *willei* to show details of exoskeleton and flagellar grooves; (d) *Mallomonas caudata*; (e) *Rhodomonas minuta* var. *nannoplanktica*; (f) *Cryptomonas* cf. *ovata*; (g) *Phacus longicauda*; (h) *Euglena* sp.; (j) *Trachelomonas hispida*. Scale bar, 10 μm. (Original photographs by Dr H. M. Canter-Lund.)

and scattering by particulate matter: there is less available *PhAR* with increasing depth. For any given autotroph in a given water mass, there is likely to be a critical depth (*the compensation point*) below which net accumulation of photosynthate is impossible. It is therefore implicit that the long-term survival of the 'alga' depends upon its ability to enter or remain in the upper, illuminated part of the water mass for at least part of its life. Many texts have emphasized this point and deduced accordingly that it is an essential characteristic of planktonic 'algae' to minimize their

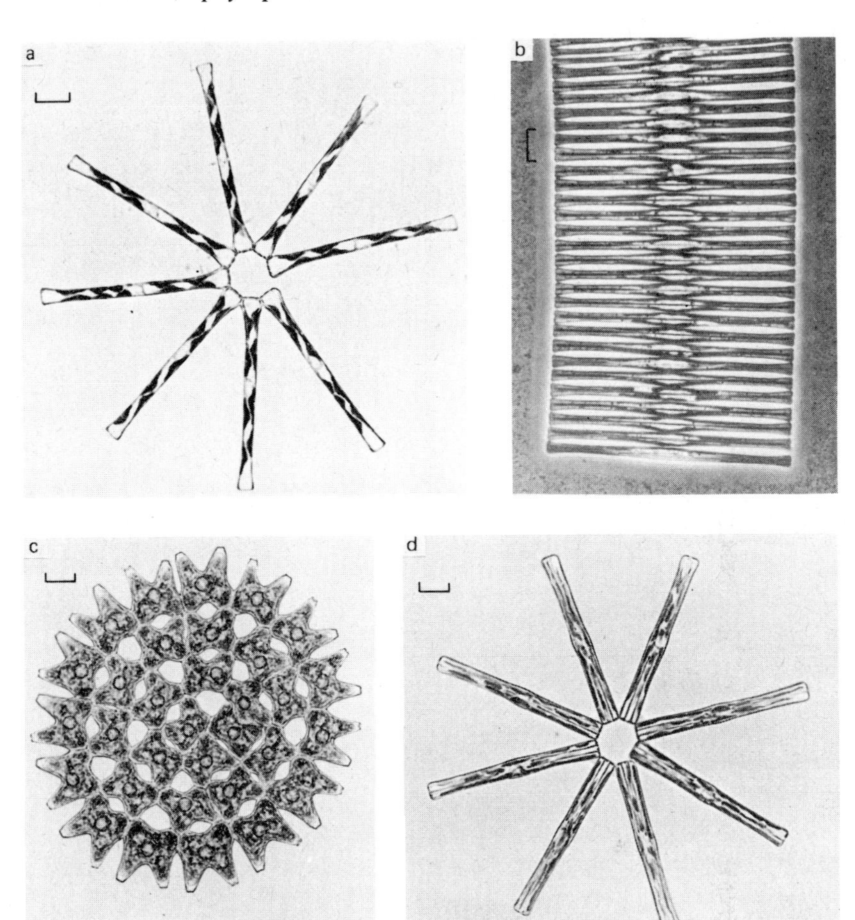

Figure 3. Coenobial phytoplankton. Colonies of the diatoms of (a) *Asterionella formosa*, (b) *Fragilaria crotonensis*, and (c) *Tabellaria flocculosa* var. *asterionelloides* and of a fenestrated-disc colony of *Pediastrum duplex*. Scale bar, 10 μm. (Original photographs by Dr H. M. Canter-Lund.)

rate of sinking. This would have to be true if water was truly static (in which case, neutral buoyancy would provide the only ideal adaptation). However, water is rarely, if ever, still. Water movements are generated by a variety of external forces, including wind, convection, gravitational flow and the rotation of the Earth (Corioli's force). Major flow patterns are accompanied by a wide spectrum of return flows and eddies, upwellings and downwellings, and progressively smaller and smaller scales of turbulent diffusion, culminating in molecular viscosity (these motions are discussed more fully in Chapter 2). To a greater or lesser degree, water movements cut across the traditional concepts of 'algal' settling, requiring a revision of the

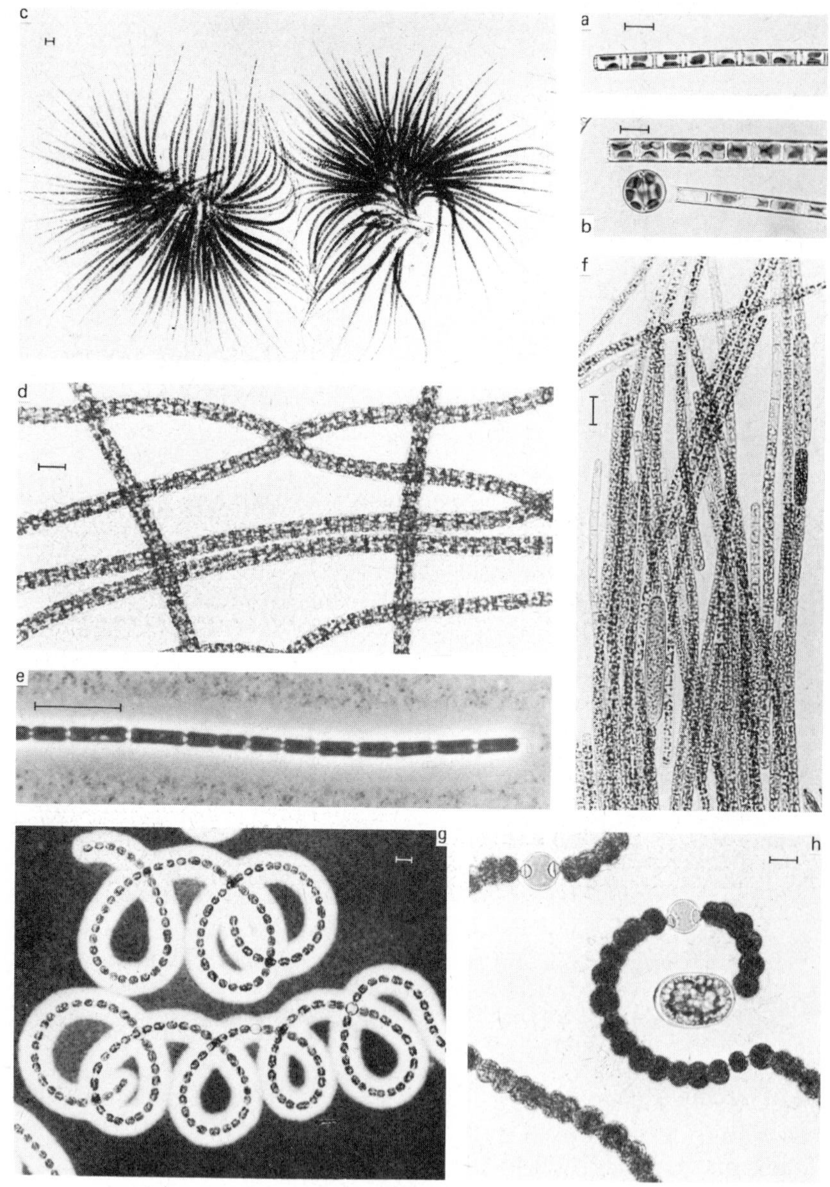

Figure 4. Filamentous phytoplankton. Filaments of the diatom *Melosira italica* subsp. *subarctica* (a, b; b also shows a spherical auxospore) and of the cyanobacteria (c) *Gloeotrichia echinulata*, (d) *Oscillatoria agardhii* var. *isothrix*, (e) *Oscillatoria redekei* (note polar gas vacuoles), (f) *Aphanizomenon flos-aquae* (with one akinete formed and another differentiating) and *Anabaena flos-aquae* (g) in India ink, to show mucilage and (h) enlarged to show two heterocysts and an akinete. Scale bar, 10 μm. (Original photographs by Dr H. M. Canter-Lund.)

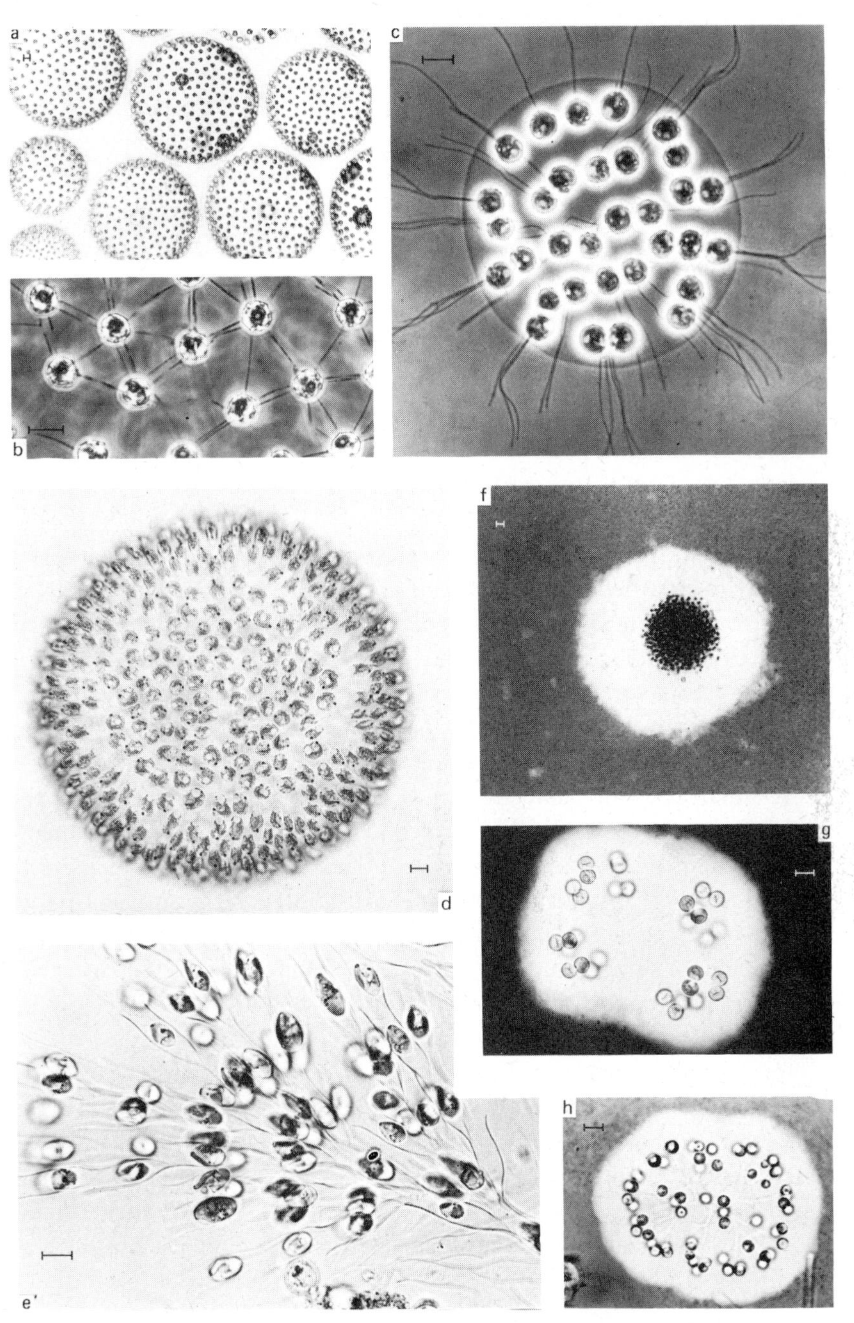

Figure 5. Colonial phytoplankton: motile colonies of (a) *Volvox aureus*, with (b) detail of cells, (c) *Eudorina elegans*, (d) *Uroglena* sp., and (e) *Dinobryon divergens*; and non-motile colonies, all in India ink to show mucilage, of (f) *Microcystis aeruginosa*, (g) *Gemellicystis neglecta* and (h) *Dictyosphaerium pulchellum*. Scale bars, 10 μm. (Original photographs by Dr H. M. Canter-Lund.)

traditional interpretation of planktonic adaptations. *The essential requirement of phytoplankton becomes one of remaining in suspension.*

The adaptive expression of this requirement is therefore towards optimizing *entrainment* within water currents. This may be achieved in a variety of ways. Those selected by most planktonic algae favour either small size, reduced excess density or devices for increasing frictional resistance with the surrounding water (or '*form resistance*'). These principles are explored further in Chapter 2 but it is important to understand how they influence phytoplankton morphology in a general sense. The only viable alternative is to attempt to overcome turbulent dispersal altogether – for instance by swimming. Even though many planktonic algae are motile, this ultimate adaptation to pelagic life is fully achieved only by certain molluscs, fish, reptiles and mammals (the nekton).

1.4.1 *Size and shape*

There is a further constraint on size. Autotrophy implies a requirement for inorganic nutrients which must be absorbed, usually in solution, from the surrounding medium. These are not often freely available, so that uptake over the cell surface against a diffusion gradient is *exothermic* (i.e. requiring the expenditure of energy). Once inside the organism, the nutrient must be translocated to the site of use. 'Algae' must rely on diffusion and transport at the molecular level to do this. These twin constraints determine that cells must remain generally small and that favourable ratio of absorptive surface to protoplasmic volume is maintained. If a linear dimension is increased then volume tends to increase as a cubic function of that dimension. Surface area increases only as its square, unless it is also convoluted so that the volume is enclosed by relatively more surface than the geometrical minimum. In this respect the 'alga's' requirements for maximizing entrainment and for enhancing the exchange of solutes across its surface coincide.

Nutrient uptake, however, is enhanced if fresh medium flows over the cell surface to displace that which may have become depleted. This is achieved if the cell has the opportunity to move relative to the adjacent water, that is, its entrainment is incomplete. It may be hypothesized that it is thus advantageous for the 'alga' to retain the ability either to float or to move relative to the water immediately surrounding it (regardless of the direction and rate of travel of the latter). Provided that such disentrainment is intermittent, the opportunity for prolonged suspension is not sacrificed.

These morphological adaptations are best illustrated by a selection of the algae themselves. The sizes, volumes and surface areas of a range of freshwater planktonic 'algal' units are presented in Table 3. The list includes examples of both unicellular algae (3A) and species in which a number of cells is grouped together to form a single larger unit (a *coenobium* of cells) taking the general shape of stellate or plate-like colonies (3B), filaments (3C) or more compact aggregations of cells embedded in a mucilaginous sheath (3D). The species selected are those which I have studied myself and have been able to make suitable measurements from which the nominal volumes and surface areas have been calculated by standard geometrical formulae. Notable previous attempts to quantify algal volumes have been made by Nauwerck (1963), Pavoni (1963), Nalewajko (1966), Besch, Ricard & Cantin (1972), Findenegg (in Vollenweider, 1974), Bellinger (1974), Willén (1976) and Trevisan (1978); data of these authors are compared in Table 3 wherever they apply to one or other of the species presented. There is not always a good agreement between observers, which is attributable more to the variation in the algae studied by the individual workers than to the methods of calculation, though these apply strictly to the nearest geometrical approximations, and not to the true values (this is especially true of the surface areas). Only Bellinger (1974) tabulated calculations of surface area (though Willén (1976) presented sufficient of her data to allow the reader to make the appropriate calculations); some of Bellinger's data are included in the collection of surface area/volume ratios shown in the extreme right-hand column of Table 3).

The species selected give, as far as possible, the full range of sizes of planktonic algae, from *Synechococcus*, a tiny unicellular cyanobacterium, to the largest single quasi-spherical colony of *Microcystis* (also a cyanobacterium) I have observed, having a diameter of 0.57 mm. The range of individual unit volumes thus covers seven orders of magnitude: if the *Synechococcus* cell was equated with the size of a hen's egg, then the equivalent *Microcystis* colony would occupy the dimensions of a small house! Nevertheless, even this 'very large' colony would scarcely cover the average pin-head: this analogy amply illustrates the characteristic small size of planktonic 'algae'.

1.4.2 *Forms with measured* SA/V

The surface areas of these same 'algae' are plotted against unit volumes in Figure 6. It is immediately clear that surface area increases with volume. In the case of the smaller unicellular 'algae' and of the larger mucilaginous colonies (the latter are identified by a different symbol), the ratio of surface area to volume remains close to the absolute geometrical minimum, represented by the lower of the two lines. This is, of course, the

Table 3. *Nominal mean maximum dimensions* (*GALD*), *volumes* (*V*) *and surface areas* (*SA*) *of some phytoplankton algae*

Notes
(*a*) Cell visualized as two frustra on elliptical bases, two cylindrical (apical) horns and two conical (lateral) horns.
(*b*) Data apply to more than one species.
(*c*) Cell visualized as two adjacent ellipsoids; area of contact ignored.
(*d*) Cell visualized as two prisms and six cuboidal arms.
(*e*) Coenobium visualized as a series of cones: areas of contact ignored.
(*f*) Coenobium visualized as eight cuboids; areas of contact ignored.
(*g*) Each cell visualized as four trapezoids; areas of contact between cells ignored.
(*h*) Coenobium visualized as 40 isolated spheres; areas of contact ignored.
(*j*) Filament visualized as a chain of spheres; areas of contact ignored.
(*k*) Colony visualized as a complete 'doughnut' ring, itself circular in cross-section (diameter: 21 μm).
(*l*) Typical habit is that of a 'raft' or flake, visualized as a bundle of filaments having an overall length of 125 μm and a diameter of 12.5 μm.
(*m*) For a 1-mm trichome length.
The volumes and surface areas given below are necessarily approximate and correspond to the nearest geometrical shapes that can be fitted to the 'algal' form; no allowance is made for detailed sculpturing of cell surfaces. Most of the data have been obtained by the author, using fresh, live material from various British lakes; data are compared with those in the literature where available. Abbreviation for shapes are: bicon, two cones fused at their bases; cyl, cylinder; ell, ellipsoid; sph, sphere.
References: (1), author, unpublished; (2), Bailey-Watts (1978); (3), Bellinger (1974); (4), Pavoni (1963); (5), Findenegg, Nauwerck in Vollenweider (1974); (6), Trevisan (1978); (7), Willén (1976) (8), Nalewajko (1966); (9), Besch *et al.* (1972).

Species	*Shape*	*GALD* (μm)	*V* (μm³)	*SA* (μm²)	*SA*/*V* (μm⁻¹)	*Reference*
A. Unicells						
Synechococcus sp.	ell	4	18	35	1.94	(1)
			5–10			(2)
Cryptomonas cf. *ovata*	ell	21	2710 (1950–3750)	1030 (820–1300)	0.381 (0.35–0.42)	(1)
			1658	942	0.568	(3)
			700–2700			(4),(5)

			3420–10000			(5),(7)
Chromulina sp.	ell	15	440	315	0.716	(1)
Chrysococcus sp.	sph	10	520	310	0.596	(1)
Kephyrion sp.	sph	5	65	78	1.20	(1)
			50			(5)
Chrysochromulina parvula	cyl	6	85	113	1.33	(1)
Cyclotella praeterissima	cyl	10	760 (540–980)	460 (380–540)	0.605 (0.551–0.704)	(1)
Cyclotella meneghiniana	cyl	15	1600	780	0.488	(1)
Stephanodiscus hantzschii	cyl	11	600	404	0.673	(1)
			800			(3)
			180–1200			(4),(5),(7)
Stephanodiscus astraea	cyl	26	5930 (2220–12010)	1980 (980–3330)	0.334 (0.277–0.441)	(1)
			18870	4410	0.234	(3)
			310–15800			(5),(7),(8)
Synedra ulna	bicon	110	7900	4100	0.519	(1)
			4900	2890	0.590	(3)
			1300–1950			(6),(8)
Monodus sp.	ell	8	105	113	1.94	(1)
Ankistrodesmus falcatus	cyl	35	30	110	3.67	(1)
var. *spirilliformis*			26	95	3.65	(3)
			30			(8)
Ankyra judayi	bicon	16	24 (3–67)	60 (12–128)	2.50	(1)
			45			
Chlorella sp.	sph	4	33	50	1.52	(1)
			30	47	1.55	(3)
			4–200 (*b*)			(4),(5),(8)
Closterium aciculare	cyl	360	4520	4550	1.01	(1)
			4000			(5)
Cosmarium depressum	(*c*)	24	7780	2770	0.356	(1)
			400–30000 (*b*)			(5),(6)
Staurastrum pingue	(*d*)	90	9450 (4920–16020)	6150 (3660–9350)	0.651 (0.584–0.744)	(1)
Staurastrum paradoxum			20000			(5)
Staurastrum sp.			20000			(6)

Table 3. (*cont.*)

Species	*Shape*	*GALD* (μm)	V(μm³)	*SA* (μm²)	*SA/V* (μm⁻¹)	*Reference*
B. coenobia						
Dinobryon sp. (10 cells)	(*e*)	145	7000 (6000–8500)	5350 (4810–6020)	0.764 (0.708–0.802)	(1)
			8000			(5),(6)
Asterionella formosa (8 cells)	(*f*)	130	5160 (4430–5890)	6690 (6130–7250)	1.30 (1.04–1.38)	(1)
			6000	6936	1.16	(3)
			2320–8000			(4)–(9)
Fragilaria crotonensis (10 cells)	(*g*)	70	6230 (4970–7490)	9190 (8010–10370)	1.48 (1.39–1.61)	(1)
			6660	10060	1.51	(3)
			6400–20000			(4),(6),(7),(9)
Tabellaria flocculosa var. *asterionelloides* (8 cells)	(*f*)	96	13800	9800	0.710	(1)
			6520–13600			(8),(9)
Dictyosphaerium pulchellum (40 cells)	(*h*)	40	900	1540	1.71	(1)
Pediastrum boryanum (32 cells)	ell	~ 100	16000	18200	1.14	(3)
Scenedesmus quadricauda	?	~ 80	1000	908	0.908	(3)
			140–1000			(5)–(8)
C. filaments						
Anabaena circinalis	(*j*)	~ 60	2040 (1310–3590)	2110 (1570–3080)	1.03 (0.858–1.20)	(1)
(20 cells)	(*k*)	~ 75	29000	6200	0.214	(1)
Aphanizomenon flos-aquae (50 cells)	cyl	125	610	990	1.62	(4)
			2316–4300			

subarctica (10 cells)					[illegible] ([illegible])	(1)
D. mucilaginous colonies						
Microcystis aeruginosa (individual cells, 30–100 μm^3)	sph	200	$4.2\ (0.034–97) \times 10^6$ $0.058–0.100 \times 10^6$	$126\ (5–1020) \times 10^3$	0.030 (0.011–0.147)	(1) (5),(6)
Uroglena cf. *lindii* (individual cells, ~ 100 μm^3)	sph	160	$2.2\ (0.12–3.43) \times 10^6$ 0.090×10^6	$81\ (12–100) \times 10^3$	0.037 (0.32–0.097)	(1) (5)
Eudorina unicocca (cells 120–1200 μm^3)	sph	130	$1.15\ (0.065–22.4) \times 10^6$	$53.1\ (7.8–385) \times 10^3$	0.046 (0.017–0.120)	(1)
Volvox globator (cells ~ 60 μm^3)	sph	450	47.7×10^6	636×10^3	0.013	(1)
Sphaerocystis schroeteri (cells 80–1200 μm^3)	sph	46	$51\ (11.5–87.1) \times 10^3$	$6.65\ (2.46–9.50) \times 10^3$	0.13 (0.109–0.214)	(1)

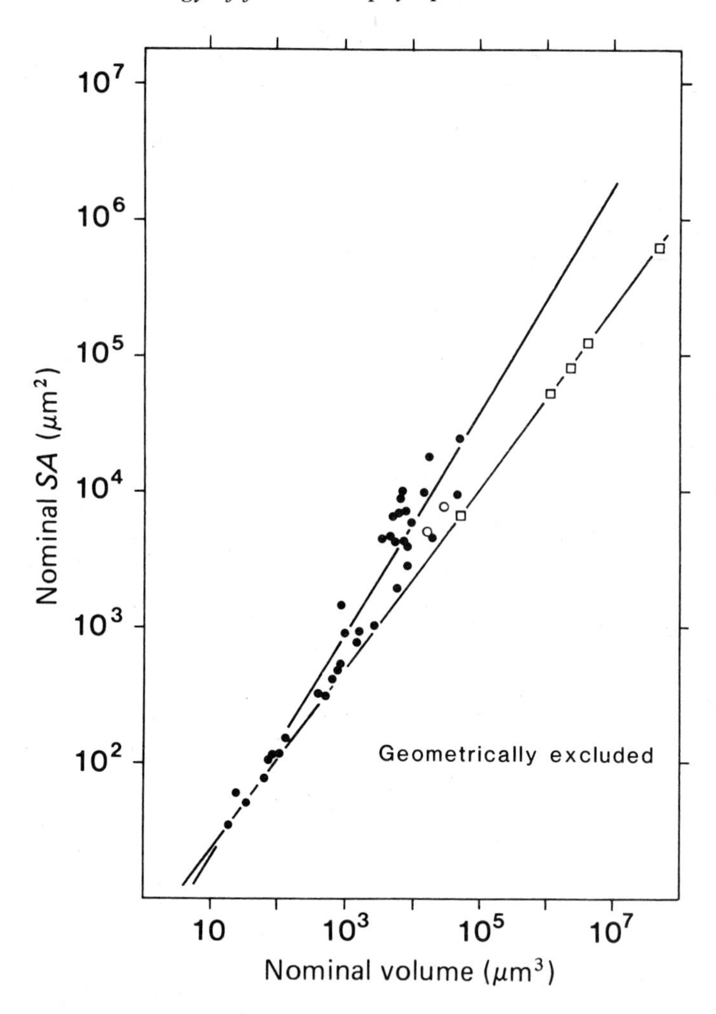

Figure 6. Log/log plots of mean surface area to volume ratio (SA/V m^{-1}) of selected 'algae' data from Table 3. The lower regression line is fitted to the points (□) representing species forming quasi-spherical mucilaginous colonies (Log $SA = 0.67 \log V + 0.70$). The upper line is the equivalent regression (Log $SA = 0.82 \log V + 0.49$) fitted to points representing all other forms (●). (Original.)

regression of the logarithm of the area of spherical surfaces (log SA) against the logarithm of volume (log V), which has a slope of $\frac{2}{3} \log V$. However, the majority of the non-mucilaginous 'algae' considered lie significantly above this line. The regression equation solved for these points (data in Table 3) has a gradient of 0.82 log V. Indeed, in several of the larger 'algae' in the series having unit volumes between 10^3 and 10^5 μm^3,

$SA > 1.0\ V$ (see Table 3 for examples). The explanation is that the volume is enclosed within a surface area greater than the theoretical minimum which is achieved in shapes which depart significantly from simple spherical and ellipsoidal forms. Thus, in spite of the manifestly wide range of unit volumes, the overall effect is that the SA/V ratio varies within only one order of magnitude.

Lewis (1976) argued that this evident conservatism of surface/volume ratio among the phytoplankton is no accident but the consequence of natural selection. Thus, however strong is the selective pressure for increased size (which may confer immunity from grazing and the opportunity for morphological division of function) the requirements for uptake and suspension are overriding. Moreover, the relatively rigid constraints imposed by maintenance of surface/volume ratio constitute the most important single factor governing the shape of planktonic algae.

Lewis developed this hypothesis through an empirical consideration of the shapes of planktonic algae in terms of their surface/volume ratios. To a greater or lesser extent, departure from the basic spherical shape is achieved by attenuation in one or (at most) two planes, so that the morphological modifications can be represented by plotting the greatest axial linear dimension (*GALD*) against surface/volume ratio. Lewis' approach is followed here (Fig. 7) but on the basis of the data included in Table 3. The plot again differentiates between the mucilaginous 'algae' and the unicellular, coenobial and filamentous forms, the latter lying within the area bounded by the SA/V values 0.22 and 3.7. Spherical forms occur along a single line bounding geometrical impossibilities (I applies to the unicells; II to the mucilaginous colonies with reduced SA/V values. The 'algal' forms that lie within the area of the plot designated III have ellipsoidal or conical shapes (e.g. *Mallomonas*, *Monodus*, *Rhodomonas*) or are squat cylinders (e.g. centric, generally unicellular, diatoms of the genera *Cyclotella* and *Stephanodiscus*: area IV). Cells having more attenuate forms (tall cylinders, long cuboids e.g. *Ankistrodesmus*, *Closterium*, *Synedra* spp.) occur in the central belt of the plot (i.e. the lower part of the area designated V) as do the coenobial forms which comprise a number of individual cells each fitting similar geometrical descriptions (e.g. *Asterionella*, *Fragilaria*, in area VI). This is also the location of the unicellular *Ceratium* and *Staurastrum* species whose surface forms are distinguished by a series of long horn-like projections. The filamentous coenobia (e.g. *Melosira*, *Oscillatoria*) represent the most extreme form of axial attenuation and occur towards the top of Area V. All these shapes represent alternative morphological variations which nevertheless preserve the supposed optimum ratio of surface area to volume. It is thus possible to conclude that

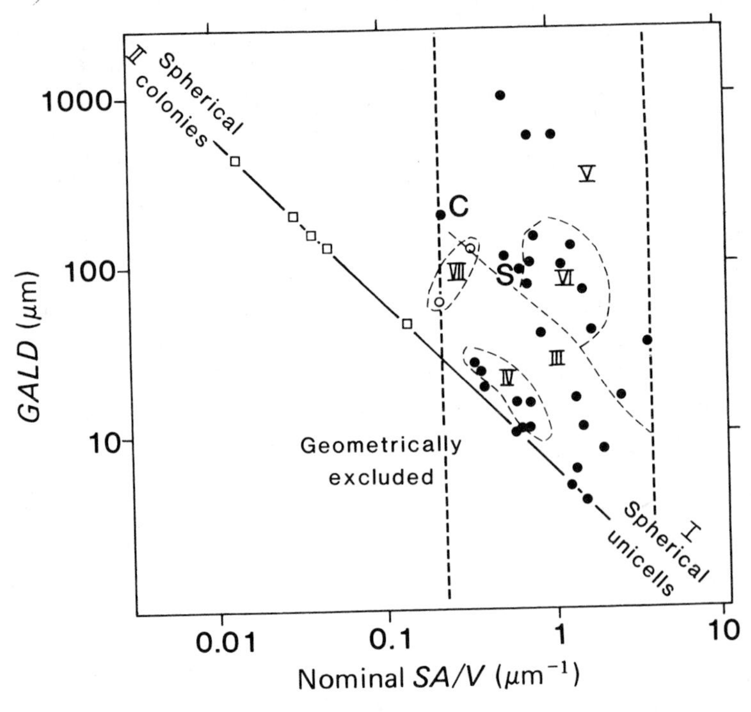

Figure 7. 'Algal' shape: log/log plot of *GALD* v. *SA*/*V* (data from Table 3). Individual points of similar morphology are grouped together (I, spherical unicells; II, spherical colonies; III, squat ellipsoids and cylinders (IV, ditto, centric diatoms); V, attenuate cells and filaments; VI, coenobia; VII, aggregates of cylinders (*Anabaena*, *Aphanizomenon*)). C (*Ceratium*) and S (*Staurastrum*) are unicells with attenuate arms. (Method of plotting after Lewis, 1976.)

the attractive and sometimes bizarre forms of phytoplankton 'algae' are functionally determined.

1.4.3 *Low* SA/V *forms*

These conclusions cannot apply to mucilaginous colonies, for which alternative explanations must be sought. The individual cells vary in number (from as few as 1, in *Gloeocystis gigas* 'colonies', to several thousands, in some *Microcystis* colonies, but in many chlorophytes the number is generally less than 100), size (typical volumes are included in Table 3) and shape (though they are most often spherical or ellipsoidal). In some cases, the cells are flagellate (e.g. in *Eudorina*, *Volvox* and *Uroglena*), their flagella passing through the surface of the mucilage to the exterior. The individual cells are rarely in direct contact (except immediately after cell division), though they may be mutually linked by protoplasmic strands (as in some *Volvox* spp.). Mucilage itself is a complex polysaccharide, which absorbs water such that, in life, as much as 99% of the volume of the mucilaginous coat *is* water. The precise function of mucilage is unknown, though a number of possibilities can be suggested. First it represents an economical way of increasing size, without incurring problems of intracellular diffusion. Increase in unit size is potentially advantageous in that it reduces susceptibility of ingestion by filter-feeding zooplankton (e.g. Burns, 1968*a*; Porter, 1973). Second, the inclusion of so much extra water within the colonial mass is likely to reduce the average density of the whole, bringing it closer to that of the surrounding water, with attendant benefits for suspension; there are limits beyond which the benefits are counteracted by the increased size of the unit (Hutchinson, 1967; see Chapter 2 for further consideration of this point). The mucilage 'bathing' the cells may also have special homeostatic properties in providing an immediate environment of a relatively steady state, though it is not clear how the efficient exchange of gases, nutrients and wastes is facilitated. Evidence of differential environmental conditions existing within the mucilage of colonial cyanobacteria has been advanced by Sirenko, Stetsenko, Arendarchuk & Kuz'menko (1968); (see also Sirenko, 1972).

A further effect of the mucilage coat is that it keeps the surface area/volume ratio close to the theoretical minimum, which reduces as the colony diameter increases (see Figure 6). This is opposite to the principle of conserving SA/V described previously, and it therefore introduces a quite separate adaptation to pelagic survival. Because the reduction in SA/V also reduces frictional drag, the colonies are equipped for relatively rapid movement through the medium. At least the larger mucilaginous colonies possess efficient means of moving through non-turbulent water (flagellar swimming in the Volvocales; buoyancy regulation in the Cyanobacteria) which enable them to influence their vertical position in still water or to regain their position in the turbulent circulation near the surface should they be transported to its lower boundary. Such movements can only be enhanced by a streamlined external form. Both groups of 'algae' are known to be capable of adjusting their vertical distribution in water columns, often within relatively short time intervals (e.g. Ganf, 1974*a*; Reynolds, 1976*a*). Provided colonies are capable of controlled migratory movements, there are good teleological grounds for supposing that the complement of a massive mucilage coat represents a discrete morphological adaptation to planktonic existence.

A comment on the two filamentous cyanobacteria included in Table 3 (*Anabaena*, *Aphanizomenon*) is also necessary. In both genera, the supposed benefits of the filamentous form, advanced above, are suppressed. In most naturally occurring *Anabaenas*, the filaments are helically coiled into more or less tight gyres, and they are also enclosed in a mucilage sheath; in *Aphanizomenon*, the filaments are loosely stacked into 'bundles' or 'flakes'. In either case, the surface area/volume ratio is reduced (see Figures 6 and 7). Both genera are also capable of regulating their buoyancy (see Chapter 3) and of making controlled vertical migrations in lakes (see Reynolds, 1978*a*; Konopka, Brock & Walsby, 1978). Moreover, it has been shown experimentally that straight *Anabaena flos-aquae* mutants apparently move less freely through water than their helically-coiled counterparts (Booker & Walsby, 1979). Thus it may be hypothesized that these 'algae' have adopted the same low-drag streamlining strategy of the mucilaginous colonies.

1.5 The composition of planktonic 'algae'

Structurally, 'algal' cells basically conform to the generalized pattern characteristic of plant cells, comprising a series of differentiated protoplasmic formations enclosed within a vital membrane, the *plasmalemma*. The membrane itself is complex, consisting of three or more separate layers; where present, the mucilage may represent the outermost of these. In the majority of 'algal' cells an additional, non-living *cell wall* is present: this may be of cellulose (as in many vegetative chlorophyte cells) or other, relatively pure, condensed carbohydrate. In some groups (especially the Chrysophyta) the wall may be more or less impregnated with inorganic substances (e.g. carbonates, 'silica'). When suitably prepared for light- or electron-microscopy these deposits can be seen to take the form of more or less continuous perforate walls (as in many diatoms) or individual scales (as in several genera of chrysomonads); both are often sufficiently distinctive to be used diagnostically in the identification of individual species.

The cytoplasm is generally a viscous gel in which the various organelles, the endoplasmic reticulum, the mitochondria, and condensed storage products are maintained. Intracellular vacuoles are to be found in most planktonic 'algae', but the large sap-filled vacuoles, characteristic of plant cells, are relatively rare, other than in diatoms. Osmoregulatory *contractile vacuoles* occur widely and are often distinctive in their number and distribution for given taxonomic groups. Many planktonic cyanobacteria potentially or actually possess proteinaceous gas-filled structures, the *gas vacuoles* (see Chapter 2).

A separate membrane-bound nucleus is characteristically present in eukaryotic algal cells. Cyanobacteria lack the membrane-bound plastids and chromatophores (see above) which are more or less prominent in planktonic eukaryotes, where they take on the intense coloration of the pigments they enclose. They are conspicuously variable, in size, shape and number, among the different taxonomic groups, ranging through solitary, cup-shaped plastids (e.g. in Volvocales), one or two broad parietal or axial plates (e.g. in Cryptophytes) or more complex shapes (e.g. in desmids, pennate diatoms), or more numerous small discoids (e.g. centric diatoms, dinoflagellates and some euglenophytes). The distribution of photosynthetic pigments among the algal groups is also variable: all the freshwater groups contain chlorophyll *a* and *β*-carotene; most also contain characteristic xanthophylls; phycobilins (especially phycocyanin) are limited to the Cyanobacteria and Rhodophyta. The relative proportions of these pigments, however, varies among the major groups. A more complete review of 'algal' cytology and pigmentation is given in Round (1965); further details will be found in the monographs dealing with individual 'algal' groups.

Stored condensates of the products of photosynthesis and normal metabolism are potentially observed in the cytoplasm of planktonic 'algae'; these too vary among groups. The chlorophytes and cryptophytes typically produce starch; the Chrysophyta produce alternative polysaccharides, such as chrysose and chrysolaminarin; the cyanobacteria store glycogen. Many 'algae' also store protein or lipids (fats and oils) in the cytoplasm. The ratios of these various substances and their absolute quantities are of course liable to considerable fluctuation as a result of cell metabolism and can be significantly altered by the ambient environmental conditions.

From the ecological point of view, the structural and compositional properties of 'algal' cells assume considerable relevance in relation to the nutritional requirements, adaptive behaviour, productivity and population dynamics. It is important therefore to establish a number of empirical criteria of 'algal' composition in order to assess the effects of environmental stresses associated with light, temperature and nutrient availability: these include methods of assessing the biomass of phytoplankton populations and the minimal nutrient levels required to support them.

1.5.1 *Dry weight*

Characteristically, the major constituent of the live 'alga' is water. If the 'alga' is air-dried to remove all the water, the residue will comprise both organic (mainly 'protoplasm' and storage products) and inorganic (e.g. the deposits impregnated into cell walls) fractions. Oxidation of the organic fraction (by further heating in air to *ca.* 500°C) yields the

Table 4. *Dry weights, ash-free dry weights, chlorophyll* a *contents and volumes determined from natural populations*

	Source	Dry weight (pg $cell^{-1}$)[a]	Ash-free dry weight (pg $cell^{-1}$)[a]	Ash (% dw)	Chl*a* (pg $cell^{-1}$)[a]	Volume (μm^3 $cell^{-1}$)[a]
Cyanobacteria						
Anabaena circinalis	(1)	45 (41–49)	—	—	0.72 (0.6–0.9)	99 (94–100)
Aphanizomenon flos-aquae	(2)	3.9 (3.2–4.2)	—	—	~ 0.04	8.2 (7–9)
Oscillatoria agardhii v. *isothrix*	(3)	28000 (15800–91200)	—	—	243 (127–813)	46600
Microcystis aeruginosa	(4)	32 (23–44)	—	—	0.36 (0.26–0.43)	73 (52–105)
Cryptophyta						
Cryptomonas ovata	(5)	2090 (1920–2250)	—	—	33 (25–40)	2710 (1950–3750)
Pyrrhophyta						
Ceratium hirundinella	(5)	18790 (12900–24000)	—	—	237 (175–260)	43740 (29080–62670)
Bacillariophyceae						
Asterionella formosa	(6)	326 (315–349)	—	—	1.80 (1.38–2.10)	—
Asterionella formosa	(7)	292	—	—	2.21	645 (554–736)
Asterionella formosa	(8)	294	—	—	2.04	630 (567–657)
Asterionella formosa	(9)	(243–291)	(104–136)	55.0	—	650
Fragilaria crotonensis	(5)	272 (196–350)	—	—	~ 2	—
Fragilaria capucina	(9)	(197–215)	(99–109)	49.7	—	350
Melosira binderana	(9)	(247–281)	(137–159)	44.0	—	1380
Melosira granulata	(5)	519 (300–764)	—	—	4–5	847
Stephanodiscus astraea	(6)	3006	—	—	121 (108–133)	8300
Stephanodiscus astraea	(7)	2770	—	—	41	5930
Stephanodiscus astraea	(9)	(115–122)	(62–66)	45.8	—	310

Stephanodiscus hantzschii	(11)	58 (47–69)	—	—	0.9 (0.8–1.8)	600
Tabellaria flocculosa v. *asterionelloides*	(9)	(383–407)	(205–210)	47.3	—	820
Mean of 11 diatoms	(9)	—	—	41.4	—	—
Chlorophyta–Volvocales						
Eudorina cf. *elegans*	(5)	273 (237–308)	—	—	5.5 (4.1–6.9)	320
Eudorina cf. *elegans*	(9)	(251–292)	(233–268)	7.9	—	320
Eudorina cf. *unicocca*	(12)	—	—	—	9.5 (8.3–10.6)	586
Volvox aureus	(5)	99.2	—	—	1.1	60
Chlorophyta–Chlorococcales						
Ankistrodesmus falcatus v. *sprilliformis*	(9)	(5.2–5.7)	(4.7–5.0)	10.4	—	30
Ankyra judayi	(12)	—	—	—	0.45	24 (3–67)
Chlorella pyrenoidosa	(9)	(5.1–6.4)	(4.5–5.7)	11.4	—	20
Scenedesmus quadricauda	(9)	(99–104)	(91–95)	8.5	—	200
Chlorophyta–Zygnematales						
Closterium aciculare	(14)	—	—	—	89	4520
Staurastrum pingue	(9)	—	—	—	57	9450 (4920–16020)
Staurastrum sp.	(9)	(4680–4940)	(4480–4620)	5.3	—	20500
Mean of 16 Chlorophytes	(9)	—	—	10.2	—	—

(*a*) Except for *Oscillatoria*, where measures quoted apply to one 1-mm filament.

Sources: (1) Crose Mere 1971; data included in Reynolds (1972).
(2) Crose Mere 1971; author, unpublished.
(3) Blelham Enclosure A, 1979; author, unpublished.
(4) From Reynolds, Jaworski, *et al.* (1981).
(5) Crose Mere, 1973; some data in Reynolds (1976*a*).
(6) Crose Mere, 1971; author, unpublished.
(7) Crose Mere, 1975; author, unpublished.
(8) Blelham Enclosures, A & B, 1978: some data in Reynolds & Wiseman (1982).
(9) From Nalewajko (1966).
(10) Crose Mere, 1968; author, unpublished.
(11) Crose Mere, 1972; author, unpublished.
(12) Blelham Enclosure A; 1978; author unpublished.
(13) Crose Mere, 1974; author unpublished.
(14) Blelham Enclosure B, 1977; some data in Reynolds & Butterwick (1979).

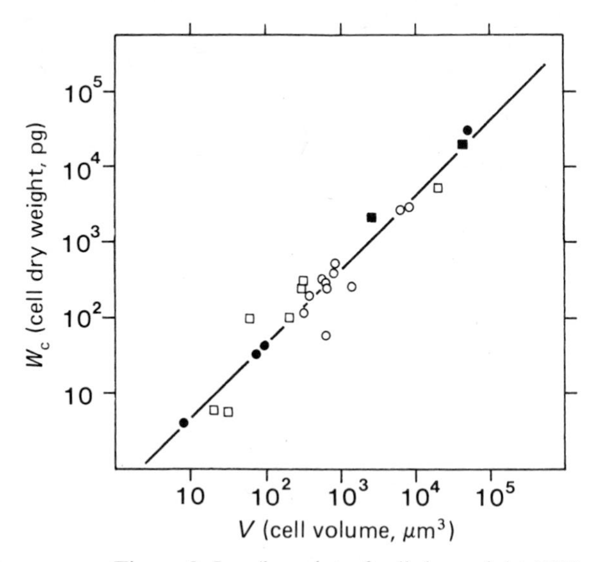

Figure 8. Log/log plot of cell dry weight (W_c) against cell volume (V) of various phytoplankton (data in Table 4: ●, cyanobacteria; ○, diatoms; □, chlorophytes; ■, others). The equation of the regression is $W_c = 0.47\ V^{0.99}$. (Original.)

inorganic 'ash' and, by back-calculation, the relative masses of the 'ash' and ash-free (i.e. organic) fractions. The dry weights and ash contents of a selection of planktonic 'algae', derived either from the literature or the author's unpublished records, are presented in Table 4. Generally, dry weight increases with the volume of the individual cell: cell dry weights (W_c) are shown against the corresponding cell volumes (V) on a log/log plot in Figure 8. The regression equation ($W_c = 0.47\ V^{0.99}$, coefficient of correlation 0.97) predicts that the dry weight of live 'algae' is between 0.41 and 0.47 pg μm^{-3}. However, the actual ratio varies between 0.10 (*Stephanodiscus hantzschii*) and 1.65 (*Volvox aureus*) but there is no consistent pattern among the groups: in the chlorophytes it ranges between 0.2 and 1.65, among the diatoms between 0.10 and 0.59 and between 0.44 and 0.62 among the cyanobacteria.

The variation in the percentage ash content of algal cells belonging to these groups varies significantly. Nalewajko (1966) found that ash accounted for between 5.3 and 19.9% (mean: 10.2%) of the dry weights of 16 planktonic chlorophytes but for between 27 and 55% (mean: 41.4%) of the dry weight of 11 diatoms, characteristically silicified.

Table 5. *The silica content of planktonic diatoms*

Species	SiO_2 (pg (cell)$^{-1}$)	% dry weight of cell
Asterionella formosa	139 (97–171)[(a)]	—
	137 (100–175)[(b,c)]	45 (32–61)
	130 (111–149)[(d)]	—
Fragilaria crotonensis	189 (170–215)[(a)]	—
	189 (188–190)[(c)]	46 (44–48)
	106 (88–119)[(e)]	—
	251 (209–304)[(d)]	—
Melosira granulata	130[(a)]	—
	620[(f)]	54
	294[(g)]	57 (38–98)
Melosira italica subsp. *subarctica*	237 (165–274)[(c)]	63 (59–66)
Stephanodiscus astraea	8500[(a)]	—
	2290[(h)]	—
	4140 (3630–4650)[(g)]	—
	2085[(g)]	69
	1600[(g)]	58
Stephanodiscus hantzschii	41 (32–47)[(c),(j)]	26 (20–33)
	35[(g)]	60 (51–74)
Tabellaria flocculosa v. *asterionelloides*	395 (370–420)[(a)]	—
	310 (250–420)[(c)]	53 (48–56)

(*a*), Einsele & Grim (1938); (*b*), Lund (1950); (*c*), Lund (1965); (*d*), Reynolds & Wiseman (1982); (*e*), Reynolds (1973*a*). (*f*), Prowse & Talling (1958); (*g*), Reynolds, unpublished, calculated from observed uptake; (*h*), Gibson *et al.* (1971); (*j*), Swale (1963).

1.5.2 *Silica*

The polymerized 'silica' which is characteristically present in the cell walls of diatoms resembles opal. It has a density of *ca.* 2.6 (2600 kg m^{-3}), so its presence adds significantly to the dry mass of the wall and, hence, of the cell. The silica contents of a number of planktonic diatoms have been derived, either by direct analysis (e.g. Mackereth & Heron, quoted by Lund, 1965; Bailey-Watts, 1974; Reynolds & Wiseman, 1982) or indirectly from the change in the dissolved reactive silica content of water in which a known number of diatoms have been produced (e.g. Lund, 1950; Reynolds, 1973*a*). A selection of these data is shown in Table 5. It is striking that the silica content consistently accounts for about half the dry-weight of the species cited (range of the mean values: 26–69%)

and, therefore, much of the diatom ash content. It is also apparent that the absolute silica content is relatively more constant in some species (e.g. *Asterionella formosa*, range 111–175 pg cell^{-1}) whereas it is conspicuously variable in others, particularly among the centric diatom species. The latter are also more variable in their absolute size, so that it is useful to assess their silica content in terms of either cell volume or surface area (see Table 6). On the limited evidence available, it can be seen that in spite of the large range in shape and size covered, the amount of silica per unit cell volume remains relatively constant. Moreover, the relationship apparently holds among the other diatom species abstracted from Table 4 and 5. For instance, if the dry weight of a *Melosira granulata* cell is taken as 519 pg (Table 4),of which silica represents 57% (Table 5: *ca.* 296 pg cell^{-1}), and its volume as 847 μm^3, a silica/cell ratio of about 0.35 pg μm^{-3} can be deduced; similarly, for *Tabellaria flocculosa* var. *asterionelloides* (dry weight: 395 pg cell^{-1}, of which 53% is silica; volume 820 μm^3) the ratio is approximately 0.26 pg μm^{-3}.

The relative constancy of this ratio is less easy to interpret. There could be some overriding biochemical or physiological control common to many species of diatoms, which determines how much silica is incorporated by the cell. However, more empirical data would be required to substantiate such a hypothesis. In any case, the silica content is more likely to be a function of the properties of the wall itself and, as has been shown, is to some extent a correlative of cell volume. A different set of results (Table 6, right-hand column) is obtained if silica content is expressed as a function of cell surface area. In the smaller cells (*S. hantzschii*, *Asterionella formosa*), the silica content is equivalent to about 0.1–0.2 pg μm^{-2}. In the larger cells of *S. astraea* (which are, perhaps of necessity, bounded by thicker walls) the ratio stands at between 0.8 and 1.0 pg μm^{-2}; according to Gibson, Wood, Dickson & Jewson (1971), the ratio may reach 2.15 pg μm^{-2}.

However, an error is apparent in the data, in that the quoted volumes and surface areas are incompatible (i.e. they fall in the 'geometrically excluded' portion of Figure 6). If the mean external volume of the cell is correct (equivalent to 8.6×10^3 μm^3 cell^{-1}), then the *minimum* surface area of such a cell (i.e. that bounding a sphere of identical volume) is 2.03×10^3 μm^2; the same silica content would then fall to a maximum equivalent of 1.13 pg μm^{-2}, that is, much closer to the values determined for other populations of the same species.

Table 6. *The silica content (as SiO_2) of some planktonic diatoms relative to cell volume and surface area*

Species	Source	*V* (μm^3 $cell^{-1}$)	*SA* (μm^2 $cell^{-1}$)	SiO_2 (pg $cell^{-1}$)	SiO_2 (pg μm^{-3})	SiO_2 (pg μm^{-2})
Asterionella formosa	(1)	630	860	130	0.21	0.15
Fragilaria crotonensis	(1)	780	1058	251	0.32	0.24
Stephanodiscus astraea	(2)	8600	1070 } (*a*)	2290	0.27	2.15
			2030 } (*a*)			1.13
	(3)	15980	4390	4140	0.26	0.94
	(3)	8300	2580	2085	0.25	0.81
	(3)	5930	1980	1600	0.27	0.81
Stephanodiscus hantzschii	(3)	600	404	35	0.06	0.09

References: (1) Reynolds & Wiseman (1982); (2) Gibson *et al.* (1971); (3) Author (unpublished: from Tables 4 and 5). (*a*) For further comment, see text.

1.5.3 *Carbon, nitrogen and phosphorus content*

Of all the elements which are universally present in living cells, carbon, nitrogen and phosphorus are regarded as being of critical importance, because exogenous sources of these elements are among the most likely to fall into short supply and hence limit 'algal' metabolic rate and, ultimately, growth (§5.1). This is especially true of phosphorus, which is available only from geochemical sources within aquatic systems; the sources of carbon (as carbon dioxide) and nitrogen are potentially augmented by solution from the atmosphere (for further elaboration on nutrient sources, please see Chapters 4 and 5). Compounds of one, two or all of these three elements are important constituents of cell protoplasm and also play an essential role in the enzymatic and energy transport systems within the cell. Their elemental proportions in living cells is thus likely to fluctuate conspicuously in relation to external conditions and internal metabolism. These elements may also be accumulated within cells in the storage products deposited during healthy metabolism; phosphates, especially, are absorbed and polymerized (polyphosphates) within cells at times when they are more freely available externally ('*luxury uptake*'). It is therefore important to distinguish the 'normal' cell contents of carbon, nitrogen and phosphorus both from the 'minimal' contents (below which cell metabolism is seriously impaired), and the concentration that can develop as a result of their accumulation within cells.

The 'normal' carbon content of 'algal' cells grown under controlled laboratory conditions of constant optimal illumination and temperature, and with an adequate supply of all nutrients provided in the culture medium, is 51–56% of the ash-free dry weight (Ketchum & Redfield, 1949; Redfield, 1958). The carbon content may fall to a minimum of about 35% of ash-free dry weight of cells deprived of light or inorganic carbon sources, or increase to around 70% if other elements limit growth (data from various sources, reviewed in Round, 1965; Fogg, 1975).

Nitrogen accounts for some 4–9% of ash-free dry weight, depending upon growth conditions (Ketchum & Redfield, 1949; Gerloff & Skoog, 1957; Round, 1965; Lund, 1970). The phosphorus content of *Asterionella formosa*, found by Mackereth (1953) to range between 0.06 and 1.42 μg P (10^6 cells)$^{-1}$ (equivalent to 0.03–0.8% ash-free dry weight) seems applicable to the majority of 'algae', though the minimum requirements of some chlorophytes may be slightly higher (see Round, 1965).

Analyses of algal tissue have given a mean 'normal' atomic ratio – the so-called Redfield Ratio – of C:N:P of approximately 106:16:1 (i.e. about

42:7:1 by weight), and it seems likely that this is also the ratio in which the same elements are ultimately required by the cells (cf. Gibson, 1971); the picture is complicated, however, by metabolic patterns and availability and, to some extent, by the rates at which these elements are recycled. At this point, the aim is simply to establish their incorporation in 'algal' cells. As these quantities are approximately related to the dry weight of the cells it follows that they are also likely to be correlatives of cell volume. The carbon content of freshwater phytoplankton is insufficiently studied to substantiate this view, though more data are available for marine phytoplankton. For example, Mullin, Sloan & Eppley (1966) established that the mean carbon/volume ratio for 14 algae was 0.012–0.26 pg μm^{-3}. Assuming the carbon content of the freshwater 'algae' to be about 50% of their dry-weights (uncorrected for ash) given in Table 4, the mean carbon:volume ratio (0.21–0.24 pg μm^{-3}) is within Mullin's range; similarly, 'normal' contents of nitrogen and phosphorus equivalent to 0.04 pg N and 0.005 pg P μm^{-3} may be predicted.

1.5.4 *Other constituents*

Definitive requirements for a further 16 or so elements have been identified in normal, healthy cells (see Chapter 5). They include hydrogen, calcium, magnesium (a constituent of chlorophyll), sodium, potassium, sulphur and chlorine and a number of so-called 'trace elements' (e.g. iron, manganese, molybdenum).

1.5.5 *Chlorophyll content*

Besides being a major photosynthetic pigment, universally distributed among the photoautotrophic 'algae', chlorophyll *a* has been widely used as a convenient correlative of biomass in estimations of phytoplankton biomass and productivity. No overview of the cellular composition of phytoplankton would be complete without some examination of the relationship between chlorophyll content and dry-weight or volume. Absolute values of chlorophyll for several freshwater 'algae' are presented in Tables 4 and 7, expressed as a function of dry-weight (or ash-free dry weight), or of cell volume. Generally, chlorophyll *a* is reckoned to account for between 0.5 and 2% of dry weight: Table 7 supports this view, though the data often apply to natural populations at, or near, their biomass maxima and may thus marginally exaggerate the typical content. Correcting approximately for ash content, the values cited lie within the range 0.9–3.9% of ash-free dry weight, with one exceptional value (for a population of *Stephanodiscus astraea*) of *ca.* 8%.

Calculations of the amounts of chlorophyll per unit specific cell volume

Table 7. *Mean chlorophyll* a *content of 'algae' expressed in terms of specific 'algal' weights and volumes*

	% Dry weight	% Ash-free dry weight[(a)]	μg chl*a* mm^{-3}
Cyanobacteria			
Anabaena circinalis	1.60	—	7.3
Aphanizomenon flos-aquae	1.02	—	4.9
Oscillatoria agardhii v. *isothrix*	0.84	—	5.2
Microcystis aeruginosa	1.12	—	4.9
Cryptophyta			
Cryptomonas ovata	1.58	—	12.2
Pyrrhophyta			
Ceratium hirundinella	1.26	—	5.4
Bacillariophyceae			
Asterionella formosa	0.68	1.52	3.1
Fragilaria crotonensis	0.74	—	—
Melosira granulata	0.87	—	5.3
Stephanodiscus astraea	4.02	—	14.6
	1.48	—	6.9
	—	—	9.3
Stephanodiscus hantzschii	1.55	—	1.5
Chlorophyta			
Eudorina cf. elegans	2.01	2.19	17.2
Volvox aureus	1.11	—	18.3
Ankyra judayi	—	—	18.8
Closterium aciculare	—	—	19.7
Staurastrum pingue	—	—	6.0

(*a*) These values are not independently analyzed, but are calculated assuming the ash content to be the same as found by Nalewajko (1966), given in Table 4.

are also given in Table 7. This particular abstraction shows some measure of difference between three major algal groups: in the chlorophytes, where chlorophyll *a* is the major photosynthetic pigment, the complement is relatively high (17–20 μg mm^{-3}, or 0.017–0.020 pg μm^{-3}) falling only among the species cited in *Staurastrum pingue* (where a large part of the cell – the arms – is unoccupied by pigment). In the diatoms, which generally have a large cell vacuole (also unoccupied by plastids) and a relatively higher content of other pigments (in this case, carotenoids and xanthophylls), the chlorophyll complement is generally lower, between 1.5 and 9.3 μg mm^{-3} (cf. the range 2.4–12.4 found by Bailey-Watts, 1978); the exceptional value, 14.6 μg mm^{-3}, for one of the *S. astraea* populations has already been mentioned. The cited chlorophyll contents of cyanobacteria (4.9–7.3 μg mm^{-3}) seem comparable with those of the diatoms, but higher

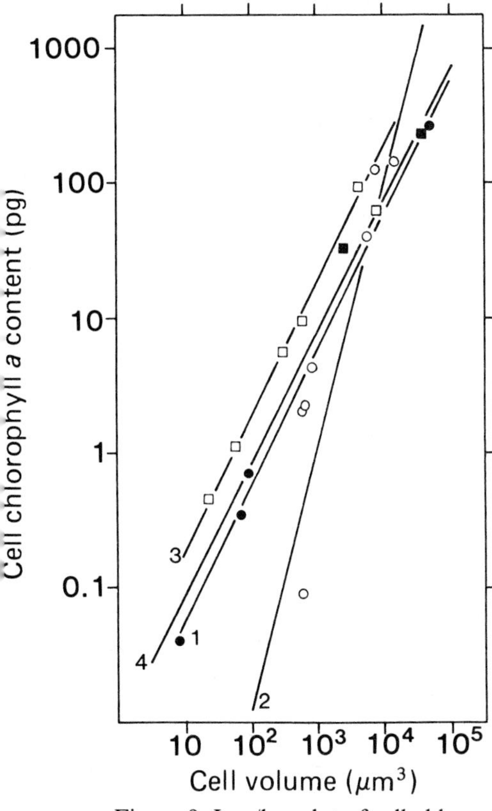

Figure 9. Log/log plot of cell chlorophyll content against cell volume of various phytoplankton (data in Table 4: ●, cyanobacteria; ○, diatoms; □, chlorophytes; ■, others). Regression equations are fitted to points for cyanobacteria (1: log chl = 1.00 log V −2.261), diatoms (2: log chl = 1.45 log V −3.77), chlorophytes (3: log chl = 0.88 log V −1.51) and *all* points (4: log chl = 0.984 log V −2.072). (Original.)

than that reported by some authors for natural populations of *Oscillatoria agardhii* and *O. redekei* (1–2 μg mm^{-3}: Gibson *et al.*, 1971; Bailey-Watts, 1978).

The regression of the available chlorophyll values against corresponding volume data (plotted on log:log scale in Figure 9) predicts a mean chlorophyll *a*:cell volume ratio of between 7.3 and 7.8 μg chl*a* mm^{-3}. The potential use of this ratio as a 'conversion' factor should be tempered by its variability with different dominant algae (see also Tolstoy, 1977).

2

Mechanisms of suspension

'I go to prepare a place for you'
John XIV, v.2

2.1 The problem of suspension

It has already been emphasized that the first requirement of phytoplankton is some means of prolonging suspension in the upper, illuminated layers of the water (the *euphotic zone*). However, it is equally important to stress that continuous residence within the euphotic zone is neither necessary nor is it always desirable (Smayda, 1970). For net autotrophic growth to occur, it is sufficient that the mean daily residence permits photosynthetic carbon fixation to exceed the immediate respiratory and secretional losses; the longer is the mean the residence time, so the potential for net increase in biomass is proportionately greater. In the longer term, cells which are lost from the euphotic zone may be able to give rise to 'resting stages' which, potentially, may allow the species to survive in the water-body and perhaps regain suspension during a later stage in the life cycle. It must also be recognized that the immediate disadvantage of sinking out of the euphotic zone is potentially offset by accelerated rates of inorganic nutrient uptake. The disadvantage will be less in turbulent water columns or in nutrient-depleted water: there is, therefore, no sinking rate which is ideal for all occasions (Walsby & Reynolds, 1980).

Nevertheless, the fact remains that without periodic access to the euphotic zone, planktonic 'algae' would become rapidly extinct. Since it has already been suggested (Chapter 1) that phytoplankton morphology represents an adaptation to pelagic survival, the relevance of structure to suspension requires examination; this topic is now pursued in greater detail. It also anticipates later chapters, in providing a basis on which sinking losses may be viewed in relation to population dynamics. The various aspects of the problem are considered to different extents: those which have been extensively reviewed by other authors (e.g. Hutchinson, 1967; Smayda, 1970; Walsby & Reynolds, 1980) are merely summarized here; relatively greater attention is given to recent findings contributing

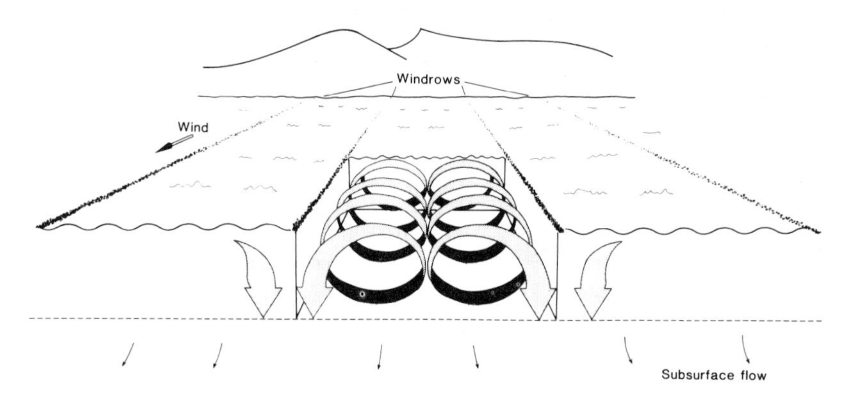

Figure 10. Diagrammatic section across the wind-induced surface flow to show Langmuir rotations. (Original.)

to the general understanding of suspension, or for which the basic theories have been scarcely developed.

Suspension of solid particles in a fluid is consequential upon the properties of the motion both of the fluid and of the particles themselves; let us first consider the relevant properties of the suspending medium.

2.2 The nature of water movements

Natural waters are almost always in motion. The energy generating movement emanates from several sources including Corioli's Force (induced by the earth's rotation), wind stress on the surface (producing waves, currents and seiche movements), and convection (caused by the alternate heating and cooling of near-surface layers). In lakes, these movements are constrained by the shores and, at least in shallow areas, by the bottom, often leading to secondary patterns of movement. In rivers the constraints are relatively greater, but the scale of motion is dominated by gravitational flow. In lakes, water masses of different mean densities, each having their own characteristics of motion, may become segregated vertically (stratification). Moreover, important characteristics of their mutual interfaces influence particle fluxes across them. The consequences of these various motions on suspension are conveniently analysed separately.

2.2.1 *Convection cells*

These reach their fullest development in the elongate, wind-induced rotations at the surfaces of lakes (represented in Figure 10), first described in detail by Langmuir (1938; see also Smayda, 1970; Smith, 1975). Such 'Langmuir cells' form under direct wind stress, at wind speeds (W)

exceeding a critical value of ~ 3 m s^{-1} (~ 11 km h^{-1}; Scott, Myer, Stewart & Walther, 1969), and there is a direct correlation between their width and depth (ratio 2.0–2.8). The velocity of downwelling (w) along lines of convergence between adjacent cells (identified by the presence of 'windrows' where bubbles and buoyant particles accumulate) also show a constant relationship with wind speed, viz. w is about 0.8 to 0.85×10^{-2} W (Sutcliffe, Baylor & Menzel, 1963; Scott *et al.*, 1969; Faller, 1971). The water removed by downwelling must be compensated by replacement of an equivalent volume, much of which is immediately derived from within the circulation cell. Thus, although the velocity of motion is neither uniform nor constant within the cell, the motion is dominated by a water current whose mean velocity is $> 2.5 \times 10^{-2}$ m s^{-1}.

2.2.2 *Turbulent flow*

The forced transport of water to one side of a basin is necessarily compensated by a reverse flow, which will occur at a depth somewhere below the surface. Such flow will be less ordered since it is driven essentially by back-pressure. Moreover, the viscous properties of the water will absorb some of the driving energy (as heat), whilst friction with the surface flow promotes the formation of intermediate 'eddies'. In this way, the ordered, laminar flow of water breaks up and becomes *turbulent.* A quantitative theory of turbulent flow has yet to be developed, but its general characteristics have been detailed by several physicists (for a good review, see Levich, 1962): in essence, turbulent eddies are superimposed upon the mean direction and velocity of the flow and can be described by a distance (L) over which their velocities (U) are significantly altered.

The interaction between changing velocities and the viscous properties of the liquid may be defined by the dimensionless Reynolds number of flow (R_f), namely:

$$R_f = \rho \bar{U} L / \eta \qquad (1)$$

where ρ is the density of the fluid and η is its viscosity. When R_f exceeds a critical value (of between 500 and 2000 for water) the flow is turbulent. In major eddies, both $\bar{U}$ and L are large, R_f is high and viscous forces are insignificant. In progressively smaller eddies, horizontal velocities and length dimensions diminish and flow becomes increasingly viscous. Thus, in fully developed turbulence there exists a complete 'spectrum' (cf. Kolmogoroff, 1941) of eddy sizes and velocities spanning the entire water circulation to molecular motion. Nevertheless, whilst the driving energy persists, the flow is dominated and characterized by the scale of the largest accelerations: water in the smallest eddies is constantly liable to be re-entrained and displaced.

The behaviour of the turbulent water mass may be expressed quantitatively, in terms either of its *eddy viscosity* (which corresponds to the shear stress of the turbulent flow) or its *eddy diffusivity* (K_z, which corresponds to vertical conductivity, across the plane of the dominant flow). Both are properties of the motion rather than of the fluid. Their derivation and mathematical expressions are given succinctly in Smith (1975); since the immediate concern here is with the vertical *transport* of particles, it is only necessary to gain some estimate of the range of the vertical velocities experienced within turbulent layers; further detail is thus avoided.

Several methods exist for evaluating K_z in water bodies, and they do not always yield identical results (for reviews, see Hesslein & Quay, 1973; Powell & Jassby, 1974). One of the simplest (that of Schmidt, 1925) has also been shown to be the most reliable and useful (Powell & Jassby, 1974). The method invokes the vertical rate of conductivity of heat through the water mass, which is greatest over freely intermixing layers but relatively lower across layers resistant to mixing because of differences in density, (ρ). Density alters with temperature, which is relatively easy to measure in space and in time.

$$K_z = S_H/(d\theta/dz) \tag{2}$$

where θ is the temperature, $d\theta/dz$ is the temperature gradient ($° \text{ cm}^{-1}$) and S_H is the heat flux rate (units: $° \text{ cm}^{-1} \text{ s}^{-1}$) which corresponds to the mean rate of change of temperature of 1 cm^3 ($\equiv$ 1 ml) of water between two points in time. Quoted values for K_z for well-mixed surface waters (that is, having minimal temperature gradients) are in the range 0.2–50 $cm^2 \text{ s}^{-1}$. As turbulent shear reduces, so viscosity becomes dominant (see above) and vertical density differences can develop. Quay, Broecker, Hesslein & Schindler (1980) have shown that no large shear occurs at values of $K_z < 1 \times 10^{-2} \text{ cm}^2 \text{ s}^{-1}$.

If zero temperature gradients are equated with turbulent mixing (in fact this is not always the case) then the vertical passage of water through 1 cm^2 in the horizontal plane will be numerically similar, that is, between 2×10^{-3} and $5 \times 10^{-1} \text{ m s}^{-1}$. A correction for the specific conductivity of water should be made, but the difference is marginal.

2.2.3 *Motion near solid boundaries*

Natural waters are constrained by solid boundaries: turbulent eddies cannot significantly pass into the bottom sediment where flow is deflected into the horizontal plane. The interaction between the surface and the flow results in a frictional damping of turbulence and the formation of viscous sub-layer (the boundary layer) in which R_f values are typically low. Depending upon the level of turbulence, the vertical profile of current

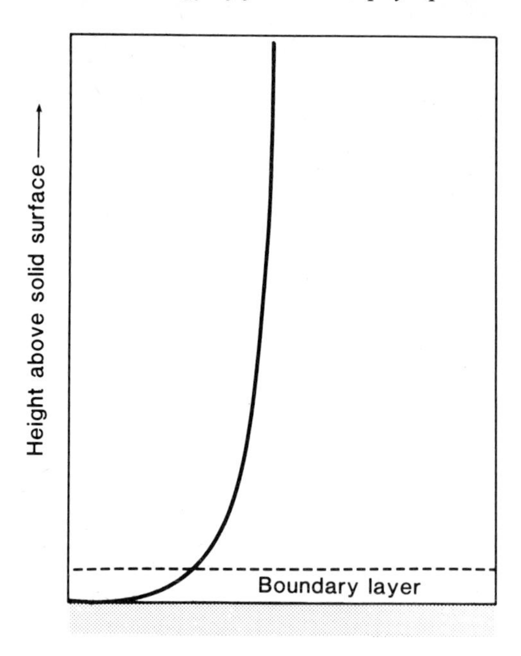

Figure 11. Vertical velocity profile (stylized) to show frictional resistance adjacent to a solid horizontal surface and the extent of the 'boundary layer' so formed. (Original.)

velocity steepens as the solid surface is approached (see Figure 11). Thus, even under conditions of fully developed turbulence, there exists a zone, of variable vertical extent, in which relatively still water conditions obtain. The presence of such boundary layers is extremely important to the consideration of algal suspension, since it both modifies particle behaviour and offers some protection to settled particles against resuspension by turbulence (see later). These remarks do not, however, apply to solid surfaces affected by wave action.

2.2.4 *Density gradients*

The input of radiation into surface layers and the concomitant increases in temperature and reductions in density potentially result in the formation of density gradients. Because the change in density for a given increment in temperature increases progressively above 4°C (Figure 12) so the amount of work necessary to overcome the density gradient increases at higher temperatures. This is the principal factor leading to the *thermal stratification* of deeper lakes under the influence of seasonal warming and, conversely, contributes to its breakdown when the surface

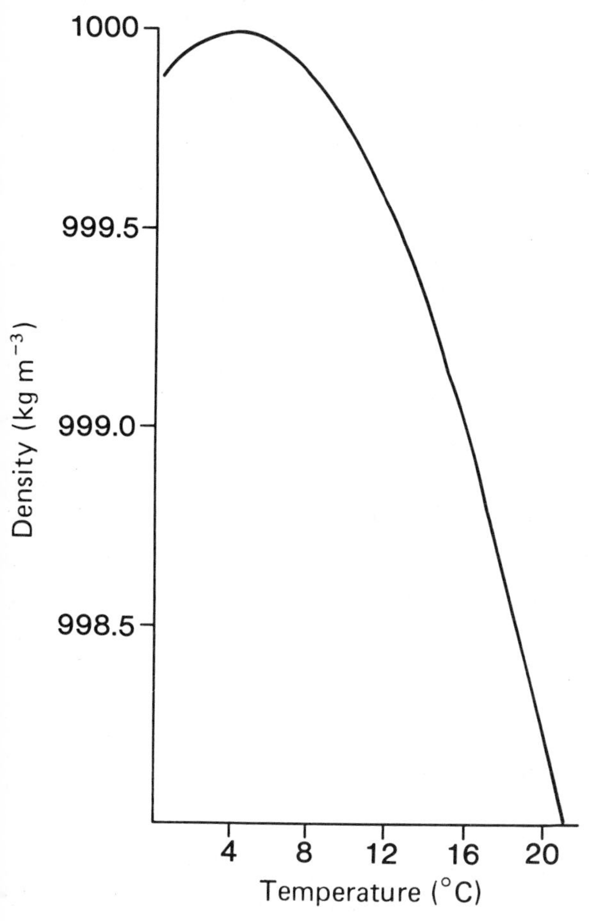

Figure 12. The relationship between density of water and temperature. (Modified after Ruttner, 1953.)

water cools. In other words, as the surface waters are warmed, so the energy required to bring about their turbulent integration with the adjacent underlying water mass is increased. If the turbulent energy is insufficient to overcome the resistance, turbulence rapidly subsides. The relative magnitude of the work required to the available energy is expressed by Richardson's Number (*Ri*)

$$Ri = g\frac{d\rho}{dz}\Big/\rho\left(\frac{du}{dz}\right)^2 \qquad (3)$$

where g is gravitational acceleration, $d\rho/dz$ is the vertical density gradient, and du/dz is the vertical gradient of horizontal velocities. Low values are indicative of turbulence; high values ($Ri > 0.25$) are equated with stability, that is, where the upper (*epilimnetic*) motion remains discrete from any deeper (*hypolimnetic*) circulation pattern. The steep density gradient between them marks the position of the *metalimnion*, or *pycnocline*, which in many cases corresponds to the band of steepest temperature gradient, the *thermocline*.

The presence of a metalimnetic layer thus acts as a barrier to vertical eddy diffusivity; accordingly the layer is characterized by a low K_z value. Determinations of vertical conductivity across actual metalimnia in stratified temperate lakes span at least two orders of magnitude ($K_z = 4 \times 10^{-6}$ cm s^{-1} to $K_z = 8 \times 10^{-4}$ cm s^{-1}: Reynolds, 1976*b*, 1979*a*; Quay *et al.*, 1980).

Under steady-state conditions, the metalimnion functions in a manner analogous to the boundary layer at the bottom of the water body. However, it is a fluid layer and can alternatively 'expand' upwards through the water column, or be 'depressed' through turbulent erosion. If epilimnetic turbulence increases, the shear stress on the metalimnion is such to set up internal waves at the interface which, just like wind-driven surface waves can become unstable and collapse ('Kelvin-Helmholtz' instability). When this happens, some of the water becomes *entrained* in an increased epilimnetic circulation, and the metalimnetic surface is depressed. In larger lake basins, the leeward build-up of water on a wind-stressed surface may also tilt the metalimnetic surface. As a result of subsequent reductions in the wind speed, the metalimnetic surface tends to return to a horizontal position, but not without some degree of 'overshoot'. Even after the wind has subsided, the metalimnion can be left 'rocking' within the basin (*seiches*). As a result of Corioli's force, seiches also rotate. These motions are extensively reviewed in Mortimer (1974): his work emphasizes the role of gross water movements in shaping the physical environment inhabited by planktonic organisms.

2.2.5 *Motion in entire water bodies*

Because the scales of motions affecting natural water bodies are variable in both time and space, it is essential to take acount of the extent of the component patterns if the impact on phytoplankton suspension is to be adequately gauged (Walsby & Reynolds, 1980). These patterns will alter substantially from season to season: Figure 13 compares the vertical temperature (density) profile in a thermally stratified lake with that of an isothermally mixed one; the diagram includes a stylized representation of the scales of motion operating in each vertical layer (the horizontal dimension is not included in the representation, but will be reconsidered in Chapter 3). It is theoretically possible to make predictive models of the vertical distribution and flux rates of solid particles in respect of wind

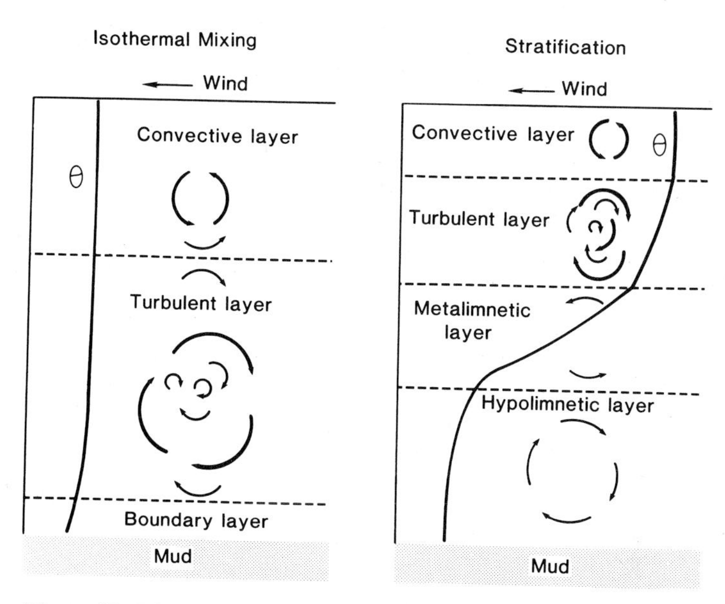

Figure 13. Diagrammatical representation of vertical circulation patterns in relation to temperature gradients in a well-mixed isothermal and a thermally stratified lake. (Modified after Walsby & Reynolds, 1980.)

stress, wave formation, current generation and the upper limit of turbulent motion, but they would require such intensive measurement of quantities, which are themselves liable to alter frequently, that they are virtually insoluble.

More practical solutions to this problem involve the consideration of synoptic integrals of eddy conductivity (K_z) or of Richardson's Number (Ri) at given depths in the profile. A simpler alternative is to characterize the vertical *stability* of the water column based upon the density difference ($\delta\rho/\delta z$) over a given depth range

$$N^2 = (g/\rho)\,(\delta\rho/\delta z) \tag{4}$$

where g is the gravitational acceleration; N^2 has the dimensions of frequency and is known as the *Brunt-Väsälä Frequency* and represents one of the resonance modes of metalimnetic oscillation.

Such evaluations (be it K, Ri or N^2) must then be compared with corresponding known distributions or flux rates of particles and potentially offer empirical solutions which may or may not be applicable to other situations. For the development of a general theory for suspension however, a knowledge of the particle behaviour is still required. This is the second part of the problem posed at the beginning of this chapter.

2.3 Particle behaviour in turbulent columns

The following consideration attempts to establish the relationship between the intrinsic sinking behaviour of particles and the superimposed fluid motion: in essence it seeks the minimum and maximum time periods during which particles can remain suspended. The development is based upon the physical literature (e.g. Dobbins, 1944; Cordoba-Molina, Hudgins & Silverston, 1978) and its application to plankton (Smith, 1982).

Let us first assume that a given particle, having a density ρ' exceeding that of water (ρ, ~ 1000 kg m^{-3}), is introduced at the surface of a completely static quiescent water column z m in depth. We must also assume that its terminal sinking velocity (v' m d^{-1}) is attained instantaneously and that it continues to sink vertically until it reaches the bottom of the column. The settling time t' is then equivalent to z/v' (units: d).

If a large number of such particles, having identical sinking rates (v') are initially distributed homogeneously through the same water column, at a concentration of N (particles m^{-3}), the individual particles will settle in times in the range $0 \rightarrow t'$; the last particle will not settle in a time significantly less than t', which continues to represent the minimum period in which the column can be cleared. At any intermediate time, t, the proportion of the original suspension which has settled is directly related to the sinking velocity, given by $N_0 v' t/z$; the quantity remaining in suspension is thus $N_0 - N_0 v' t/z$.

Let us now assume that at time t, the column is instantaneously and homogeneously mixed, such that the particles remaining in suspension are redistributed throughout the column but that those having settled remain protected by the bottom boundary layer and are not resuspended. Let us call the new concentration of particles N_t. Now,

$$N_t = N_0(1 - v' t/z) \quad (5)$$

Because particles are now reintroduced towards the top of the water column, the time taken for complete settling will be proportionately extended by a factor, not exceeding 2.

Let us now suppose that more mixings are accommodated within the period t'. If the number of mixings is m, then the period of quiescence is t'/m. The population remaining in suspension immediately after the first mixing is:

$$N_{t'/m} = N_0(1 - v' t'/m\ z) \quad (6)$$

After the second, it will be

$$N_0(1 - v' t'/mz)(1 - v' t'/mz)$$

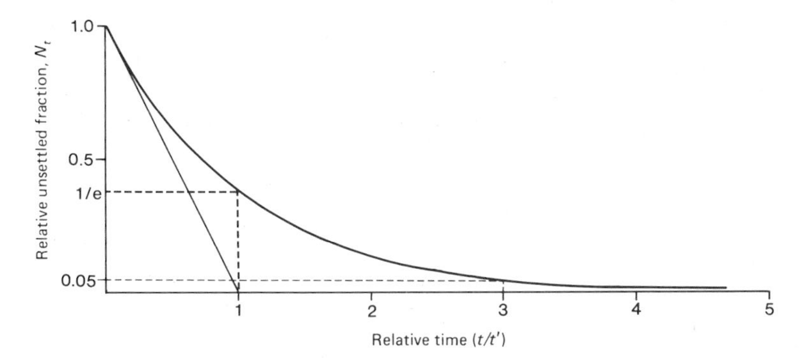

Figure 14. The number of particles retained in suspension in a continuously-mixed suspension compared with the retention of the same particles from a static column of identical height, which they clear completely in time t'. (Original, based on Fig. 1 of Smith, 1982.)

After the mth mixing, it will be

$$N_{t'} = N_0(1 - v't'/mz)^m \tag{7}$$

Since $t' = z/v'$, Equation (7) simplifies to

$$N_{t'} = N_0(1 - 1/m)^m \tag{8}$$

Moreover, as m becomes large ($m > 5$), so the equation tends toward

$$N_{t'} = N_0(1/e) \tag{9}$$

where e is the natural logarithmic base. Solving empirically,

$$N_{t'} = 0.368\, N_0 \tag{10}$$

This solution is instructive in several respects. It shows that at an infinite number of mixings (fully developed turbulence), the fraction of the original material remaining suspended at t' is independent of the intensity of turbulence. Thus, it follows that the time for complete settling under persistent mixing is similarly independent, although it cannot be estimated from equation (10), which is asymptotic to zero. However the time taken (t_e) to achieve 95 to 99% elimination (i.e. when 0.05 or $0.01 N_0$ remains in suspension) is given from:

$$t_e/t' = \log_e 0.05/\log_e 0.368$$
$$= 3.0$$

or

$$t_e/t' = \log_e 0.01/\log_e 0.368$$
$$= 4.61$$

In this way, an upper limit on the settling time for the particles can be defined (see Figure 14): for 99% elimination, the period is approximately

4.6 times the minimum predicted from the intrinsic, still-water settling velocity.

It may also be noted that the number of intermediate mixings need not be large before the effect is approached. Thus, substituting in equation (8), if $m = 2$, $N_{t'} = 0.25\ N_0$; if $m = 5$, $N_{t'} = 0.33\ N_0$; if $m = 20$, $N_{t'} = 0.36\ N_0$. In other words, neither the frequency nor the duration of full mixing is necessarily large to prolong particle suspension. This point is particularly relevant to the consideration of natural water columns which frequently oscillate between periods of active vertical turbulence and of more-ordered laminar flow, at least within certain depth ranges. Particle settling times will vary between those derived for quiescent and fully-developed turbulent conditions, with a tendency towards the latter.

The relationship also permits the calculation of mean settling velocity from the observed changes in population density, as well as the assessment of the true growth rate of live algae for an observed rate of change (see Chapter 7).

That the assumption of a uniform sinking rate is unlikely ever to be realized in natural populations does not materially affect these considerations, since the result is independent of sinking rate. It is less simple to dismiss the effect of particle resuspension which may further prolong the presence of particles in turbulent water columns. It does seem, however, that most resuspension in lakes occurs mainly from relatively shallow sediments whose surfaces tend to be rough and liable to physical movement as a result of wave action. A further factor affecting the outcome of the solution which has not been considered is *entrainment*. Particles which are entrained more or less completely in the turbulent flow will not escape easily from currents deflected by boundary layers: hence, they may be returned to the water column from deep water without intermediate settling.

Nevertheless, the mixing model predicts that particles are eventually lost from suspension, regardless of turbulence, which delays their settling by a factor directly related to the quiescent settling velocity. The implication for planktonic 'algae' is that a slow quiescent settling velocity is an essential prerequisite for true suspension. The mechanisms influencing 'algal' sinking rates are considered in the next section.

2.4 Phytoplankton settling velocities

The settling properties of planktonic cells and colonies are determined by the same forces which govern the movements of inert bodies in viscous fluids. As the body moves, it displaces some of the suspending fluid. Provided that the movement of the displaced fluid past the particle is laminar, the terminal velocity of a spherical body is given (in m s^{-1}) by the well-known *Stokes equation*, namely

$$v_s = 2\,gr^2(\rho' - \rho)/9\eta \qquad (11)$$

where g is the gravitational acceleration (units: m s^{-2}), η is the coefficient of viscosity of the fluid medium (units: kg m^{-1} s^{-1}), ρ is its density and ρ' is the density of the particle (units: kg m^{-3}) and r is its radius (units: m). The term $(\rho' - \rho)$ is known as the excess density: when it is negative, the body *floats upward* at a velocity of $-v_s$. Although v_s(or $-v_s$) is a terminal velocity, it is attained almost instantaneously.

The Stokes equation has been verified empirically in a series of experiments conducted by McNown & Malaika (1950), who measured the velocities of machined metal shapes sinking in viscous oils. The equation thus provides a convenient basis for a consideration of quiescent 'algal' settling velocities. It is, however, essential to take account of the factors which lead to non-conformity with the equation. One of these concerns the condition of laminar flow.

Laminar flow occurs provided the viscous forces are not overwhelmed by inertia. The ratio of these two forces is expressed by the Reynolds Number, *Re*. Though calculated on similar premises, R_e is not equivalent to the Reynolds number of flow (R_f), discussed above.

$$Re = 2\,r\,v'\,\rho/\eta \qquad (12)$$

McNown & Malaika (1950) showed that, for calculated *Re* values of < 0.1, there is very little departure from the Stokes equation and that the error is $< 10\%$ for $Re < 0.5$. The maximum reported velocity for any unicellular alga is 6 mm s^{-1}, observed for empty frustules of the large marine diatom *Ethmodiscus rex* having diameters of *ca.* 1 mm (Smayda, 1970). Thus, putting $\rho = 10^3$ kg m^{-3} and $\eta = 10^{-3}$ kg m^{-1} s^{-1} in equation (12), $Re \simeq 6$, indicating that the observed settling velocity departs significantly from the value predicted by the Stokes equation. However, the example is an extreme one, for the settling velocities of most other live diatoms considered by Smayda (1970) were 1–2 orders of magnitude less than that of *E. rex*. Walsby & Reynolds (1980) calculated *Re* for *Coscinodiscus wailesii*, another large marine diatom ($r \sim 75$ μm; $v' \sim 0.1$ mm s^{-1}), to be 0.0156; for one of the largest freshwater diatoms, *Stephanodiscus astraea* ($r = 25$ μm; $v' \sim 25$ μm s^{-1}: Reynolds, 1973*a*), *Re* is ~ 0.00125. Thus, it may be assumed that the Reynolds numbers for most planktonic 'algae' lie within the limit for laminar flow ($Re < 0.1$) and, hence, do not detract significantly from the Stokes equation.

A second source of departure from the predicted settling behaviour concerns shape: most 'algae' are *not* spherical (Figures 1–5). The settling rate of the sphere can be changed by deforming its shape whilst its density and volume remain constant. Such deformation inevitably results in a greater surface area and, under most circumstances, a greater frictional resistance ('drag') is exerted by the medium. This effect is difficult to quantify, other than for relatively simple shapes (ellipsoids), but in terms of settling, it may be expressed by a correctional term, ϕ_r, *the coefficient of form resistance*:

$$\phi_r = v_s/v' \quad (13)$$

The form resistance may be embodied into the modified Stokes equation (14):

$$v' = 2\,g\,r^2(\rho' - \rho)/(9\eta \,.\, \phi_r) \quad (14)$$

There is a third component influencing departure from the predicted settling velocity which is based purely on a simple observation: that suspended 'algal' populations are usually dominated by live cells, whereas dead cells are rapidly eliminated. There are many reasons for this, the most obvious being that cells decompose rapidly after death, as a result of bacterial action, or that they are consumed by a detritus-feeding zooplankton. Neither explanation applies in all cases. There are sufficient 'large' species, with conservative cell components (e.g. siliceous walls) to suppose that their cells are removed from suspension relatively more rapidly than the living ones. By implication, 'suspension' includes a certain vital element. Indeed, it has been demonstrated by a number of workers that dead diatoms or even living senescent ones, sink faster than viable cells (literature reviewed in Smayda, 1970), by factors of three to five (Smayda & Boleyn, 1965; Eppley, Holmes & Strickland, 1967; Smayda, 1974; Reynolds, 1973*a*; Titman & Kilham, 1976; Wiseman & Reynolds, 1981) but without visible alteration in shape, size or form-resistant structures. The vital component is therefore essential to the consideration of suspension mechanisms. Without any simple means of distinguishing the vital, 'physiological' contribution to settling rate from that afforded by 'form resistance', it is advantageous to treat the vital component as a contribution to ϕ_r and then to investigate ϕ_r on killed cells.

2.5 Adaptive mechanisms for depressing v'

Four components of equation (14) are properties of the 'alga', as opposed to those of the medium, and hence available to adaptive regulation: v', size (r), density (ρ') and 'form-resistance' (ϕ_r), in the wider sense discussed in the previous paragraph. It is also possible that metabolites produced by the algae could affect the viscosity of the suspending medium. So far as is possible, these 'algal' suspension mechanisms are considered separately below.

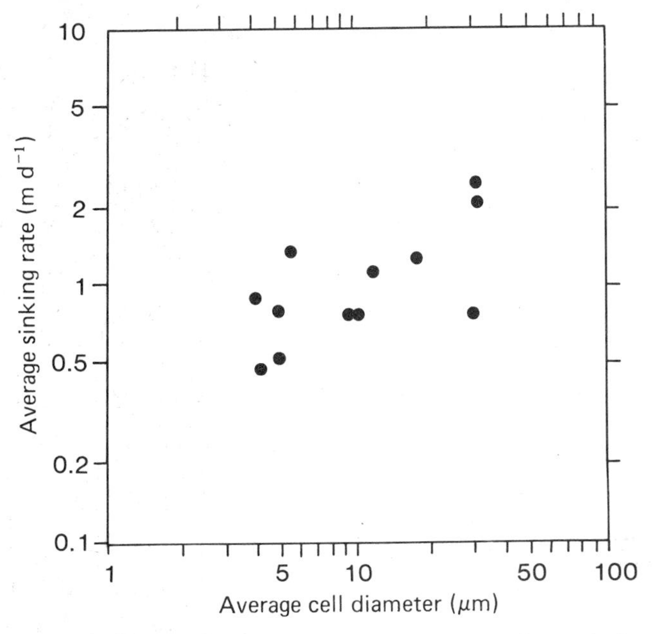

Figure 15. The relationship between maximum observed mean sinking rates and the mean cell diameters of chain-forming marine diatoms. (Redrawn from collected data presented by Smayda, 1970.)

2.5.1 *Size*

The relatively small size of most planktonic algae has been alluded to in Chapter 1. The radii (r) of spheres having respectively identical volumes to the 'algae' listed in Table 3 are in the range 1–285 μm, although many species potentially present a larger maximal dimension (*GALD*) to the direction of movement. The effect of size on sinking velocity has been demonstrated by Smayda (1970), who compared the settling velocities of several marine species of centric diatoms, which share a generally similar morphology and construction. Some of his data are redrawn as Figure 15. Subject to the qualification (above) concerning the condition and age there is a straight-line relationship between the logarithm of the mean sinking velocity and the logarithm of the mean cell diameter. The slope of the regression is much closer to 1 (i.e. $v' \propto r$) than to 2 ($v' \propto r^2$) as might be expected if r were the only variable in the Stokes equation (14). This suggests that larger size is in fact partly compensated by decreased density or increased form resistance. This effect may be vital to large diatoms: as Smayda (1970) showed, the large cells of *Coscinodiscus wailesii* would sink at about 40 m d^{-1}, had they the same density as the much smaller *Cyclotella nana*, as against the observed 7 m d^{-1}.

A decrease in density might be explained by the fact that the silica in

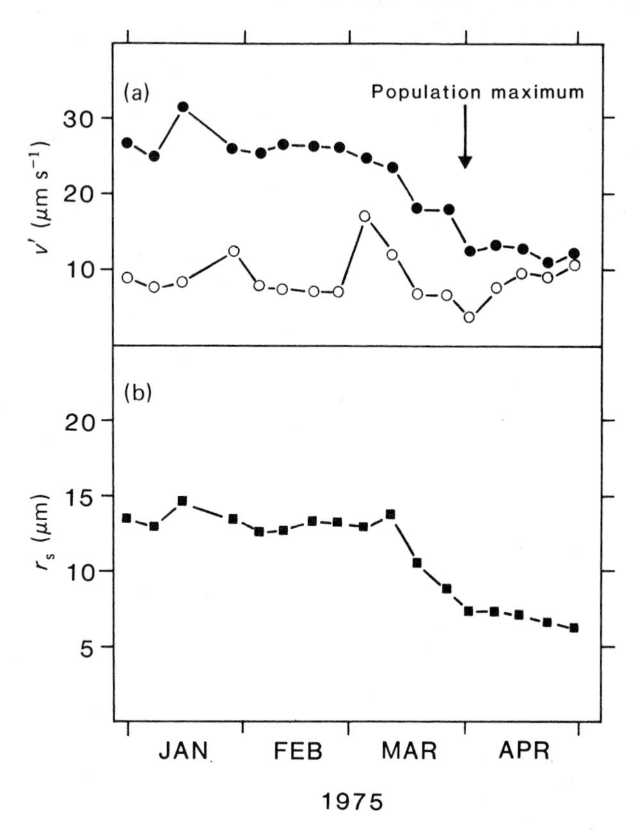

Figure 16. Changes with time in (a) the mean intrinsic settling velocities (v') of *Stephanodiscus astraea* cells, freshly collected from Crose Mere, Shropshire (○, live; ●, killed cells), and (b) the radius of a sphere of identical volume to the mean cell volume (r_s). (Original.)

the cell wall accounts for much of the diatom's excess density, and that the volume of wall decreases as a proportion of total volume as size is increased (Hutchinson, 1967; see also Walsby & Reynolds, 1980). However, the limited data available, presented in Table 6, suggest that silica content may be directly correlated with cell volume. An alternative explanation, advanced by Walsby & Reynolds (1980), is that the relative volume of the sap vacuole increases with increasing size; the sap is likely to have a lower density than that of the cytoplasm. It is also known that some marine diatoms, at least, can also regulate their sap density by selective ion accumulation (Gross & Zeuthen, 1948; Anderson & Sweeney, 1978). This mechanism is not available to freshwater diatoms (see § 2.5.2.2).

Nevertheless, some previously unpublished data indicate that the relationship between v' and r may hold for freshwater diatoms. In Figure 16, a series of *in vitro* sinking rate determinations on freshly collected

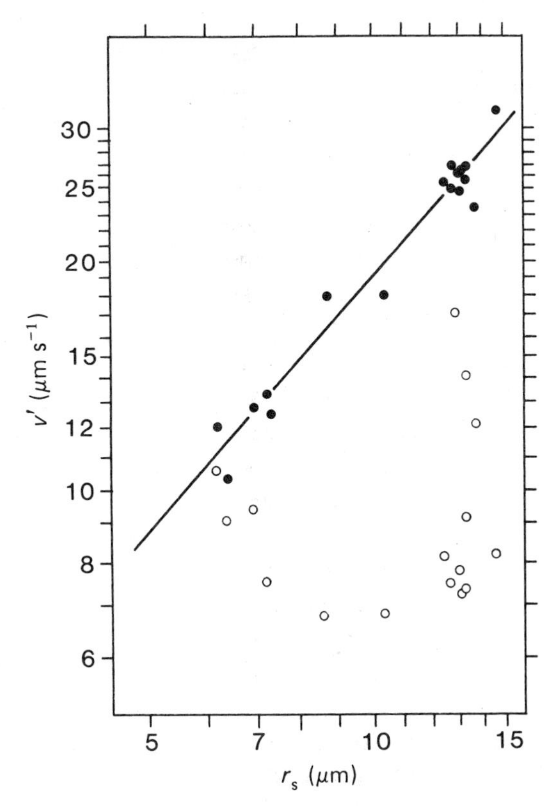

Figure 17. Log/log plot of v' against corresponding r_s values (from Figure 16; ○, live cells; ●, killed cells). There is no correlation between v' and r_s for live cells, but a strong positive correlation for killed cells (coefficient of correlation = +0.980). The equation of the regression fitted to points for killed algae is: $\log v'_k = 1.127 \log r_s + 0.152$.

Stephanodiscus astraea cells and shortly after killing (by boiling, followed by re-equilibration of the temperature) is plotted against the dates of collection, together with corresponding determination of the radius of a sphere (r_s) having the identical volume to the mean algal volume. Although the mean settling velocities of the untreated algae (v') fluctuate over the course of the population development, the mean sinking rates of killed cells (v'_k) and the cell size (r_s) steadily diminish. Indeed, these quantities are significantly correlated when plotted on a log/log scale (Figure 17; coefficient of correlation = +0.980). The regression coefficient (1.127) again indicates that $v'_k \propto r$. That there is no significant correlation between v' and r (correlation coefficient = +0.241) may be attributable to 'vital interference'. This cannot apply to the killed diatoms, which are then more likely to conform to a predictable sinking behaviour. The implication is that, whilst a generally small size is advantageous to prolonged suspension, increased dimensions are not necessarily as counter-productive as suggested by the Stokes equation, whereas they may offer positive advantages for other reasons, such as resistance to grazing by zooplankton.

The effective size of many planktonic 'algae' is also increased by colony formation (Table 3). However, this usually involves a change of particle shape, and is therefore considered separately (§2.5.4).

2.5.2 *Density*

Most of the components of living protoplasm have densities greater than that of water: carbohydrates have a density of ~ 1500 kg m^{-3}; proteins, ~ 1300; nucleic acids, ~ 1700. Many condensed storage products commonly found in healthy cells are also relatively dense, notably polyphosphate bodies (~ 2500 kg m^{-3}), whilst the opaline silica deposits in diatom walls have a density of about 2600 kg m^{-3}. Only lipids are characteristically less dense than water (the lightest having a density of ~ 860 kg m^{-3}: Sargent, 1976) but these rarely account for more than 10% of the cell dry weight. It follows that phytoplankton cells are normally heavier than water and will therefore sink (cf. Hensen's definition of plankton!).

Unfortunately, few quantitative determinations of 'algal' densities have been made, largely because of inherent methodological problems (for a review of the methods currently available, consult Walsby & Reynolds, 1980). Some quoted and hitherto unpublished densities are presented in Table 8. Most of these data apply to diatoms, wherein the siliceous walls contribute to the overall densities, which fall in the range $\rho' = 1078 - 1263$ kg m^{-3} (apart from values calculated by Einsele & Grim, 1938). The density of one of the non-siliceous 'algae', *Microcystis aeruginosa*, includes the contribution of mucilage which generally makes up between 75 and 98% of the colony volume. Reynolds, Jaworski, Cmiech & Leedale (1981) used simultaneous equations to distinguish the density of the mucilage (ρ'_m, which was scarcely different from that of the suspending water) from that of non-vacuolate cells (ρ'_c, about 1016 kg m^{-3}). The latter value may be more representative of non-siliceous 'algal' cells which also lack large sap vacuoles, than any of the data cited in Table 8.

Many phytoplankton 'algae' have evolved mechanisms which directly, or incidentally, reduce average density. These include (i) the storage of relatively light lipids, (ii) the regulation of ions, (iii) the secretion of mucilage and, in the cyanobacteria, (iv) the provision of gas vacuoles.

Table 8. *The densities of a marine and some freshwater phytoplankton 'algae'*

Species[a]	Number of determinations	Density ($kg\ m^{-3}$)	Method[b]	Reference
Cyclotella praeterissima	1	1196 (±13)	gradient centrifugation	(1)
Stephanodiscus astraea	5	1091 (±13)	gravimetric	(1)
Synedra acus[c]	1	1100 (±33)	gravimetric	(1)
S. acus[c]	1	1115 (±23)	gradient centrifugation	(1)
Asterionella formosa	1	1130	gravimetric	(1)
Fragilaria crotonensis	6	1198 (±82)	gravimetric	(1)
F. crotonensis[c]	1	1196(±13)	gradient centriguation	(1)
F. crotonensis[d]	2	1165 (±65)	gravimetric	(1)
F. crotonensis	?	1100–1465	calculation	(2)
Melosira italica subsp. *subarctica*[e]	1	1250 (±13)	gradient centrifugation	(1)
M. italica subarctica[f]	1	1169 (±14)	gradient centrifugation	(1)
Tabellaria flocculosa v. *asterionelloides*		1142 (±14) 1142 (±14)	gradient centrifugation	(1)
Cyclotella meneghiniana	1	1020–1200	gradient centrifugation	(3)
Chlorella vulgaris	2	1095 (±7)	gradient centrifugation	(3)
Chlorococcum sp.[g]	3	1074 (±16)	gradient centrifugation	(3)
Chlorococcum sp.[h]		1089 (±51)		
Chlorococcum sp. [i]		1044 (±24)		
Thalassiosira fluviatilis[j]	?	1121	gravimetric	(4)
Microcystis aeruginosa[k]	72	999.4–1004.0	calculation	(5)

References: (1), author, unpublished; (2), Einsele & Grim (1938); (3), Oliver *et al.* (1981); (4), Walsby & Xypolyta (1977); (5), Reynolds, Jaworski *et al.* (1981).
(*a*) Natural populations unless otherwise stated.
(*b*) For full description of methods, see Walsby & Reynolds (1980).
(*c*) Culture.
(*d*) Dead algae, killed by boiling.
(*e*) Pre-auxospore population.
(*f*) Post-auxospore filaments.
(*g*) Early log-phase culture.
(*h*) Late log-phase culture.
(*i*) Senescent culture.
(*j*) *T. fluviatilis* is a marine diatom (syn. *T. weisflogii*).
(*k*) After pressure treatment to destroy gas-vacuoles.

2.5.2.1 *Lipid accumulation.* Fats and oils normally account for some 2–20% of the dry weight of 'algae'; under exceptional circumstances it may rise to 40% (Grøntved, 1952, quoted by Smayda, 1970), though such levels are associated with cellular senescence (see also Fogg, 1975). Generally, the lipids accumulated are lighter than water and inevitably their presence in cells reduces excess density. The idea that lipids would adequately compensate the excess density to the point that cells were rendered neutrally buoyant and that their presence could be construed as a buoyancy aid is an old one (Hensen, 1887). However, it has long been recognized that many non-planktonic 'algae' may contain similar lipid complements so that the benefit to flotation may be incidental (e.g. Gessner, 1955). Moreover, Smayda (1970), using data of Strickland (1960, 1965), has calculated that an increase in the normal lipid content of the marine diatom *Coscinodiscus concinnus* from 9 to 40% would reduce the cell density from 1190 to 1150 kg m^{-3} (an absolute reduction of 3.5%). Assuming the density of the suspending sea water to be $\sim$ 1027 kg m^{-3}, the density reduction would give a 25% reduction in sinking rate, though this is far from sufficient to overcome it altogether. Similarly it can be argued that if the entire internal volume of an otherwise empty cell of *Asterionella formosa* were to be completely filled with the lightest known lipid (density 860 kg m^{-3}), its density (1005 kg m^{-3}) would still exceed that of the suspending medium. The only attested case of a freshwater alga becoming buoyant as a direct result of lipid accumulation under conditions of extreme nitrogen deficiency, is that of *Botryococcus braunii* (Belcher, 1968). The carotenoid content of the cells often increases simultaneously; the appearance of orange surface accumulations of this alga on certain lakes has been attributed to these phenomena (Fogg, 1975).

In each of these cases, the reduction in density consequential upon intracellular lipid accumulation is not questioned; however, it is to be doubted that this accumulation has any adaptive significance, nor is it used to regulate buoyancy (Walsby & Reynolds, 1980).

2.5.2.2 *Ionic regulation.* The exploitation of significant inherent differences in the densities of equimolar solutions of different organic ions has been invoked to propose selective retention of 'light' ions by 'algae' as a buoyancy-regulating mechanism. In a classical paper, Gross & Zeuthen (1948) calculated the density of the cell sap of the marine diatom *Ditylum brightwellii* to be $\sim$ 1020 kg m^{-3} (i.e. less than that of the suspending sea water) and to be sufficient to reduce the cell density to neutral buoyancy.

They showed that the sap density could be explained if the composition was identical with sea water but for the replacement of divalent ions (e.g. Ca^{2+}, Mg^{2+}) by monovalent ones (Na^{+}, K^{+}). Anderson & Sweeney (1978) have since determined the composition of the sap of *Ditylum brightwellii* grown under alternating light-dark periods, and have shown that its density may be indeed altered by up to ± 15 kg m^{-3}, by selectively accumulating either sodium (Na^{+}) or potassium (K^{+}) ions. The difference is insufficient to overcome the negative buoyancy of the cells, but the mechanism clearly provides a means of regulating sinking rate. Kahn & Swift (1978), however, have shown that by reducing the content of Ca^{2+}, Mg^{2+} and SO_4^{2-} in its sap, the marine dinoflagellate *Pyrocystis noctiluca* is able to become positively buoyant.

The effectiveness of ionic regulation in reducing buoyancy is also dependent upon the alga maintaining a relatively large sap vacuole. Equally, the scope of the reduction in its density is limited by the density of isotonic solution of the lightest ions. In the more dilute freshwaters ($\rho < 1002$ kg m^{-3}), there is extremely little scope for regulating sap density (maximum ± 2 kg m^{-3}) and, hence, for making any significant impact upon overall cell density. The effectiveness of ionic regulation in freshwater 'algae', if it occurs at all, would lend no selective advantage to suspension (cf. Lund, 1959).

2.5.2.3 *Mucilage secretion.* The provision of a mucilaginous external sheath, characteristic in many freshwater 'algae' (e.g. many cyanobacteria, chrysophytes and especially among the chlorophytes) has for long been advanced as a mechanism for reducing sinking rate. Mucilages are gels formed by a network of hydrophilic polysaccharides which, though themselves of high density (~ 1500 kg m^{-3}), are able to hold large volumes of water such that their average density (ρ'_m) is very close to that of the suspending medium: Reynolds *et al.* (1981) estimated the density of the mucilage of *Microcystis* to average ($\rho + 0.7$) kg m^{-3} (its density cannot, of course, be less than ρ, so it cannot provide positive buoyancy). The effect of a mucilage sheath in reducing the overall 'algal' density of the cell is mitigated by the consequent increase in the size of the 'algal unit'. According to the Stokes equation (14), the latter inevitably contributes to an increase in sinking rate. For mucilage to be effective in reducing buoyancy, the advantage must always outweigh the disadvantage.

The critical conditions for this relationship have been defined by Hutchinson (1967). If a spherical cell, of density ρ'_c, is enclosed in mucilage of density ρ'_m, such that its overall radius is increased by a factor a, then the sinking rate of the cell with mucilage will be less than that of the cell alone, *provided that*:

$$\frac{\rho'_{c}-\rho'_{m}}{\rho'_{m}-\rho} > a(a+1) \qquad (15)$$

Because a is always $\geqslant 1$, the density difference between cell and mucilage must be at least double that between mucilage and water if the provision is to hold. If we take the respective values already suggested as being 'typical' for 'algal' cells and mucilage and assume the density of the suspending water to be 999 kg m^{-3}, equation (15) may be resolved as follows:

$$\frac{1016-999.7}{999.7-999} > a(a+1)$$

$$\frac{16.3}{0.7} > a(a+1)$$

$$23.3 > a(a+1) \qquad (16)$$

In this instance, the maximum value for a is ~ 4.3. The optimal value of a, that is, the solution which gives the lowest sinking velocity of the alga, may be gauged from Figure 18, which plots the density ρ'_{c+m} and the settling velocity (v') of the 'alga' relative to the settling velocity of the cell without mucilage. It can be seen that the secretion of mucilage reduces settling velocity up to a maximum value of a (in this example, $a \sim 2.3$, $v' = 0.63$), above which settling velocity begins to increase again. Thus, it is possible to hypothesize that if mucilage is to be advantageous in reducing sinking, its volume relative to that of the cell is governed by physical constraints imposed by Stokes' Law. The precise value of a will vary from alga to alga according to its density (for a diatom cell whose density is 1200 kg m^{-3}, the maximum value of a might be as high as 16.4, with an optimum value of about 8.7).

In order to compare the typical values for a in many of the 'algae' which produce mucilage, which are either non-spherical or colonial, it is necessary first to express the cell volume to total unit volume (V_c/V_{c+m}: a conversion scale is included in Figure 18). Observed relative mucilage and cell volumes of some selected algae are included in Table 9. In each case, the radius of a sphere equivalent to the total colony volume (V_{c+m}) is expressed as a ratio of the radius of a sphere equivalent to the total cell volume (V_c). The available data appear to conform with the suggested range of 'benefit' from mucilage presence, though in *Gemellicystis* and some *Sphaerocystis* colonies the ratio appears to be unfavourable, at least if the assumptions concerning the density of cells continue to apply in these cases.

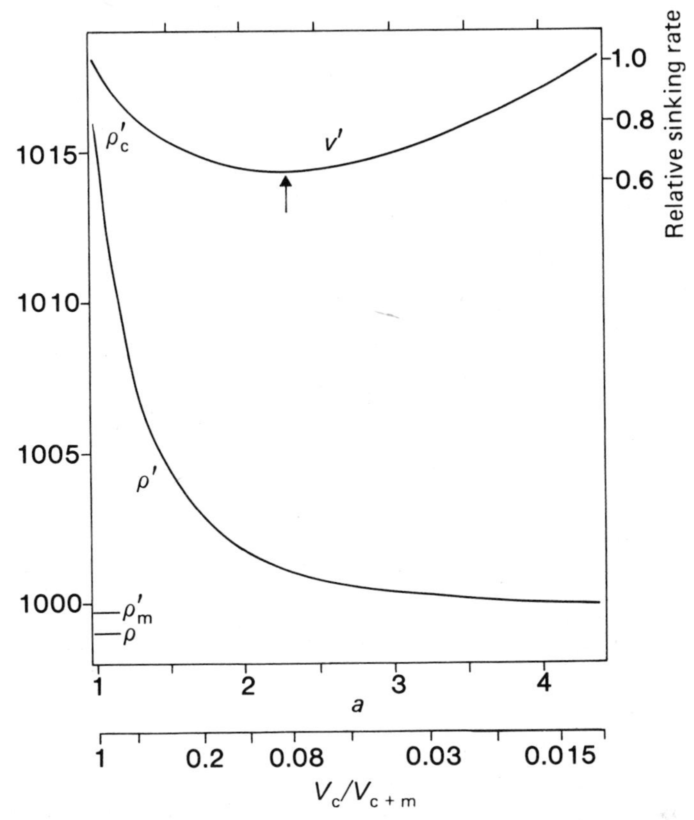

Figure 18. The effect of increasing mucilage thickness (a, as a proportion of r) on the density ($\rho' = \rho'_c + \rho'_m$) and the relative sinking rate of a spherical algal cell, of constant radius (r) and density ($\rho'_c = 1016$ kg m^{-3}). The arrow on the velocity plot indicates the point at which the extent of the mucilage secretion is most advantageous ($a = 2.25$). Based on Figure 10.2 of Walsby & Reynolds (1980).

2.5.2.4 *Gas vacuoles.* From the time that their existence was first established in the cells of *Gloeotrichia* (by Klebahn, 1895), gas vacuoles have been assumed to impart buoyancy to the cells of planktonic cyanobacteria. Although this may have been neither their original nor their only function (cf. Lemmermann, 1910; Porter & Jost 1973, 1976), these uniquely prokaryotic structures certainly do reduce the average unit density, frequently to the extent that $\rho' < \rho$: gas-vacuoles are characteristically present in the cyanobacteria forming surface scums ('water blooms') on lakes during quiet weather. Because gases have much lower densities than solids or liquids, maintenance of gas-filled structures clearly offers the most effective of all buoyancy mechanisms. The simplest such structures might be bubbles (having densities of zero) but there are several physical and

Table 9. *Relative volumes of cells in mucilage-producing planktonic algae* (V_c/V_{c+m}), *expressed in terms of a.*

'Alga'	Source	V_c/V_{c+m}	a
Microcystis aeruginosa	(1)	0.03–0.05	1.59–3.22
Anabaena circinalis	(2)	0.045–0.124	2.00–2.81
Eudorina unicocca	(2)	0.055–0.262	0.64–2.63
Gloeocystis sp.	(2)	0.126	1.99
Gemellicystis neglecta	(2)	0.008–0.013	4.30–4.92
Sphaerocystis schroeteri	(2)	0.005–0.532	1.24–9.52
Staurastrum brevispinum	(3)	—	2.2–2.3
Fragilaria crotonensis	(4)	0.032–0.047	2.78–3.14

Sources: (1), Reynolds, *et al.* (1981); (2), Reynolds, unpublished measurements on natural populations; (3), From Figure 27 of Ruttner (1953), by direct measurement of a; (4), From Plates 1c, 1d of Canter & Jaworski (1978), by direct measurement of linear dimensions and approximation of V_c, V_{c+m}.

physiological objections against their formation at a size small enough to be accommodated within an 'algal' cell (Walsby, 1972). The gas vacuoles of cyanobacteria are complex structures comprising numerous hollow submicroscopic (diameter: 70 nm) cylinders, called *gas vesicles*. Their rigid proteinaceous walls are fully permeable to gases but a hydrophobic inner surface prevents the influx of liquid water (for a recent review of gas vacuole structure, see Walsby, 1978*a*). In this way, the vesicles collectively maintain a finite volume of metabolic gases at ambient pressure: the buoyancy that they provide is thus more a consequence of the space rather than of the gas which fills it.

The role of gas vacuoles in providing buoyancy is evident in Figure 19, which traces the *in vitro* velocity rates of natural *Anabaena circinalis* filaments against the relative volume of the cell occupied by gas. At the temperatures at which the experiments were conducted (15 ± 1°C, see Reynolds, 1972) the volume of gas required to give neutral buoyancy ($\rho' = \rho$) was 1.1% of the original volume, suggesting that the density of the cell without vacuoles is ~ 1010 kg m^{-3}. Cell density may vary independently of vacuole content, of course, through the accumulation of condensed phytosynthates and storage products; in addition, gas vacuoles must also compensate for the mucilage in the cyanobacteria which produce it in large quantities. There will thus be no unique level of relative gas vacuole content which will always provide neutral buoyancy. Actual values determined for natural cyanobacteria vary between 0.7 and 2.3% of cell volume (Reynolds & Walsby, 1975). Nevertheless, Reynolds *et al.* (1981) have shown that in *Microcystis* colonies where the volume of the healthy

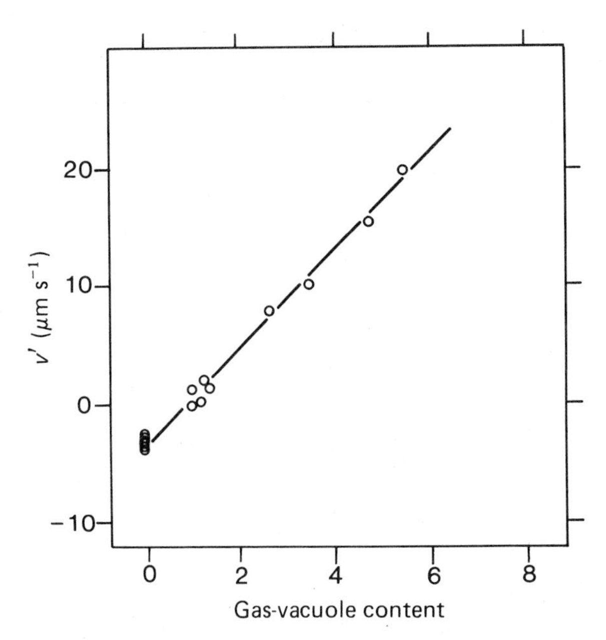

Figure 19. The relationship between sinking or floating velocity (v') and relative gas-vacuole content of cells of *Anabaena circinalis*. (Modified after Reynolds, 1972.)

cell fraction falls to a very low fraction of total colony volume (< 0.5%), the attainable relative gas-vacuole volume of those cells is insufficient to render the whole colony buoyant.

However, the gas vacuole is more than a mere buoyancy aid: by dynamically regulating the number of vesicles in existence at any given time, the alga is able to *control* its buoyancy. In an admirably thorough series of studies, A. E. Walsby and his co-workers (e.g. Walsby, 1971, 1978*b*; Dinsdale & Walsby, 1972; Walsby & Klemer, 1974; Grant & Walsby, 1977) have been able to show that though individual gas vesicles are capable of withstanding external pressures of between 400 and 700 kPa (4–7 atmospheres; those of the marine *Trichodesmium thiebautii* will withstand as much as 35 atmospheres: Walsby 1978*b*), they nevertheless collapse irreversibly when the 'critical pressure' is exceeded. The intracellular osmotic pressure, which may increase during photosynthesis as soluble photosynthates are elaborated, may well rise to within the range 350–500 kPa, that is sufficiently high to eliminate the weaker vesicles. At the same time, new vesicles can be constantly assembled; the relative dynamics of collapse and assembly thus provide one potential mechanism for buoyancy regulation: under conditions favouring rapid photosynthesis, sufficient gas vesicles may be collapsed for cells to lose buoyancy; at lower light intensities the reverse might apply. It also seems highly likely that, as the cells divide, the existing vesicles will be similarly divided between the daughters: if the relative rate of vesicle assembly is slower than that of the elaboration of other cell components, vesicles will become progressively 'diluted' out as a consequence of growth. This provides an alternative method of buoyancy control (for more complete reviews of gas-vacuole function, see Walsby 1972, 1978*a*).

Both mechanisms appear to function in natural populations (e.g. Reynolds 1972, 1973*b*; Walsby & Klemer, 1974; Konopka *et al.*, 1978), although the specific responses vary. Reynolds & Walsby (1975) identified three quite discrete behaviour patterns adopted by planktonic cyanobacteria all of which invoked the mediation of buoyancy control. The first of these accounted for the stratification of *Oscillatoria* populations in the metalimnia of transparent, stable lakes where depth maxima persist for weeks on end (see also Walsby & Klemer, 1974); the second related to the more or less continuous suspension within the epilimnia of relatively more turbid, less-stably-stratified basins, where buoyancy responded on a day-to-day basis (see also Reynolds 1972, 1973*b*); the third alluded to the rapid buoyancy accommodation of *Microcystis* populations in the shallow tropical basin of Lake George, Uganda, to diel alternations between a stratified and fully-mixed water column (see also Ganf, 1974*a*). In the latter two examples, especially, and in the first to a limited extent, failure of the buoyancy-regulating mechanism to adapt sufficiently rapidly to fluctuating environmental conditions imposed by light penetration and water turbulence potentially result in the formation of surface scums. These 'blooms' are given added poignancy when compounded by the presence of relatively large populations and by subsequent drifting towards lee shores. These features, perhaps more than any other, have contributed to the notoriety of cyanobacterial blooms and to their objectionable association with 'eutrophication' (Reynolds, 1979*b*; see also Chapter 5).

2.5.3 *Form resistance*

The sinking rate of a spherical particle is altered by deforming its shape, even though its volume and density remain constant. Certain distortions can result in increased relative settling rate (e.g. the tear-drop shape) provided that they contribute to the streamlining of the body. At low Reynolds numbers, however, the effect is negligible. Most other shape changes lead to a reduced rate of sinking. The reduction is attributed to 'form resistance' which is expressed by the dimensionless coefficient ϕ_r (see equation 13, above). At the present time, its value can be predicted only in general terms, except in the case of regular ellipsoids for which theoretical derivations have been verified experimentally (McNown & Malaika, 1950). The classical studies of McNown & Malaika (1950), however, showed that, at low Reynolds numbers ($Re < 0.05$), equiaxial bicones and prisms had similar sinking rates to spheres of similar volume (i.e. $\phi_r \sim 1$) and that the presence of sharp edges and angles make little impact on form resistance. The form resistance of more elongate forms (four times longer than broad) was nevertheless greater ($\phi_r \sim 1.3$). Extrapolating these findings to the generalized shapes of phytoplankton cells, it is probable that small projections and irregularities on cell surfaces do not greatly reduce v'.

As yet, there have been few empirical attempts to suggest that such extrapolations have any validity. One approach, which is developed here, is to evaluate all the separate components of the Stokes equation (11) and to then compare the calculated v_s with the measured v'; the factor by which v_s exceeds v' is then equivalent to ϕ_r. Some of the data currently available to the author (some previously unpublished) have been used to calculate ϕ_r values, which are presented in Table 10. These show that the sinking velocities of spherical cells and the squat cylindrical forms of centric diatoms apparently conform well with that predicted by Stokes equation for the sphere of identical volume and density difference. Distortions of the spherical form, whether as cylinders, plates or other more elaborate forms, result in 2–5-fold reduction in the sinking rate with respect to the equivalent sphere. The most extreme departure among the unicells is in the case of the *Synedra*, a needle-like diatom whose dynamic shape approximates to an attenuate cylinder (the mean length of the cultured cells used in the experiment was 127.9 ($\pm$11.2) μm, having a mean diameter ($\bar{d}$) of 8.87 μm, i.e. the greatest axial linear dimension exceeded d by a factor of 13–16). The single cells of *Asterionella* (*GALD* $\sim$ 66 μm; $\bar{d} = 3.5$ μm), *F. crotonensis* (*GALD* $\sim$ 70 μm; $\bar{d} = 3.4$ μm) and *Melosira italica* (*GALD* of single cell: 19.0 μm; $d = 6.3$ μm) also approximate to cylinders yielding v_s/v' ratios of 2.3–2.8.

Filament formation in *Melosira* has the effect of directly extending the length of the cylinder. The data set from which the entry in Table 10 is extracted is represented graphically in Figure 20: sinking rates of killed filaments were measured from initially mixed suspensions (method 10.7.1 of Walsby & Reynolds, 1980) in which the data were logged for chains of 1–2, 3–4...11–12 and 12+ cells. In Figure 20, the range of results from a series of such experiments (but using the same suspension throughout) is plotted against the corresponding chain length (in terms of cells per filament). The figure also shows the curves describing v_s, for spheres of the same density and volume as *Melosira* colonies of increasing length, and the ratio v_s/v'.

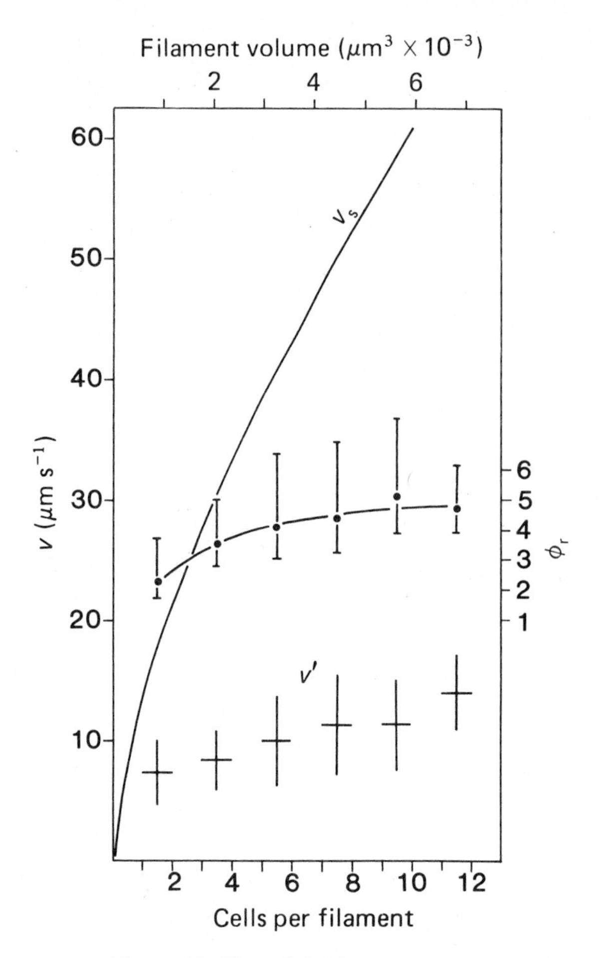

Figure 20. Plot of sinking rates (v') against length (cells per filament) of killed *Melosira italica* filaments and calculated sinking velocities of spheres (v_s) of equivalent volume and density, and the ratio v_s/v' ($= \phi_r$). (Original.)

The information impinges upon the subject of colony formation which is discussed below (see §2.5.4) but, in the present context, it helps to demonstrate the principle by which progressive attenuation of the basic cylindrical form of centric diatoms increases form resistance. As size increases, so does v' (while all other factors remain constant) and this seems contrary to the principle of decreasing sinking rate. Yet by extending in one plane only (i.e. attenuation) the surface area/volume ratio is conserved: instead of increasing hyperbolically against filament length, as does v_s, v' rapidly becomes asymptotic to a value of about 15 $\mu m\ s^{-1}$. This is the calculated velocity of a sphere of a volume equivalent to 1.2 cells

($\sim$ 710 μm^3) whose diameter is then just over 11 μm, nearly twice that of the cylinder. The length of such a filament is 12 cells, or $\sim$ 228 μm, that is $\sim$ 36 times its diameter, whilst its volume is $\sim$ 7120 μm^3. Thus, attenuation of the cylinder permits a 10-fold increase in volume with no increase in sinking rate! Equally, the proportionate increases in sinking rate become smaller for given increases in chain length once it has reached the equivalent of 3–4 cells (length: 57–76 μm; volume: 1780–2370 μm^3); here, sinking rate is reduced (from about 30 to 8 $\mu m\ s^{-1}$) by a factor of $\sim$ 3.8 with respect to the sphere of equivalent volume (diameter 15.0–16.5 μm), while the chain length is 9–12 times greater than the diameter. These deductions support Hutchinson's (1967) worked examples, derived on theoretical grounds.

Many larger planktonic 'algae' increase surface area through major morphological protuberances (e.g. *Staurastrum* spp., *Ceratium*; see Figures 1,2,7), though their relative effect on sinking rates is unclear. Several 'algal' genera characteristically possess narrow projecting spines that have been shown to increase form resistance. The observations of Conway & Trainor (1972) on different *Scenedesmus* strains, some of which bore spines on the end cells of the four-celled coenobia, implied that spined coenobia had a higher form resistance than spineless forms. Similarly, Smayda & Boleyn (1966) showed that spineless pre-auxospore cells of the marine diatom *Rhizosolenia setigera* generally sank faster than those with spines. However, the most conclusive evidence of the contribution of fibrous protuberances comes from Walsby & Xypolyta (1977): they were able to remove the long chitinous fibres from *Thalassiosira weisflogii* cells, using chitinase, and to compare their sinking rates with those of untreated cells. Those without fibres sank nearly twice as fast. In that density of the fibres (1500 kg m^{-3}) was found to be greater than that of 'naked' cell (1120 kg m^{-3}), their presence could scarcely contribute to a reduced mean density.

2.5.4 *Colony formation*

Colony formation among planktonic organisms is an independent characteristic in many taxonomic groups, the colonial structures being formed in many ways. The adpositioning of cells to form discrete new units must always result in a direct increase in volume (and hence of r), whilst surface area is sacrificed at the points of mutual contact between the component cells and so always makes for increases in sinking rate. If other selective pressures favour increased size (e.g. grazing), however, increased sinking rate may be relatively offset if the cell arrangement preserves form resistance. Some colonial forms, for instance those formed by compaction of cells into more or less globular bodies (e.g. in *Eudorina*, *Coelastrum*,

Microcystis), with or without mucilaginous encapsulation, do not conform with this condition.

Many colonies take the form of chains. The simplest structure is that of spherical cells forming filaments resembling a string of beads, as typified by many *Anabaena* spp. Booker & Walsby (1979) investigated the sinking rates of two non-gas-vacuolate mutants of *A. flos-aquae* one of which was characterized by straight trichomes, the other being helically coiled. They showed that mean sinking rate was scarcely increased in chains of 20–50 cells over those having 10–20 cells (4.5 against 3.9 $\mu m\ s^{-1}$) provided that the filament was straight; single gyres of the coiled mutant (~ 10 cells in length) sank at 4.6 $\mu m\ s^{-1}$ but chains of 4.5 gyres (i.e. up to 50 cells) sank at *ca.* 8.1 $\mu m\ s^{-1}$. Booker & Walsby (1979) argued that the evident lower relative form resistance of helical coiling was an adaptation facilitating more rapid buoyancy-regulated movements, a feature shared by other bloom-forming cyanobacteria (Reynolds & Walsby, 1975; see also Chapter 1).

Melosira spp. typically form chains by the mutual adhesion of cells by their valve surfaces, effectively elongating the cylinder. The effect of chain formation on form resistance and reducing the potential increase in sinking rate attending increased size has been discussed above (see also Figure 20). The most beneficial effects of chain formation in terms of reducing sinking rate occur in colonies of up to five or six cells. It may not be entirely coincidental that the modal length of filaments in natural *Melosira italica* and *M. granulata* populations rarely exceeds six to nine cells (though individual filaments may often be considerably larger).

Other genera do not form either chains or globular colonies but their colonies may adopt new shapes such as discoids (e.g. *Pediastrum*) and flat plates, ribbons (e.g. *Fragilaria*) or stellate colonies (e.g. *Actinastrum*, *Asterionella* and certain varieties of *Tabellaria*). In these cases, no theoretical consideration of their form resistance has been adequately elaborated, although it is possible to compare known sinking rates of two species of diatom (*Fragilaria crotonensis* and *Asterionella formosa*) for which density measurements and cell dimensions have been made.

The individual cell of *Fragilaria crotonensis* is essentially cuboid in shape (though cell width varies along its length) and, hydrodynamically, may be expected to behave like a cylinder: the typical cell may have a mean cross-sectional area equivalent to that of a circle 3–4 μm in diameter and a length of 60–80 μm (i.e. 15–27 times the diameter). Comparison of the sinking rates of killed cells with those calculated for spheres of identical volume and density (Table 10) confirms the anticipated effect of cylindrical attenuation ($\phi_r = 2.75$). Colonies are typically formed by parallel

Table 10. *Comparison of measured v′ and values of* v_s *(units:* μms^{-1}*) calculated from Equation 11 for a selection of freshwater algae; the approximate shapes of the algae are also noted*

Alga	Dynamic shape	v'	v_s	ϕ_r	Reference
Chlorella vulgaris	± spherical	—	—	0.98–1.07	(1)
Chlorococcum sp.	± spherical?	—	—	1.02–1.04	(1)
Cyclotella meneghiniana	squat cylinder	—	—	1.03	(1)
Stephanodiscus astraea	squat cylinder:				
	$r_s = 6$–$7\ \mu m$	11.52 ± 0.81	13 ± 1	1.06	(2)
	$r_s = 12$–$14\ \mu m$	27.62 ± 2.64	26 ± 2	0.94	(2)
Cyclotella praeterissima	squat cyliner	9.52 ± 1.04	10.14	1.07	(3)
Synedra acus (cultured)	attenuate cylinder ($GALD = 17d$)	7.31 ± 1.2	29.8	4.08	(3)
Melosira italica					
(1–2 cells)	cylinder	7.43 ± 2.75	17.17	2.31	(3)
(7–8 cells)	attenuate cylinder	11.40 ± 4.11	50.09	4.39	(3)
Asterionella formosa	cylinder	2.98 ± 0.43	7.23	2.46	(3)
(4 cells)	stellate	5.78 ± 0.22	18.21	3.15	(3)
(8 cells)	colonies	7.33 ± 0.57	28.91	3.94	(3),(4)
(16 cells)		10.73 ± 1.23	45.89	4.28	(3)
Tabellaria flocculosa var. *asterionelloides*	stellate 8-celled colonies	10.26 ± 0.97	56.3	5.49	(3)
Fragilaria crotonensis					
(1 cell)	cylinder	3.86 ± 0.21	10.62	2.75	(3)
(11–12 cells)	plate	11.19 ± 0.45	54.08	4.83	(3)

References: (1), Oliver *et al.* (1981); (2), From data in Figure 16; (3), Unpublished data of Reynolds; all components corrected to 15°C from determinations at 15 or 20°C on killed algae; (4), Includes data from Wiseman & Reynolds (1981) for killed *Asterionella* colonies.

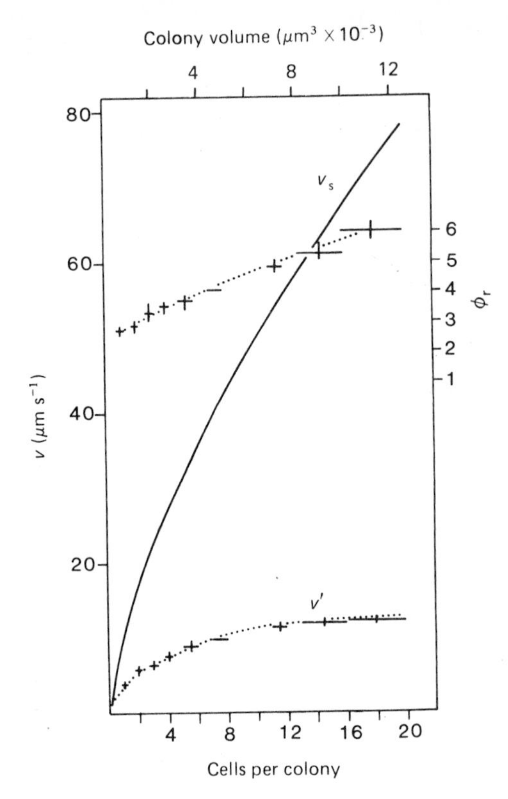

Figure 21. Plot of sinking rates (v') against the size of killed ribbon-forming *Fragilaria crotonensis* colonies (as cells per colony) and calculated velocities (v_s) of spheres of equivalent volume and density. The ratio v_s/v' ($= \phi_r$) is also shown. (Original.)

arrangement of cells adhering laterally over a relatively small area in their central regions. As more cells are added, the cylindrical shape of the colony is progressively transformed to that of rectangular plate, barely more than 3 μm in thickness yet retaining a fenestration occupied by water (if the passage of water through these fenestrations when the colony is sinking is impeded, as seems theoretically possible, then it will have an additional effect of reducing the average density of the plate). A further tendency, observable in some natural populations, is for the plates to be spirally twisted, with between 140 and 200 cells per complete rotation; the effect of this corkscrew formation on sinking rate is unquantified. In the majority of populations observed by the author, however, filaments were either flat or too short, on average, to reveal significant twisting. This was true of the cultured strain used to determine ϕ_r presented in Figure 21, all the filaments being < 20 cells in length. Again, increasing the number of cells

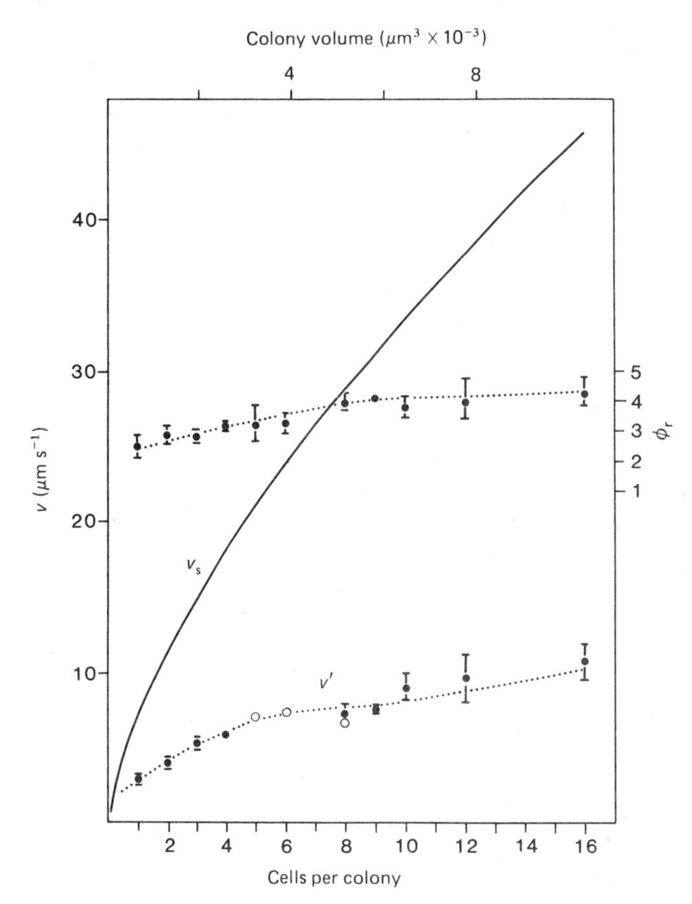

Figure 22. Plot of sinking rates (v') against the number of cells comprising single killed stellate colonies of *Asterionella formosa* and the calculated velocities (v_s) of spheres of equivalent volume and density. (●, collected unpublished data of the author; ○, Wiseman & Reynolds (1981)). The ratio v_s/v_m ($= \phi_r$) is also shown. (Original.)

per colony also increased the sinking velocity but the latter quickly became asymptotic to about 12–13 $\mu m\ s^{-1}$, once the chain length exceeded about 10 cells (when the shape of the plate is nearly square). Above this limit, the curve of ϕ_r approaches a straight line, yielding values up to six in the range considered.

The stellate arrangement of cells to make a flattened spiral, adopted by *Asterionella formosa*, provides a quite different effect upon form resistance (Figure 22). Again, the individual cell is cuboid and may be treated as a cylinder, the length of which (50–70 μm) exceeds the equivalent mean diameter (3.5–4.0 μm) by a factor > 12. The determined coefficient of form resistance (Table 10: $\phi_r \sim 2.3$) is similar to that of the single *Fragilaria* cell.

Up to about eight or nine cells per colony, the sinking velocity of colonies is increased in an asymptotic manner to about 6–8 μm s^{-1}, with a commensurate increase in ϕ_r. Once this point is passed, however, sinking rate begins to increase again, with ϕ_r apparently stabilizing around four. In other words, the supposed advantage, in terms of increased form resistance, reaches a peak when the colonies contain eight or nine cells. This result suggests that the change in overall unit shape from a cylinder to a spoked disc is initially beneficial but that as the gaps are progressively 'plugged' and the disc becomes more 'solid', with almost no increase in its overall diameter, the advantage is lost. That the maximum should occur at about eight cells per colony, a modal ratio commonly observed in both natural and cultured populations of this diatom, furnishes a functional explanation of the usual form of *Asterionella* colonies. If true, the argument would presumably hold for the relatively more massive cells of *Tabellaria flocculosa* var. *asterionelloides* which, unlike the more typical habits adopted by this genus (see Knudsen, 1953), forms eight-armed stellate colonies. Frequently, up to 16 cells can be simultaneously present but the cells remain in mutually adherent pairs, thus preserving the eight-radiate form.

2.5.5 *The significance of form resistance*

The arguments concerning the increase of surface area with volume, either by distortion of shape or the formation of what are, at first sight, bizarre colonial structures, can be advanced as evidence that increased form resistance is subject to powerful natural selection. However, the inescapable fact remains that increased size inevitably increases sinking rate, whereas the hypothetical ideal is that planktonic organisms should reduce their sinking potential. We must therefore presume that there are other pressures that act on size and form. Two possibilities, which are by no means mutually exclusive, are advanced here. One of these is that the distortions of form serve an alternative function, such as the resistance to ingestion by planktonic animals or the improvement of photosynthetic efficiency; the other is that increased form resistance acquires an additional significance in relation to hydrodynamic flow. Although neither of these possibilities has yet been adequately researched, data exist which enable a preliminary discussion.

2.5.5.1 *Resistance to grazing.* The argument in favour of protuberances serving a defensive function invokes the existence of non-planktonic species (e.g. of *Micrasterias*) whose surface projections are highly irregular.

Were they planktonic[1], no doubt they could be cited as extreme examples of the adaptive increase of form resistance! As it is, the significance of the shape of *Micrasterias*, at least, is uncertain, though the possibility that it permits an increase in overall dimensions, whilst still maintaining a favourable surface area: volume ratio, and the attainment of a form that renders it unpalatable by benthic grazers, remains a plausible explanation. Extension of this argument to planktonic species could be justified if it had been shown that the forms possessing high coefficients of form resistance were also proof against grazers. Investigations into the dietary preferences and feeding behaviour of the freshwater zooplankton are still in their infancy, although there is now a body of evidence accruing that the major consumers in lakes are filter-feeding Cladocera (especially among Daphniidae), calanoids (e.g. *Eudiaptomus*) and Rotifera. Many of the species investigated (Gliwicz, 1969; Porter, 1973, 1977; Pourriot, 1977; Nadin-Hurley & Duncan, 1976; Ferguson, Thompson & Reynolds, 1982) definitely show dietary preferences but the selectivity is largely dependent upon size and shape (Burns, 1968*a*, *b*; Ferguson *et al.*, 1982). Most filter-feeding rotifers are restricted to food particles $< 15\ \mu m$ *GALD* (Gliwicz, 1969), whereas individuals of *Daphnia hyalina* can ingest particles of 50–60 μm, or more in the case of flexible filaments of suitably narrow diameter (Nadin-Hurley & Duncan, 1976). Ferguson *et al.* (1982) found that particles $>$ 50–60 μm in two or more planes were unavailable to the same species; they suggested that chains of *Fragilaria crotonensis* of > 13 cells were barred to *Daphnia* which could, however, ingest shorter filaments and also feed relatively freely on the smaller *Asterionella* colonies. The point is that the particular mode of colony formation in these two diatom genera is not a guarantee against grazing animals. On the other hand, increasing size is an advantageous adaptation against consumption by common filter feeders but because it also reduces the relative adsorptive surface area as well as increasing sinking rate, some adaptation to maintain a favourable surface area : volume ratio is necessary if the advantage is not to be lost. That many species of smaller algae (and hence available to the filter feeders) show no apparent morphological adaptation against consumption does suggest that distortion among larger forms is essentially directed towards the conservation of an optimal SA/V ratio (see Figure 7).

There are other possible advantages of attenuation or other distortion of form which should not be ignored. The relatively elongate, curved or digitate horns of certain *Ceratium* species inhabiting warm tropical oceans

[1] There are planktonic species in tropical lakes.

(and which were formerly regarded as adaptations for suspension in less-dense, less-viscous water) may assist the efficient distribution of the photosynthetic pigment within the cell (Taylor, 1980). This might be especially true of freshwater phytoplankton species inhabiting lower levels of the light gradient (see also Chapter 4).

2.5.5.2 *Orientation and entrainment.* The second alternative (2.5.5, above) is that disortions affect the orientation of the alga in relation to hydrodynamic flow. It is well-known that non-spherical bodies moving through liquids tend to adopt a position with the greatest area of projection perpendicular to the motion (Hutchinson, 1967); at the low particle Reynolds numbers ($Re < 0.01$) characteristic of planktonic algae, this principle does not apply, as the experiments of McNown & Malaika (1950) clearly showed. Provided the sinking body is symmetrical, it should maintain any orientation at which it is set. Although there are reports in the literature alluding to the preferred orientation of planktonic algae, Walsby & Reynolds (1980) suggested that these may often have been the artefactual consequence of convection currents. Asymmetrically weighted cells, however, would be expected to re-orientate themselves during fall (cf. Smayda & Boleyn, 1966; Taylor, 1980). Duthie (1965) observed that long-armed *Staurastrum* species persistently re-orientate themselves during descent to the extent that they rotate and tend away from the vertical path; again, it is not clear how much of this effect is due to currents set up within the observation chambers.

However, the behaviour of algae in relation to currents counter to the direction of fall does provide a valuable clue as to their behaviour in natural, open waters; surface layers are beset with motions which, both in scale and direction, greatly exceed the gravitation of algal particles. The imposition of a lateral or upwards current upon a sedimenting alga will be expected not only to re-orientate it with respect to the vertical but, if its velocity significantly exceeds v', to transport it in the direction of flow as well. That is, the particle becomes partially entrained in that flow. It is 'partial' because the particle will continue to settle downwards in relation to the immediately-adjacent water, even if that water is itself travelling upwards at a greater velocity. It is the frequent changes of direction and velocity of the water itself which prolong the suspension of the particle; the more complete the entrainment, the greater is the delay.

In this way, we can speculate upon the crucial role of a high coefficient of form resistance. By resisting the passage of water across its surface, the 'alga' is more likely to remain within a given flow and to be transported by it. The results of some preliminary laboratory experiments on the

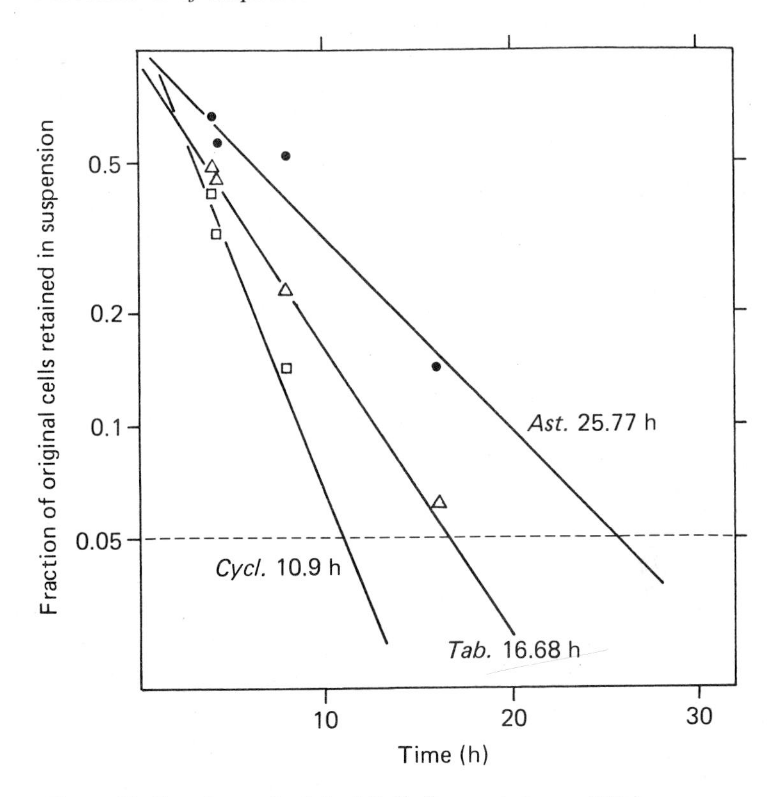

Figure 23. Fractions of original (killed) populations of (●) *Asterionella formosa*, (△) *Tabellaria flocculosa* f. *asterionelloides* and (□) *Cyclotella praeterissima* remaining in suspension in a continuously-mixed vessel (but with intact basal boundary layer), shown on a $\log_{10}$ scale against time. The equations of the fitted regressions were used to calculate the 95% clearance times, which are compared with clearance rates in still water and in mixed water (predicted by equation 10) in Table 11. For *Asterionella*, $F_N = 0.013 - 0.051\ t$; t when $F_N = \bar{2}.699$ is 25.77 h. For *Tabellaria*, $F_N = 1.967 - 0.076\ t$; t when $F_N = \bar{2}.699$ is 16.68 h. For *Cyclotella*, $F_N = 0.073 - 0.126\ t$; t when $F_N = \bar{2}.699$ is 10.91 h. (Original.)

simultaneous behaviour of algae of different form resistance in a mixed water column can be invoked in support of this idea.

The experiments were briefly reported in Walsby & Reynolds (1980). A thick suspension of killed diatoms was pipetted into a 7-l flat-bottomed vessel of water mixed by a continuous jet of compressed air discharged about 50 mm above the bottom. The bubbling was adequate to keep the water constantly mixed and to disperse the diatoms uniformly yet the boundary layer at the bottom of the vessel was kept sufficiently intact to allow diatoms to accumulate visibly there during the course of each

Table 11. *Times taken* (t_e) *for 95% elimination* ($F_N = 0.05$) *from suspensions of killed* Asterionella, Tabellaria *and* Cyclotella *mixed in a cylindrical vessel* ($z = 180$ mm), *predicted from Figure* 23, *compared with the expected elimination time* (z/v', *in h*) *in still water and that predicted for* $t_e = 0.05$ *from Eqn* (10)

	t when $F_N = 0.05$ (h)	v' (from Table 10) ($\mu m\ s^{-1}$)	z/v' (h)	Predicted t_c (h)
Asterionella	25.77	6.76–7.90	6.33–7.34	18.97–21.99
Tabellaria	16.68	9.29–11.23	4.45–5.38	13.33–16.12
Cyclotella	10.91	8.48–10.56	4.74–5.90	14.20–17.67

experiment. Samples were taken periodically in a glass tube (diameter 8 mm) lowered slowly into the water to a predetermined depth (10 mm above the bottom) to enclose a vertical column of the mixed water. With the index finger closing the upper end, the intact column was withdrawn. Loss of diatoms was followed by monitoring the fraction (F_N) of each population still in suspension at each population. Values of F_N are plotted against time in Figure 23, together with the appropriate regression equations. From these, the times taken to achieve 95% specific elimination (i.e. $F_N = 0.05$; log $F_N = \bar{2}.699$) have been calculated and are presented in Table 11. The equivalent times calculated from equation (10), putting $N_0 = 1$, $N_t = 0.05$ and interpolating values of v', abstracted from Table 10, are included.

The results in Table 11 show that the prolongation of suspension by mixing is generally as predicted from the theoretical model developed in §2.3, though the delay in settling seems relatively greater in *Asterionella* and *Tabellaria* than in *Cyclotella*; although the *Tabellaria* and *Cyclotella* had similar mean still-water settling velocities, *Cyclotella* was apparently eliminated more rapidly from a mixed water column (cf. Walsby & Reynolds, 1980). More experiments are required before the conclusion may be justified statistically. Nevertheless, the evident tendency of the species having a high form resistance to take longer to settle than predicted by the model equation ($t_e = 3.0t'$) requires an explanation. Either numbers of colonies are persistently resuspended having once already settled or, if the supposition of an intact, protective boundary layer is correct, some additional factor resists the penetration of this boundary layer by diatoms still in suspension. Since the energy during the imposed turbulence is continuous, currents directed towards the boundary zone must be diverted therefrom. Those particles which are entrained more completely in the flow

Table 12. *Comparison of sinking rates (in* μm s^{-1}*) in healthy* (H) *diatoms with those of killed* (D) *or senescent cells* (S) *from the same population*

	(H)	(S or D)	(S or D)/H	Reference
Asterionella formosa	2.89	D:6.83	2.36	(1)
(5- to 8-celled colonies)	2.31 ± 0.69	S:17.13 ± 12.15	7.42	(2)
	1.97 ± 0.81	D:6.48 ± 3.01	3.29	(3)
	3.12 ± 0.45	S:11.36 ± 2.90	3.64	(4)
	3.04 ± 0.59	D:7.33 ± 0.57	2.41	(5)
Fragilaria crotonensis (11–12 cells)	5.16	D:11.19 ± 0.45	2.17	(5)
Stephanodiscus astraea	7.06 ± 0.98	D:22.35 ± 2.94	3.17	(6)
(diameter unspecified)	8.43 ± 0.98	D:25.29 ± 2.35	3.00	(6)
($r_s \equiv$ 6–7 μm)[a]	9.25 ± 1.36	D:11.52 ± 0.81	1.25	(5)
($r_s \equiv$ 12–14 μm)[a]	9.91 ± 3.33	D:27.62 ± 2.64	2.79	(5)

References: (1), Smayda (1974); (2), Titman & Kilham (1976); (3), Wiseman & Reynolds (1981); (4), Jaworksi *et al.* (1981); (5), Reynolds (unpublished); (6), Reynolds (1973*a*).
(*a*) From data in Figure 16 and Table 10.

will sink from it less easily and, thus, will be more likely to be returned in the general circulation.

2.6 Vital regulation of sinking rate

In addition to morphological features which are arguably adaptations to planktonic existence, short-term fluctuations in buoyancy occur in living cells. The physiological regulation of the numbers of gas vesicles in existence within the cells of planktonic cyanobacteria provides an outstanding example of density control (see §2.5.2.4). Indeed the mechanism can be sufficiently fine-tuned to avert the loss of cells from the photic zone or from the mixed layer, whichever is the deeper. By analogy, flagellate organisms which can swim sufficiently to overcome gravitation may be able to prolong residence within the mixed layer or migrate towards some preferred depth beneath it, through their own controlled movements.

There is considerable evidence that non-motile organisms, especially planktonic diatoms, are also able to regulate their buoyancy, or at least to reduce their sinking rate, and that this control is effected physiologically. Allusion has already been made (§2.4) to the significant increases in natural sinking rates that accompany death, or merely senescence, of diatoms. Some reported instances are presented in Table 12.

Indeed, there is also some evidence that, under certain circumstances, similar cells will have sinking rates intermediate between those of healthy or dead ones (e.g. Wiseman & Reynolds, 1981), and that maximal sinking rates can be induced in otherwise healthy cells very rapidly, in the presence of sub-lethal concentrations of fixatives, such as propanol (Smayda, 1974). It is these characteristics which necessitated the use of suspensions of killed diatoms in the experiments on form resistance, described in §2.5.4, 2.5.5, in order to distinguish the morphological contribution to ϕ_r *per se* from the vital component.

The precise mechanisms determining these changes have yet to be explained. Taking the Stokes equation once again as the basis, it may be argued that killing diatoms or even natural senescence, if it is not accompanied immediately by cell death or cell loss, should not appreciably alter the size, shape, or, by implication, the form resistance of the diatoms in question; certainly these would not account for the observed 3- and 5-fold increases in sinking rate. (Distinction is made at this point between the floccular aggregation of dead or moribund cells which occur spontaneously in the decline phases of natural populations; the 'new' particles will have increased size and, while other components of the Stokes equation remain unaltered, a higher settling velocity than the component parts. Faecal pellets of compacted diatoms also have greatly enhanced settling rates (e.g. Ferrante & Parker, 1977) over their live counterparts.)

Density changes in healthy diatoms have not been adequately investigated. It is known, however, among planktonic algae, that the internal concentrations of condensed photosynthate (e.g. starch, glycogen) and of other storage products do fluctuate in relation to growth rate and nutritional state of the cells (e.g. Fogg, 1975) and to their immediate photic history (e.g. Gibson, 1978). Since carbohydrates, polyphosphate bodies and proteins themselves have high densities (see §2.5.2) it may be presumed that their presence will contribute to the mean overall density of the cell. A simple calculation shows how much of these materials would be required to account for a three-fold increase in sinking rate. If a single *Asterionella* cell ($V = 645\ \mu m^3$) is visualized as comprising only pure water ($\rho = 1000$ kg m^{-3}) enclosed by a wall of opaline silica ($\rho = 2600$ kg m^{-3}) having a total weight of 140 pg (i.e. occupying 54 μm^3), then its overall density is equivalent to:

$$\begin{aligned}\rho_c' &= [(591 \times 10^{-18}\ m^3 \times 1000\ kg\ m^{-3}) + (54 \times 10^{-18}\ m^3 \\ &\quad \times 2600\ kg\ m^{-3})]\ /645 \times 10^{-18}\ m^3 \\ &= 1134\ kg\ m^{-3} \qquad (17)\end{aligned}$$

Its density difference $(\rho_c' - \rho)$ will be 134 kg m^{-3}. For it to treble its sinking rate, by adjustment of density alone, $(\rho_c' - \rho)$ is required to increase to 402 kg m^{-3}, that is, for ρ_c' to increase to 1402 kg m^{-3}. If the silica cell wall is intransmutable, then the contents must increase in weight to 764×10^{-15} kg, whilst the volume they occupy is unchanged (591×10^{-18} m^3). This would require the displacement of 344×10^{-18} m^3 (58%) of the original water by carbohydrate of density 1500 kg m^{-3} (or 155×10^{-18} m^3 by polyphosphate of density 2500 kg m^{-3}, or by 577×10^{-18} m^3 of protein, 1300 kg m^{-3}) to achieve the desired effect. Whilst these solutions are hydrodynamically acceptable, their achievement raises biological objections, not least that the substances are stored intracellularly, yet healthy frustules are normally occupied in part by vacuole sap. Second, the more rapid adjustments in sinking rate (for instance, when cells are killed) are unlikely to result in such condensations of carbohydrate. Third, there is no evidence that the sinking velocities of killed cells are greatly influenced by their pre-treated physiological condition: in other words, density changes in life are unlikely to produce large fluctuations in sinking rate. Fourth, there is the important point that diatoms tend to store fats and oils in their senescent stages, which would be expected not to increase density but, if anything, to reduce it.

It is also of interest that the calculated density of the *Asterionella* cell (1134 kg m^{-3}) is close to the measured value given in Table 8. To invoke density changes as the means of reducing sinking rate by a factor of three, we require a method of reducing mean density to about 1045 kg m^{-3}. One possibility is that live diatoms maintain gas-filled structures analogous to the gas vesicles of the cyanobacteria: following the argument above, for the weight of the entire diatom cell to fall to 674×10^{-15} kg the non-siliceous weight of the *Asterionella* cell would be required to fall to 534×10^{-15} kg, i.e. for 57 μm^3 of the hypothesized water content to be replaced by gas-filled space. No such structure has been observed within the protoplast of diatom cells, though it is not inconceivable that the vacuities in the silica wall might maintain gas-filled spaces. However, with this in mind, I have on several occasions subjected suspensions of healthy diatoms to pressures of 1.3 MPa (the limit of the apparatus available to me) without producing any significant effect on their sinking rates. Yet, these structures, if they exist, have to be sensitive to killing and to ageing of the cell.

The remaining possibility is that diatom cells are somehow able to alter the viscosity of the adjacent medium: for instance, if viscosity was increased, sinking rate would be decreased. Although phytoplankton cells release measurable quantities of organic substances into the water (see Chapter 4), some of which might potentially alter its physical properties (a distinct reduction in surface tension, manifest as an increased tendency to produce detergent-like foam on the water surface, can be associated with waters in which *Eudorina* is abundant), it must be doubted that natural concentrations of diatoms are able to make any measurable impact on the general viscosity. However, if water was immobilized in the immediate vicinity of the cell wall, its effect would be to create a 'new' particle, of diatom + water which, though larger than the diatom itself, would have a reduced average density, in a completely analogous manner to the presence of a hygroscopic mucilage secretion. This theory of 'structural' viscosity is attributed to Margalef (1957), who proposed that the effective viscosity of the adjacent medium is due to the electrical properties of the outer cell wall. By varying the electrical charge (electrokinetic- or zeta-potential) on its outer surface, the cell might regulate its sinking rate. Apart from Margalef's own observations (he detected differences in both polarity and zeta potential between *Scenedesmus* coenobia drawn from actively-growing and darkened cultures), there are many fragments of information which could be invoked to support Margalef's (1957) theory. In the first instance, it must be emphasized that like any other small particles dispersed in an electrolyte (albeit a weak one, lake water) 'algal' cells carry a surface charge, in any case. This is determined, in part, by the ionic strength of the medium. Several workers (e.g. Ives, 1956; Grünberg, 1968; Hegewald, 1972; Zhuravleva & Matskevich, 1974) have used electrophoretic procedures to measure the charge on various algal cells. Ives (1956) showed that each of the 12 freshwater species he investigated carried a consistently negative zeta potential, which fluctuated within comparatively narrow limits and independently of cell size and colonial configuration. The negatively-charged cell wall will attract positively-charged metal ions which are naturally complexed by *y* co-ordinated water molecules, having the general configuration $M(H_2O)_y^{z+}$ (e.g. Davison, 1980). In this way, the conditions for structural viscosity can be met.

In order to *regulate* structural viscosity, however, the planktonic cell must be presumed actively to enhance or to reduce its surface charge in relation to external conditions. Increase in sinking rate, as a consequence of addition of algicides (Ives, 1956) or sublethal doses of butanol (Margalef, 1957) or propanol (Smayda, 1974) to algal suspensions, has been cited as evidence that such controls exist. Moreover, since the development of an enhanced charge density requires the expenditure of energy, it might be argued that it is likely to be a characteristic of physiologically healthy and active cells, rather than of senescent or physiologically damaged ones. Weakening of surface charge might then account for the increases in sinking potential that accompany ageing. In much the same way, short-term fluctuations in buoyancy would be expected to be coupled with photosynthetic rate. Skabiachevskiĭ's (1960) observation that diatom cells are 'lighter' when photosynthesizing has been supported recently by the findings of Jaworski, Talling & Heaney (1981) that increases in the sinking rates of *Asterionella* and *Fragilaria* populations are consequential upon carbon limitation of photosynthetic rate.

Although these observations can be satisfactorily compounded to explain how the phenomenon of a biologically-mediated zeta potential might regulate sinking rate, there is as yet no clear indication that the phenomenon actually exists. With this in mind, Wiseman & Reynolds (1981) set about making simultaneous measurements of electrophoretic mobility (a direct correlative of surface charge) and sinking rate in individually-observed *Asterionella* colonies. Each set of observations was repeated using killed colonies. Although the expected differences in sinking rate, between live and dead cells and between colonies comprising different numbers of cells, were readily realized, in no experiment was any significant change in electrophoretic mobility detected. Indeed, electrophoretic mobilities remained in a similar order throughout the experiments and were unaltered by colony size or by four-fold change in ionic strength. In other words, the changes in sinking rate associated with death, senescence or simply a reduced photosynthetic rate, have *not* been related to altered surface charge, nor to any accountable alteration in structural viscosity. The statement made in the fourth paragraph of this section is still substantially true: the vital mechanisms which control autogenic changes in sinking rate have yet to be explained; further investigation is required.

2.7 Conclusions

The discussion in this chapter has established that the generalized supposition that a low overall sinking rate is an essential pre-requisite for prolonged residence in the plankton is substantially correct. This requirement emanates not from a delayed vertical passage *per se*, but from minimizing the opportunity of escape from the turbulently-mixed layer into lower, non-turbulent layers. The various groups represented in the freshwater phytoplankton achieve sinking-rate reductions in strikingly different ways. Microscopic size is perhaps the most widespread adaptation; many algae have also evolved mechanisms for reducing overall density.

Flagellated algae can use their motility to return to mixed layers or to offset permanent settlement in non-turbulent water. The gas-vacuolate cyanobacteria have developed along a different evolutionary line in which, in some genera, buoyancy regulation is (at first sight, paradoxically) coupled with increased size, compactness and streamlining. These organisms are able to move relatively rapidly through water and, hence, are better equipped to accommodate to fluctuations in turbulent mixing. Non-motile organisms, especially the diatoms and representatives of some chlorophyte orders, appear to be more directly reliant upon entrainment within flow-lines and for which developed form-resistance may be directly advantageous.

Many planktonic organisms, however, retain a behavioural or physiological control over their sinking behaviour. This can be advantageous in two respects: the lower is the sinking rate, the more suspension within a given mixed depth will be prolonged. Equally, when conditions obtaining within that layer are unfavourable, either because of excessive insolation or nutrient depletion, an enhanced rate of sinking can be positively advantageous in contributing to the survival of a specific inoculum within the water body in question. The vital mechanisms of sinking rate regulation by diatoms are imperfectly understood.

3

Spatial and temporal distribution of phytoplankton

'From out our bourne of Time and Place, the flood may bear me far.'
TENNYSON, *Crossing the Bar*

3.1 Distribution patterns: the problem of scale

In order to gain adequate understanding of the life of planktonic organisms and the functioning of pelagic ecosystems, it is desirable to have some insight into the patterns of distribution and the scales of spatial and temporal heterogeneity within the fluid environment, as well as some of the phenomena that contribute to their existence.

The environments inhabited by plankton are manifestly heterogeneous. Temporal changes in mean temperature, irradiance, hydraulic throughput and nutrient availability are among the more obvious variables. Here, the characteristic oscillations can be recognized to occur on different frequency scales – either annually, seasonally or dielly. They also contain smaller, more-frequent oscillations on much shorter scales, of hours and minutes (consider, for instance, the effect of sunshine and flash floods in the examples cited). Vertical segregation of limnetic space is no less apparent: the establishment of thermal stratification is both cause and effect of the physical differentiation of water layers, each having their own identifiable characteristics of motion, temperature, density and salinity. Equally, these various layers are separated by relatively abrupt discontinuities. Gradients of underwater light penetration and hydrostatic pressure also exist with depth and cross from one layer to the next. Under such conditions, discontinuous vertical distributions of planktonic organisms '. . . are therefore to be expected and are commonly encountered' (Harris & Smith, 1977). 'Feedback mechanisms' involving planktonic organisms may accentuate vertical structure: for instance, uptake of nutrients near the surface by organisms that subsequently settle into, and decompose within, deeper layers, contributes directly to superficial depletion and profundal enrichment of nutrients; absorbance and scatter of light by suspended organisms directly enhance light attenuation with depth and the steepening of the light gradient.

In the horizontal plane, striking differences between the physical and chemical characteristics of major current systems and the phytoplankton assemblages they transport can often be detected: these are best known from the oceans (Hardy, 1936; Bainbridge, 1957). The development of methods for continuous sampling devices and techniques for *in situ* measurement of phytoplankton abundance from chlorophyll-fluorescence (Lorenzen, 1966; George, 1976; Herman & Denman, 1977) has enabled a considerable expansion of the study of horizontal 'patchiness'. Spectral techniques have also been evolved for partitioning the total variance among specific contributions occurring at different characteristic length scales (Platt, Dickie & Trites, 1970; Platt, 1972; Denman & Platt, 1976; Platt & Denman, 1975, 1980; Fasham & Pugh, 1976). Sufficient analyses of this kind have been carried out for it to be supposed that horizontal patchiness occurs as the rule rather than as the exception (Harris & Smith, 1977). Large-scale discontinuities in distribution have been identified in coastal seas and embayments (e.g. Bernhard & Rampi, 1965; Platt, 1972, 1975; Horwood, 1978; Lekan & Wilson, 1978) especially where these are influenced by localized upwellings of deep, cold currents (Lorenzen, 1971) or by tidal displacement of stratified layers (e.g. Pingree, Holligan, Mardell & Head, 1976). Comparable discontinuities, both in space and time, have been recognized in comparatively large lakes (> 100 km^2) including Erie (Verduin, 1951; Glooschenko, Moore & Vollenweider, 1974), Ontario (Munawar & Munawar, 1975), St Clair (Leach, 1972), Kinneret (Pollingher & Berman, 1975), Tahoe (Richerson, Powell, Leigh-Abbott & Coil, 1978) and Memphrémagog (Watson & Kalff, 1981). Often, these relate to identifiable limnological characteristics of the lakes concerned, such as point sources of nutrient enrichment and seasonally-dominant circulation patterns, that may be effective over tens of kilometres and over several consecutive months of each year.

In smaller lakes, significant variations in the horizontal distribution of planktonic species are observable over shorter periods (see for instance, Colebrook, 1960; Small, 1963; Horne & Wrigley, 1975; Powell *et al.*, 1975; Heaney, 1976; George & Edwards, 1976; George & Heaney, 1978; Sandusky & Horne, 1978), the discontinuities being more continuously related to the interaction of wind speed and direction with the thermal stability of the water column; nearshore upwellings, downwellings and seiches contribute to this effect (e.g. George, 1981*a*, *b*). The downwind drifting of buoyant organisms (e.g. planktonic cyanobacteria, as well as some dinoflagellates and *Botryococcus*, under certain conditions) can give a particularly spectacular impression of localized abundance of their biomass (e.g. Small, 1963; Reynolds, 1971, 1978*b*; Yamagishi & Aoyama, 1972; Nakamoto, 1975; Baker & Baker, 1976; George & Edwards, 1976).

There is evidence that horizontal clustering also occurs on much finer scales, of about 1–12 m (e.g. McAlice, 1970; Sandusky & Horne, 1978; Therriault & Platt, 1978; Therriault, Lawrence & Platt, 1978) but that these are relatively transient and patterns may be frequently modified or suppressed by wind-stress. In each of these examples, the temporal persistence of the discontinuity is directly related to the spatial dimension across which it fluctuates.

Although the causes of patchiness are still incompletely understood (see §3.3), there is little doubt that much of the variance relates to water movements, in the broadest sense. The important points to be stressed are that (i) the fluid environment inhabited by plankton should not be regarded as being homogeneous and (ii) the spatial and temporal scales of variability are interlinked (Bowden, 1970; McNaught, 1979; Harris, 1980*a*). It follows that environmental variability has a considerable impact upon phytoplankton ecology, though the effects alter with scale. At one extreme (large-scale variation), the fluctuations may be climatically determined: here we may cite major changes in the planktonic flora of lakes occurring within post-glacial time or in response to long-term changes in external nutrient-loading (e.g. Haworth, 1969, 1980). Oscillations with a wavelength of one year distinguish seasonal changes in community composition and productivity that are, to some extent, recapitulated in successive years. At shorter wavelengths, day-to-day fluctuations in the physical environment (thermal structure, irradiance input) affect populations already in existence, perhaps selectively influencing growth rates by alternately providing and overriding localized microhabitats offering specifically preferred conditions. Still finer scales will distinguish between the rates and directions in which existing cells are transported; these will alter in a matter of metres and minutes. Thus, we can envisage that cells being carried rapidly across the full depth of the illuminated layer will be exposed to an irradiance spectrum ranging from the maximum available to zero, all within a matter of minutes (cf. Farmer & Takahashi, 1982). Here, the response will be essentially physiological (Harris, 1980*a*): intracellular metabolism (e.g. photosynthesis, respiration, excretion) will alter, but the full range of conditions can be experienced by the same organism.

To a greater or lesser extent, these various fluctuations characterize the habitats of freshwater phytoplankton. The most relevant scales depend upon the particular aspect under study (e.g. production, growth, seasonality, floristic change), although the scales inevitably interact and are never mutually independent. Since this book is largely directed towards the ecology of populations, upon which all scales impinge, it is important to select a time unit in which terms the effects of environmental fluctuations may be judged. Here, I shall adopt *cell generation time*, that is, the mean time period in which a newly-formed cell grows up and itself divides into daughters; this may vary between several hours and several days, according to species and conditions. Then, the question that may be asked is 'which scales are *perceived* by the intact cell?'

Resolution of this question may be analogized to Hutchinson's (1967) deliberations on structural diversity in aquatic habitats. It is possible to subdivide a habitat (e.g. a lake) into smaller (e.g. its littoral zone) and smaller (e.g. beds of submerged macrophytes) component biocenoses and, eventually into microhabitats (say, the leaf axils as opposed to their laminae or petioles) which, like a mosaic of fragments, together contribute to the single, whole lake. Hutchinson visualized a stretch of shallow water with a sparse cover of rooted, submerged macrophytes growing in isolated clumps. To the fish swimming near these plants, the area appears to be '*homogeneously diverse*', that is, it is perceived as being uniform, the mosaic elements being small relative to the fish's range. To the rotifer living its entire life on one or two leaves of one of the plants, slowly probing its way between surface irregularities and a forest of attached algal filaments, the same area is '*heterogeneously diverse*': its range ends at the leaf edge beyond which lies the totally hostile and unfamiliar environment of an adjacent-mosaic fragment. The presence of other plants nearby, possibly supporting other rotifers of the identical species, is almost irrelevant to the rotifer. The leaf in question is its perceived world.

In the same way, the environmental fluctuations perceived by the phytoplankton cell during its brief life-span as a separate entity will be those determining whether it will obtain sufficient materials from which to assemble into additional biomass; whether it will receive adequate light, carbon and other nutrients within the flow pattern it finds itself; whether it is likely to be irretrievably lost from suspension and settle out to the depths below. Large-scale fluctuations (i.e. those occurring on scales exceeding two to three months) will be perceived as being temporally heterogeneous, of relevance only to past and future generations. Equally, the fine, high-frequency variations, emanating from turbulent circulation will be perceived as being temporally homogeneous: these are features of the whole environment of the cell which it must tolerate or to which it must adapt within its life span. The response will be manifest over a longer time scale; the scale perceived will be that of the ambient conditions. The critical scales to which the rate of population increase will respond will be those on which the ambient 'equilibrium' conditions are shifted or perturbed. These will include alterations of the imposed pattern of flow, of thermal structure, of the underwater light field and the pools of biologically-available nutrients therein. Such scales, to which the existing pattern of population development will react, will be measurable in terms of, at least, hours and, at most, days. This scale range is uppermost in the following consideration of spatial and temporal distribution patterns.

3.2 Vertical distribution

Discontinuities in spatial distribution of phytoplankton that persist for hours or days are probably best known in the vertical plane (i.e. with depth), largely because they have a longer history of study. Several early researches – now regarded as classical – followed changing specific vertical distributions through time (e.g. Ruttner, 1938; Findenegg, 1943); some later studies pursued a similar approach to great advantage, epitomized by Nauwerck's (1963) monograph on the plankton of Erken, Sweden. The methods of data presentation have evolved over the years, the 'cylindrical curves' (where the data are plotted as two-dimensional representations of conceptualized three-dimensional models) having largely given way to 'contour maps' on time v. depth plots, the isopleths linking points of identical concentrations (e.g. Lund, 1959; Lund, Mackereth & Mortimer, 1963) or to histogrammic figures blocking data for individually sampled depth ranges (e.g. Irish, 1980). These plots nevertheless depend upon a large number of individual samples being sequentially collected (in, for instance, Friedinger, Ruttner or van Dorn samplers) and analysed, each requiring tedious enumerations of all or selected species represented. The revolution in direct ship-borne recording of vertical profiles of horizontal fixed light beams (transmissometry, turbidometry: Baker & Baker, 1976; Jones, 1977; Talling, 1981) has alleviated much of this effort, yielding effective and reliable results for monospecific populations or ones overwhelmingly dominated by a single species. For mixed populations, however, where the specific proportions may alter with depth and, in any case, may have different properties of light-scattering, there is no substitute for the earlier methods. These various methods of presentation are exemplified in the instances illustrated below.

Specific vertical distributions can vary substantially with depth, with time and in relation to physical segregation of the water column. Moreover, the observable patterns are, to a greater or lesser degree, isolated instances in a dynamic process of change, like still frames from a movie. The examples given here have been selected to illustrate the various interactions occurring between organisms (that are typically either negatively, positively or neutrally buoyant) and diffusive motions in water layers whose vertical velocities are either much greater than or comparable to the intrinsic velocities of the specific particles in question. The patterns become increasingly complex when gradients of light attenuation or nutrient availability are superimposed: that the water is sufficiently quiescent for intrinsic movements to manifest themselves may or may not indicate that the focus of a particular organism's movements represents the depth range within which the optimal combination of its specific growth requirements obtains. However, it is recognized that the observed behaviour will be consequential upon specific adaptations to meet the scale and frequency with which these events are likely to occur.

3.2.1 *Non-motile, negatively buoyant algae* ($\rho' > \rho$)

From our consideration of the factors involved in the suspension of phytoplankton (Chapter 2), it will be clear that planktonic diatoms furnish the prime example within this category. Not only are they usually (if not always) negatively buoyant, but they may spontaneously increase their mean intrinsic rate of sinking under certain, supposedly unfavourable, conditions. Many published examples may be cited to illustrate diatom sedimentation that is apparently accelerated by the onset of thermal stratification (e.g. Lund, 1959; Lund *et al.*, 1963; Reynolds, 1973*a*; Reynolds & Wiseman, 1982). Species of *Melosira* are particularly prone to sinking losses (Lund, 1966*a*, 1971), even in quite large lakes (Lund, 1954). Sinking loss is by no means confined to old, non-growing populations: Reynolds & Wiseman (1982) showed that net increase of *Fragilaria* can occur in small lakes provided that the losses from the surface layer do not exceed the recruitment of new cells (see also Chapter 7). Only one instance of diatom sinking is depicted here (see Figure 24): it traces changes in the vertical distribution of *Asterionella* cells in a Lund Tube (Blelham Enclosure A: see Lund & Reynolds, 1982) over a five-week period in which thermal stratification first intensified (as a result of surface warming and low vertical down mixing) and then became depressed by increased epilimnetic mixing; the latter came too late, however, to resurrect the sinking population. The concentrations within successive one- or two-metre layers were determined from one-metre Friedinger-collected samples; the data are plotted on a normal scale. Whilst the surface layer remained isothermally-mixed to a depth of about 2.5 m (5 May), the concentrations in the first three one-metre layers were similar. Between 12 May and 19 May, the mixed depth decreased, with thermal layering extending almost to the surface. The surface water was rapidly voided, the

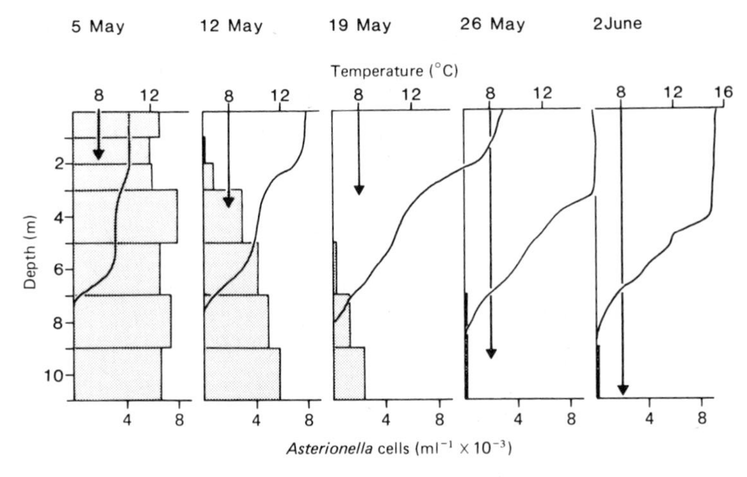

Figure 24. Changes in the vertical distribution of *Asterionella formosa* cells (represented as histograms) in Blelham Enclosure A, in relation to temperature gradient (curved lines) and Secchi disc extinction depth (vertical arrows), during May 1980. (Original, based on author's unpublished data.)

downward-moving 'boundary' of diatoms suggesting a minimum *in situ* sinking rate of 0.3–0.6 m d^{-1}. Just over half the cells present on the 5 May (integrating areally, *ca.* 7.59×10^6 cm^{-2}) had been lost from the column by 12 May. Assuming uniformity of sinking rates and initial vertical homogeneity of cell distribution, a mean effective sinking rate of not less than 0.81 m d^{-1} can be deduced. Both solutions are commensurate with *in vitro* sinking rates determined experimentally (see Chapter 2). That they were apparently manifest in the natural population is, in part, due to the absence of turbulent vertical transfer and insufficient diffusive upwelling through all but the top few centimetres of the stratified column. In other words, the turbulent layer (as evidenced by minimal gradients of temperature and, hence, of density, against depth) is so relatively shallow that the opportunities for particles to escape from it are proportionately increased. Moreover, the behaviour of the particles will be related principally to whether the flow is turbulent or laminar, rather than to its actual velocity.

This hypothesis may be supported by comparing the observed changes in the vertical distribution of inert, conservative (non-living) and approximately uniform particles (*Lycopodium* spores), introduced into the surface flow of a Lund Tube on three occasions (each characterized by differing thermal stabilities), with the distributions predicted from the simultaneous application of the appropriate model solutions (Smith, 1982) describing the fractions of material remaining in suspension in the mixed and stratified

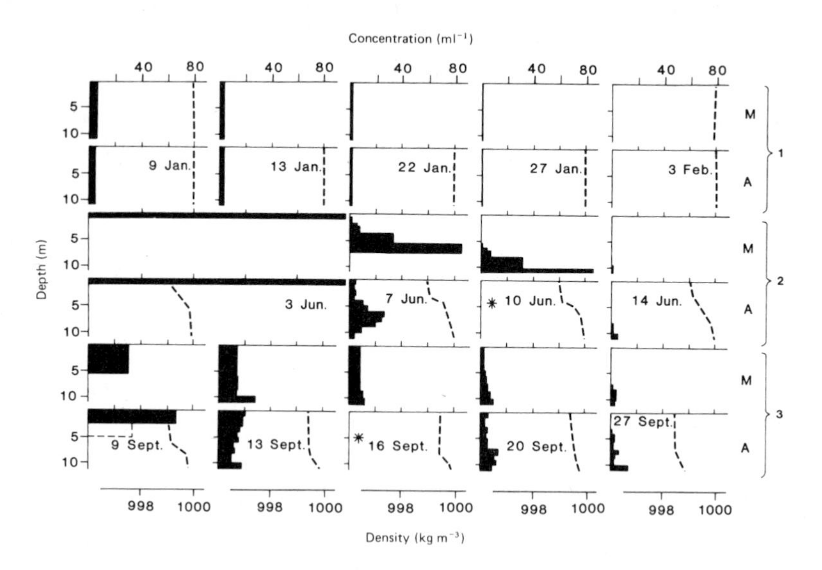

Figure 25. Modelled (M) and actual (A) depth-time distributions of *Lycopodium* spores (of predetermined sinking characteristics) introduced at the water surface of Blelham Enclosure B on three occasions (1: 9 January; 2: 3 June; 3: 9 September) during 1976, under differing conditions of thermal stability. *Lycopodium* concentrations plotted as *Asterionella* in Figure 24; density gradients shown by dashed lines; asterisk indicates that no field data are available. (Original, based on data in Reynolds, 1979*a*.)

layers, of given depths. Predetermined intrinsic sinking velocities, corrected for density and viscosity of the water, were interpolated (Reynolds, 1979*a*). Excerpts of the predicted distributions are directly compared with the corresponding actual distributions in Figure 25: they show a good general agreement. Detail differences are evident: these may be attributed to the incorrect assumption that actual sinking rates are uniform and to compounded errors in sampling and counting. Though lacking refinement, the model solution represents an acceptable description of particle dispersion: ordered sinking in non-turbulent layers; randomized distribution within the turbulent layers, from which particles are nevertheless lost, by first-order reaction, in the proportion v'/z_m.

The critical point at which ordered sinking gives way to turbulent 'dilution' from a mixed system can be approximated in physical terms: as turbulence, when the Reynolds number of flow (R_f) exceeds 500–2000; as stability, when Richardson's number (Ri) is < 0.25; as vertical eddy diffusivity, when K_z exceeds 10^{-2} to 10^{-3} cm^2 s^{-1}. At these critical limits, the density gradient ($\delta\rho/\delta z$) can be scacely greater than 0.02 kg m^{-3} m^{-1} (or 0.02 kg m^{-4}, corresponding to a temperature gradient of

$\sim 0.64°C\ m^{-1}$ in the range 5–10°C, $0.11°C\ m^{-1}$ between 15 and 20°C, and $0.07°C\ m^{-1}$ between 25 and 30°C). If the density gradient is $\gg 0.02\ kg\ m^{-4}$, then the water will be sufficiently stable to permit effective settling velocities to approach intrinsic values at the same temperature. The duration of the stability is equally important, especially when the stratification is periodically subject to increased wind-driven horizontal currents during parts of the day, as has been observed in some tropical lakes (e.g. Ganf, 1974*a*) and in temperate lakes during periods of anticyclonic weather (George & Heaney, 1978).

Richardson's number relates the gradients of horizontal velocity and density; the stability condition (when the energy available is insufficient to meet the work required to overcome the layering) is that $Ri > 0.25$. Rewriting equation (3) and interpolating $Ri = 0.25$, $(\delta\rho/\delta z) = 0.02\ kg\ m^{-4}$ $g = 9.81\ m\ s^{-1}$ and $\rho = 999\ kg\ m^{-3}$, we can deduce that

$$(\delta u/\delta z)^2 = 9.81 \times 0.02/999 \times 0.25\,(m\ s^{-1}\ m^{-1})^2 \qquad (18)$$

whence, the critical velocity gradient that can be withstood:

$$\delta u/\delta z \sim 0.028\ m\ s^{-1}\ m^{-1}$$

The mean horizontal current velocity in a stably-stratified layer is likely to be generally within the range 0 to $0.02\ m\ s^{-1}$. Clearly, a surface horizontal current velocity of $0.028\ m\ s^{-1}$ would be insufficient to overcome a stability at one metre depth, whereas wind-driven surface flows of $0.04–0.05\ m\ s^{-1}$ would generate the required energy to overcome the density gradient. Equally, if the stability occurred at 2 m depth, it would be liable to be overcome when surface flows exceeded $0.06–0.08\ m\ s^{-1}$.

The ratio of surface current speed (u_0) to wind speed (W) is not constant (e.g. Haines & Bryson, 1961; see also George & Edwards, 1976). At wind speeds between 1 and $4\ m\ s^{-1}$, u_0 increases little, from $\sim 0.025\ m\ s^{-1}$ ($0.025\ W$, when $W = 1\ m\ s^{-1}$) to $\sim 0.036\ m\ s^{-1}$ ($0.009\ W$ when $W = 4\ m\ s^{-1}$). When $W > 4\ m\ s^{-1}$ (which wind speed is sufficient to generate Langmuir rotations), u_0 remains approximately in direct proportion wind speed ($0.008–0.009\ W$). Thus a wind of $< 4\ m\ s^{-1}$ is insufficient to break down a pre-existing density gradient of $0.02\ kg\ m^{-4}$ at 1 m, or to prevent one developing under suitable irradiance conditions; when $W = 8–9\ m\ s^{-1}$, they may survive only at depths of > 2 m. Clearly, the work required to overcome a given density gradient increases with its depth beneath the surface; ultimately, the breakdown of limnetic density gradients depends proportionately more upon the spontaneous weakening of the gradient itself (for instance, through cooling of the upper mixed layer) as depth is increased. Thus, it is possible to distinguish between major segregations of water masses into epilimnia, metalimnia and hypolimnia, which may survive as mutually separate entities for months, and the formation of layers ('microstratification') nearer the surface, which may survive only for hours, or, at most, days on end. The depth range in which short-term microstratification (recognized here by a density gradient exceeding 0.02 kg m^{-4}) can persist will often depend largely upon the wind speeds experienced and changes of heat content in the upper layers. The stronger the gradient, so the wind speed required to overcome it is greater, in accordance with the Richardson equation.

To the non-motile particle, it is the persistence of the microstratification as much as its vertical position which can be critical. If the intrinsic settling rate (v') of a diatom is 0.3 m d^{-1}, then a density gradient of > 0.02 kg m^{-4} through the upper one metre need persist for only 3.1 d before the layer is 95% cleared of that diatom. On the other hand if the gradient is broken down overnight (by cooling) and reforms by day such that the same layer is mixed at least once every 24 h, it will take more than eight days to achieve 95% elimination. If the layer is continuously mixed, by turbulence, to a depth of one metre, so that the 'dilution' model applies throughout, it will take 9.9 days. These times will be shorter if the intrinsic sinking rate of the diatoms is physiologically enhanced (see Chapter 2). Cell division will not significantly alter the time taken to clear the 1-m layer if it is continuously microstratified, but the survival of diatoms in the 1-m layer will be prolonged should mixing occur. For no net change of diatoms to occur in a continuously mixed 1-m layer the cells must increase by a mean factor of 1.35 d^{-1} (see also §7.3)

In the light of these considerations, it is not difficult to understand the distribution of the *Asterionella* population represented in Figure 24. Given the formation of a persistent density gradient at *ca.* 2.0 m, an intrinsic settling rate of 0.81 m d^{-1} and zero growth, it would take just seven days before the upper 2-m layer was effectively depleted of 95% of the *Asterionella* cells; in fact, the data indicate a decline from 1253×10^3 to 14.8×10^3 cells cm^{-2} (i.e. 98.8% elimination) between 5 and 12 May.

We may also conclude that prolonged suspension of non-motile, non-buoyant cells will always be resisted by shallow microstratification; it is no coincidence that diatoms tend to be favoured in well-mixed water columns (although other factors are involved in determining how well they can grow) where microstratified density gradients are transient. If the latter are persistent (such as in the metalimnion of a seasonally stratified lake), then the depths at which they are located may become critical. One of the most convincing demonstrations of this principle is provided by the depth-time survey of the vertical distribution of *Asterionella* in Windermere carried out by Lund *et al.* (1963), reproduced here as Figure 26. Verticality of the isopleths implies uniform vertical distribution; their frequency

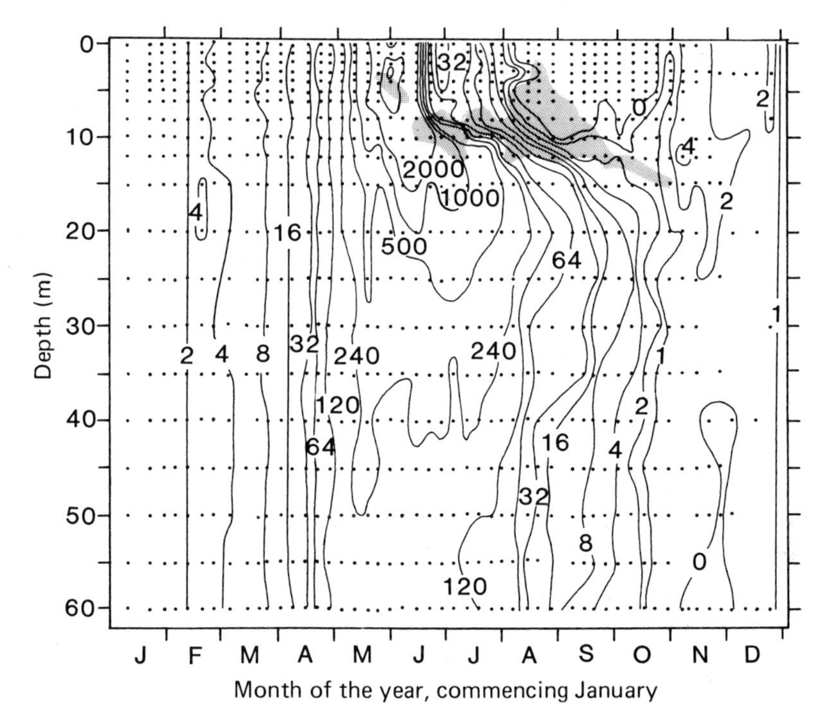

Figure 26. The vertical distribution of *Asterionella formosa* cells in Windermere (North Basin) through 1947. Isopleths in live cells ml^{-1}. The shaded area represents the extent of the metalimnion. (Redrawn from Fig. 2 of Lund *et al.*, 1963.)

represents the exponential rate of change. Horizontality during the stratified period indicates increasing differences in the concentration of cells with depth. The final elimination of cells (to $< 1\ ml^{-1}$) from the upper 10 m occurred in August when the microstratified layer came within three metres of the water surface.

3.2.2 *Positively buoyant algae* ($\rho' < \rho$)

The vertical distribution of buoyant organisms, which include many of the planktonic cyanobacteria during at least some stages of their development (usually, though not exclusively, their later stages), is likely to be subject to similar constraints to those experienced by non-buoyant forms, save that they will float, rather than sink, through stable layers. A further difference is that, potentially, they are likely to remain in the water column, rather than to settle irretrievably from it. The qualification 'potentially' is used since buoyant organisms at or near the surface are liable to removal from suspension by shoreward wind-drift (see §3.3).

It has long been appreciated that the formation of surface blooms

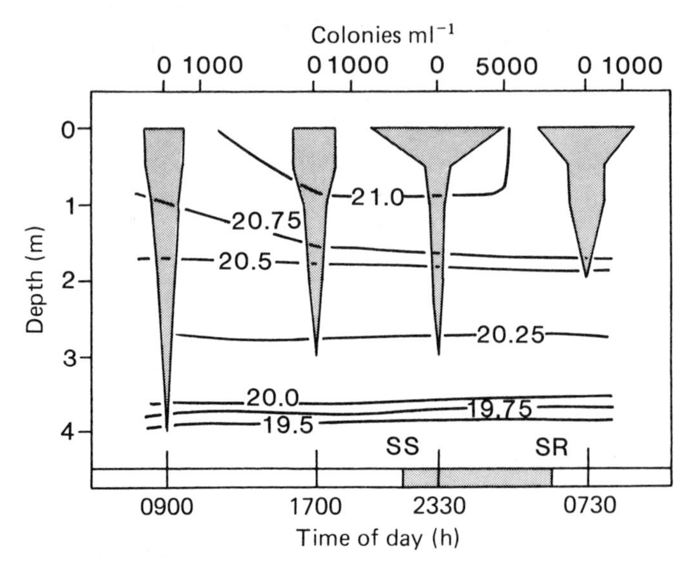

Figure 27. Changes in the vertical distribution of *Microcystis aeruginosa* colonies, shown as 'cylindrical curves' (equated with laterally-viewed solid figures, whose diameters are proportional to the cube root of the concentration, in colonies ml^{-1}) in relation to thermal stratification (isopleths in °C) during 28 and 29 July, 1971 (SS, sunset; SR, sunrise). (Original, based on data in Reynolds, 1973*b*.)

involving species of *Microcystis*, *Anabaena*, *Aphanizomenon*, *Gloeotrichia*, *Coelosphaerium* and *Gomphosphaeria* are prone to form in still, windless conditions (e.g. Griffiths, 1939). The mechanisms of bloom-formation are not straightforward; rather, they depend upon the coincidence of three preconditions – a pre-existing population, a significant proportion of the organisms containing sufficient gas vesicles to render them buoyant, and stability of the water column (for a full review, see Reynolds & Walsby, 1975). The present discussion assumes that the first and second of these conditions are satisfied and that bloom formation depends upon suitable hydrological stability. This third condition will be met by an abrupt reduction in near-surface turbulent flow, i.e. that $Ri > 0.25$. One example is presented in Figure 27, which traces (by means of a series of cylindrical curves) the changing vertical distribution of *Microcystis* colonies in a small lake during one 24-h period in July, 1971. The *Microcystis* colonies were, on average, buoyant throughout this period, having a mean flotation rate of $\sim 9\ \mu m\ s^{-1}$ (or $1\ m\ d^{-1}$) increasing almost two-fold during the hours of darkness (Reynolds, 1973*b*). The metalimnion was located between four and 6 metres depth but the epilimnion was secondarily microstratified with a maximum density gradient of $\sim 0.1\ kg\ m^{-4}$ developing between one and

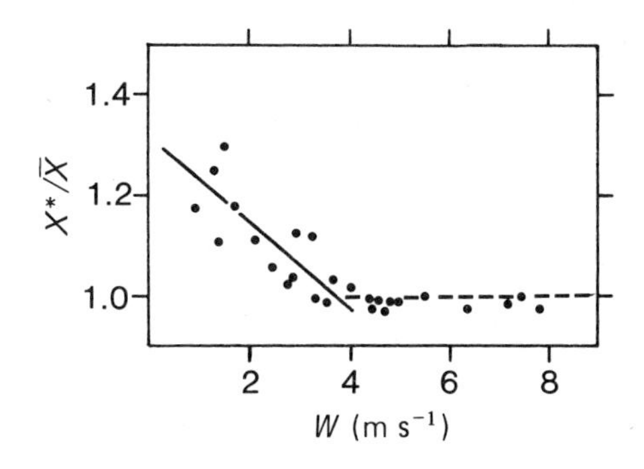

Figure 28. The relationship between vertical patchiness of *Microcystis* in the Eglwys Nynydd reservoir and wind speed (W). (Redrawn from Figure 8 of George & Edwards, 1976.)

two metres depth during the day. A light breeze (estimated to have been 1–2 m s^{-1}) was blowing until early evening on 28 July after which it became completely windless. The third precondition to bloom formation, namely a reduction in the velocity gradient such that $Ri > 0.25$, was thus fulfilled. *Microcystis* colonies, hitherto distributed uniformly within the top 0.5 m (mean concentration: $\sim$ 160 colonies ml^{-1}) rapidly began to accumulate at the surface, as the profile at 23.30 hours shows. The 37-fold increase in surface concentration (to 5940 colonies ml^{-1}) was presumably sustained by recruitment of colonies from below, nominally requiring the complete flotation of all the colonies within the top 0.37 m at 17.00 h; for this to have occurred within 6.5 h requires a minimum effective flotation rate of 15.8 μm s^{-1}. The available data may be invoked to explain the observation. It is of additional interest that, presumably as a result of nocturnal cooling and localized 'overturn', the buoyant population was partially redispersed within the upper one metre before morning.

In their analysis of a series of discontinuities in the vertical distribution of *Microcystis* in a small, shallow reservoir (Eglwys Nynydd; mean depth: 3.5 m; area: 1.01 km^2) over a two-year period, George & Edwards (1976) plotted the ratio of mean crowding ($\overset{*}{x}$) to mean density ($\bar{x}$), after Lloyd (1967), where $\bar{x}$ is the mean concentration in a three-metre standard water column and $\overset{*}{x} = (\bar{x}+s^2/\bar{x}-1)$, ($s^2$ being the variance between individual samples from intermediate depths), against contemporaneous wind-speed recordings. They showed (see Figure 28) that significant vertical patchiness occurred only when $W \leqslant 3.7$ m s^{-1} (i.e. when $u_0 < 0.043$ m s^{-1}). This abstraction agrees with the current speeds critical to sinking behaviour deduced above (§3.2.1).

3.2.3 *Neutrally buoyant ($\rho' \sim \rho$) and motile 'algae'*

The use of the term 'neutral' in this section is relative, merely serving to distinguish those organisms whose effective density is more nearly similar to that of the suspending water than in the instances so far considered, or whose behaviour enables them to regulate their own vertical position. Apart from one or two striking exceptions, it is inconceivable that phytoplankton cells, composed of materials whose aggregate density exceeds that of most natural freshwaters, could continuously maintain a density which was always exactly equal to that of the adjacent water (whose density varies with temperature and solute content). Nor, indeed, is this desirable: the ability to move relative to the adjacent medium (regardless of the velocity and direction of the particular parcel of water) is itself advantageous to the maintenance of diffusion gradients, conducive to the uptake of dissolved nutrients and gases and to the expulsion of waste. Most non-motile algae, then, strictly fall within the previous two categories though they may transiently approach isopycny (equal density) against strong external density gradients. The present category is taken to include those non-motile representatives of the Chlorococcales (especially), whose relatively small size (r) determines that their intrinsic settling (v') is low, and those larger colonial forms endowed with mucilaginous sheaths whose presence reduces average density of the whole body. It also includes (motile) flagellate organisms (from among the green algae, the chrysophytes, the dinoflagellates and the cryptomonads), whose controlled movements permit the effects of gravity to be countered, enhanced or negated so that they can migrate towards, or maintain station within, discrete layers in the water column. Vertical migrations and station maintenance are also achieved by gas-vacuolate prokaryotes (including cyanobacteria). These organisms, alone, genuinely regulate their overall density, by means of a delicate control of their intracellular gas-vesicle content (see §2.5.2.4).

Sequences in the vertical distributions of two non-motile green algae (the small unicellular *Ankyra* and the palmelloid colonies of *Sphaerocystis*) are represented in Figure 29. In both cases, the algae apparently remain homogeneously dispersed within the epilimnion. Though some sinking across the metalimnetic density gradient is evident, the loss of cells from the epilimnion is evidently offset by increases in the population. Nominal sinking rates of 0.1–0.3 m d^{-1} may be deduced from the downward spread of the higher specific concentrations, though these may well be erroneous if the algae either increased their numbers by growth at depth or were eliminated from the water by animals (as food) or through death and bacterial decomposition. Neither species was well represented on the surface deposits (Reynolds & Wiseman, 1982, and unpublished), possibly

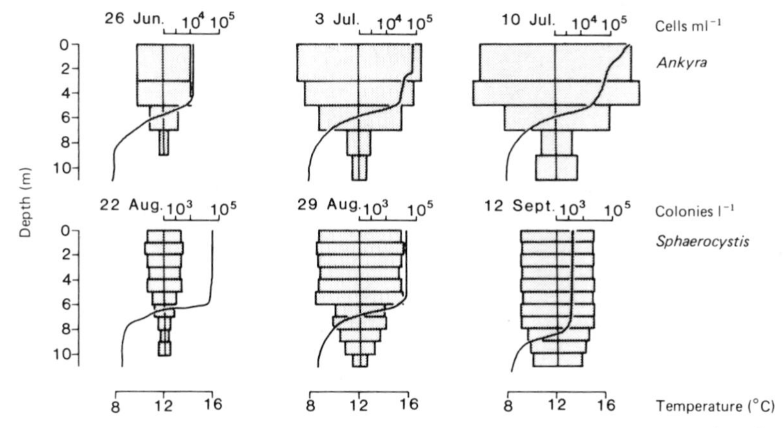

Figure 29. Instances in the vertical distribution (shown as stacks of 'cylindrical curves', one for each 1-m layer), relative to stratification, of (above) a small, non-motile unicellular chlorophyte, *Ankyra judayi*, in Blelham Enclosure A, during 1978 and (below) a palmelloid colonial chlorophyte, *Sphaerocystis schroeteri*, in the same enclosure during 1977. (Original, based on data included in Reynolds & Butterwick, 1979 and Reynolds & Wiseman, 1982.)

because they rapidly disintegrate (Reynolds, Morison & Butterwick, 1982*a*). Both species are always likely to become freely distributed within water layers liable to intermittent turbulent mixing and to settle therefrom only slowly.

The swimming movements of motile organisms can appear very impressive (in terms of unit lengths transversed per unit time) when observed in life under a microscope. In reality, the rates of progress are, at best, in the order of 0.1–1.0 mm s^{-1}. These movements are too feeble to overcome wind-driven current speeds having velocities an order or two greater. Nevertheless, the vertical direction of the movements will be important, for if the intrinsic movements were (say) all in the downward direction, the predicted effect would be analogous to the sinking of non-motile particles. Vertical migrations would always be more effective in non-turbulent layers and the latter are essential if vertical station is to be even approximately maintained.

Many descriptions of the vertical distribution of flagellates are available in the literature (e.g. Nauwerck, 1963; Moss, 1967; Baker, 1970; Fee, 1976); a further four examples (all drawn from observations on populations in the Blelham Enclosures) are presented in Figure 30. The first of these shows a typical diurnal distribution of *Eudorina elegans*[1] colonies during

[1] The specific epithet is used in the wide sense of Ehrenberg's description; the alga in question should be called *E. unicocca* following the revised nomenclature of Goldstein (1964).

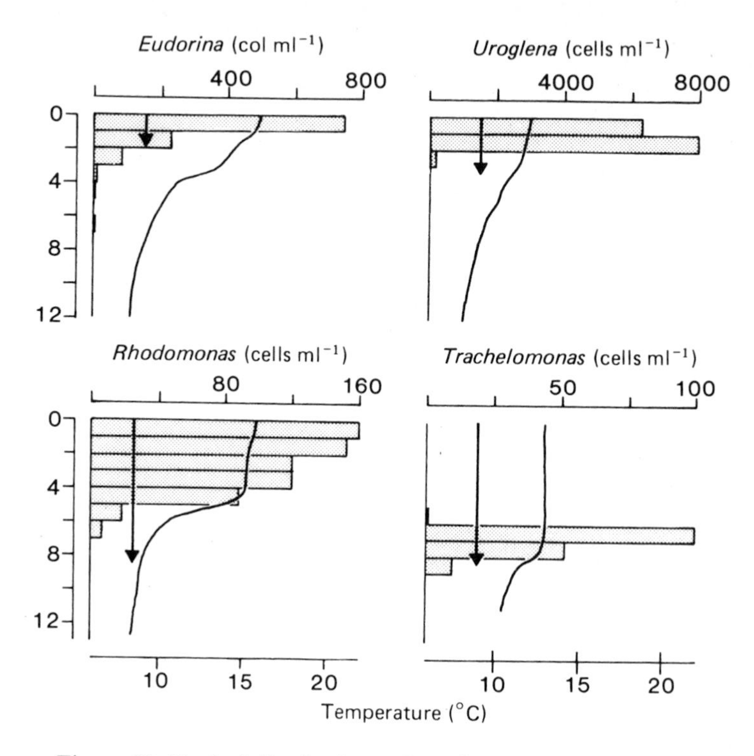

Figure 30. Vertical distributions of motile algae in Blelham enclosures (shaded histograms) in relation to thermal gradients (superimposed curves) and Secchi-disc extinction depth (arrows): *Eudorina elegans* (Encl. B: 23 May 1977); *Uroglena* sp. (A: 16 May 1977); *Rhodomonas minuta* (B: 20 June 1977); *Trachelomonas* sp. (B: 3 October 1972). (Original, based on data in Lund & Reynolds (1982), for *Trachelomonas* only, or author's unpublished records.)

its late-spring maximum: here, a majority of the population is distributed within the top metre under conditions of calm weather and high surface irradiance. It is likely (from unpublished preliminary observations) that the alga becomes more widely dispersed with depth during the night, in the manner described for other members of the Volvocales (e.g. Smith, 1917). The second shows a similar distribution of *Uroglena* colonies against a temperature gradient not exceeding 0.3°C m^{-1} ($\sim$ 0.04 kg m^{-4}). *Rhodomonas* is a small unicellular flagellate which is nevertheless capable of maintaining itself in the epilimnion. *Trachelomonas* is a unicellular euglenoid often found in the hypolimnia of transparent lakes; Figure 30 depicts a metalimnetic population towards the end of the seasonal stratification in 1972.

Among the freshwater flagellates, probably few exceed the ability of

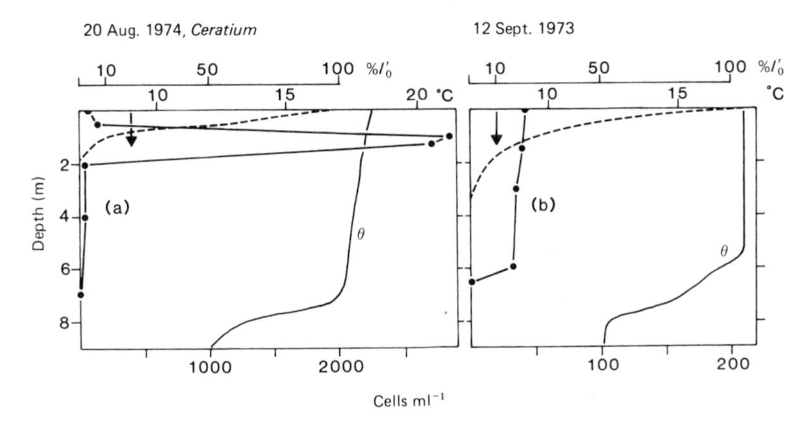

Figure 31. Vertical distributions of *Ceratium hirundinella* in Crose Mere in relation to thermal gradients (θ), light penetration (pecked lines, as a percentage of I'_0, the immediate subsurface irradiance intensity) and Secchi-disc extinction depths (vertical arrows), under (a) calm conditions following windy weather, and (b) during strong wind. (Original, drawn from data in Reynolds 1976*a*,*c*.)

Ceratium hirundinella to regulate its vertical position and to perform conspicuous vertical migrations. The behaviour of this relatively large unicellular dinoflagellate is complex and is governed by a variety of environmental factors which have been carefully elucidated by Heaney and his co-workers (e.g. Heaney, 1976; Heaney & Talling, 1980*a*, *b*; Heaney & Furnass, 1980; George & Heaney, 1978; Harris, Heaney & Talling, 1979). Under certain conditions, *Ceratium* cells will migrate towards the surface by day and redisperse by night (e.g. Talling, 1971) yet at other times, remain continuously stratified at depth (Heaney & Talling, 1980*a*). Movements appear at all times to avoid anoxic hypolimnia and, to some extent, high irradiances obtaining immediately beneath the lake surface. Diel migratory patterns are further complicated by the effects of nutrient limitation (e.g. Heaney & Eppley, 1981). The present consideration seeks only to establish the scale of the density gradients upon which the behaviour of *Ceratium* is likely to be manifest.

Owing to the inherent mobility of the organism, the response may occur within a matter of hours so that quite striking vertical discontinuities can develop and disappear on diel cycles. The right-hand profile in Figure 31 contrasts the vertical distribution of *Ceratium* in Crose Mere during windy weather (zero temperature gradient from the surface to 5.5 m depth) with that during a calm interlude (no wind), following several days of wind-mixing to 6.5 m. In the latter instance (Figure 31a), the remarkable aggregation of cells between 0.5 and 2.0 m depth coincides with a temperature gradient of 0.25°C in 1.5 m (representing a mean density gradient of 0.035 kg m^{-4}). The differences between the concentrations of *Ceratium* at 0 and 6 m in Figure 31b are insignificant (at the 95% confidence level) but at ⩾ 6.5 m, (i.e. beneath the epilimnion) concentrations were not observed to exceed 1 ml^{-1}, in spite of a metalimnetic density gradient of 0.26 kg m^{-4}. Thus while the *Ceratium* cells were unable to regulate their vertical station within the epilimnion, they were apparently able to resist elimination from it.

True buoyancy regulation is the preserve of planktonic cyanobacteria. The ability to form stable metalimnetic depth maxima of approximately isopycnic individuals is exemplified particularly well by *Oscillatoria* species belonging to the *O. agardhii–O. prolifica–O. rubescens* complex (e.g. Eberley, 1959; Zimmermann, 1969; Reynolds & Walsby, 1975; Klemer, 1976). This behaviour differs from that of other *Oscillatoria* spp., whether gas-vacuolate (e.g. *O. redekei*) or not (e.g. the *O. limnetica* group), which characteristically remain dispersed within the mixed layers of exposed or turbid lakes, where the depth of mixing typically considerably exceeds the depth of light penetration (see §8.1.2.1). Stratification of *Oscillatoria* appears to depend upon two factors. The first is the existence of stable density gradients within the euphotic zone (i.e. above the depth to which 1% of the immediate subsurface irradiance can penetrate – see also Chapter 4), which persist for weeks or months on end; this condition can apply regularly in some sheltered, clear-water alpine and small continental lakes. The second combines the ability of *Oscillatoria* to overcome its density difference with the adjacent medium (i.e. $\rho' - \rho \simeq 0$) with the solitary habit of its trichomes (i.e. r_s remains relatively small) which, in accordance with the Stokes equation (14), restricts both the sinking or floating velocity to < 0.5 μm s^{-1} (Walsby & Klemer, 1974; Reynolds & Walsby, 1975). This means that individual trichomes travel only a short vertical distance before the sinking or floating movement can be reversed. However, populations can respond to sustained and marked changes in light penetration by migrating upwards or downwards as appropriate (thus preserving the maximum at a new depth) provided always that the movements are confined to the stable layers. It has also been shown (Walsby & Klemer, 1974; Klemer, 1976) that these migrations can be modified by nutrient availability (trichomes may float relatively higher in the light gradient if more nutrients are available to support growth). However, Lund & Reynolds (1982) were unable to discern any material difference between the depth of *Oscillatoria* stratification with respect to light penetration in Lund Tubes, regardless of nutrient availability; neither were they able to detect significant net growth of stratified

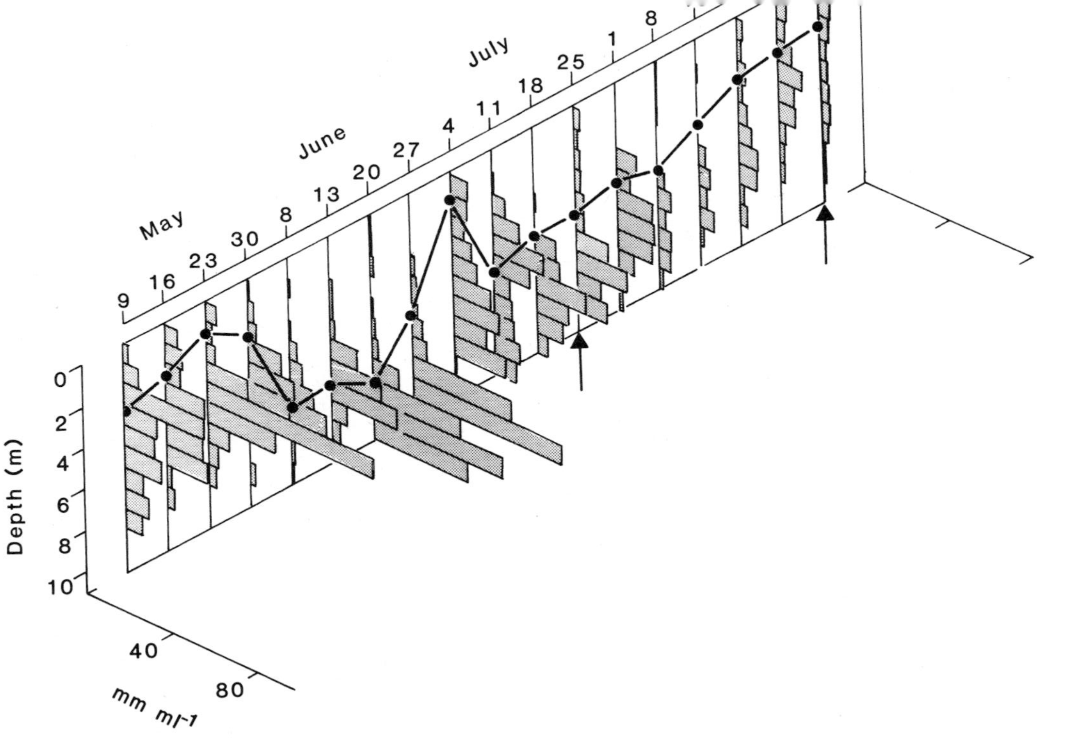

Figure 32. Isometric plot of seasonal changes in the vertical distribution of *Oscillatoria agardhii* var. *isothrix* (shown as concentrations of total trichome length) in Blelham Enclosure B, 1977. Secchi-disk extinction (z_s) is shown as a solid line; the arrows indicate dates on which the epilimnetic deepening entrained part of the stratified *Oscillatoria*. (Original presentation of the data used in constructing Figure 17b of Lund & Reynolds 1982.)

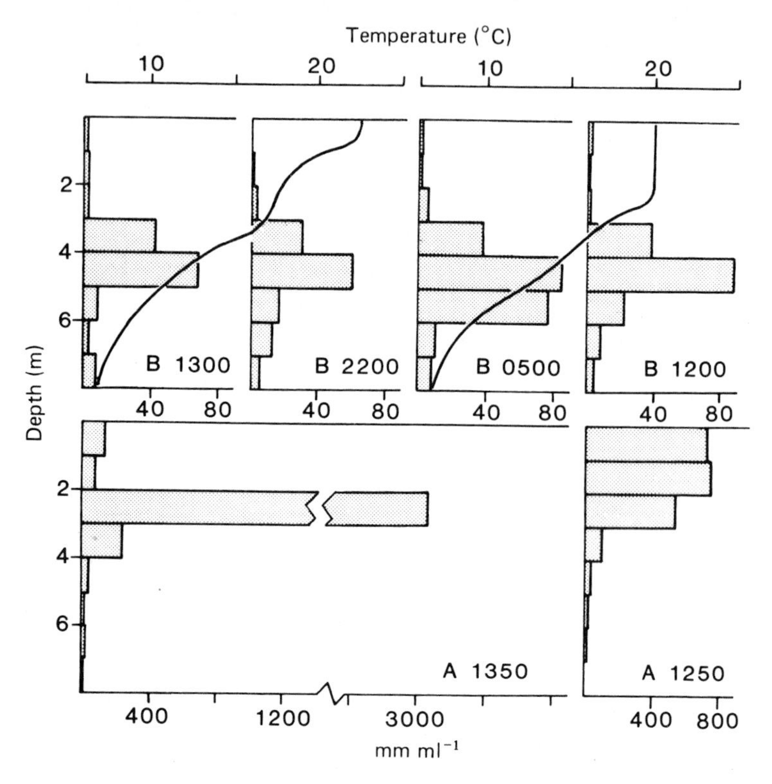

Figure 33. Four instances in the vertical stratification of *Oscillatoria agardhii* v. *isothrix* (as mm ml^{-1}) in Blelham Enclosure B (above) and in Enclosure A (below) through 24 h on 20–21 June, 1979. Two temperature profiles (1300 h 20 June and 0500 h 21 June) are included to show that the nocturnal mixing to ~ 2.5 m was sufficiently deep to entrain the *Oscillatoria* depth maximum in Enclosure A but not to affect the population stratified at 4–5 m in Enclosure B. (Original, drawn from author's unpublished data.)

populations. What was apparent was a well-defined pattern of responses to the changes in light penetration over the summer season. In Figure 32, changes in vertical distribution through the summer of 1977 are presented as isometric projections. Most of the population persistently avoided the surface layer, variously forming its peak concentration at depths between 1.5 and 2 times the Secchi-disk extinction (z_S) but moving up or down in response to changes in z_S, at apparent mean rates of 0.25–0.35 m d^{-1} (3–4 $\mu m\ s^{-1}$). Exceptions to this statement occurred during periods of high epilimnetic turbidity (generated largely by other algal species), when *Oscillatoria* was forced to rise into the epilimnion and to be randomized by turbulent flow (Figure 32: 4 July, 22 August) or when increased epilimnetic mixing entrained erstwhile stratified trichomes from depth

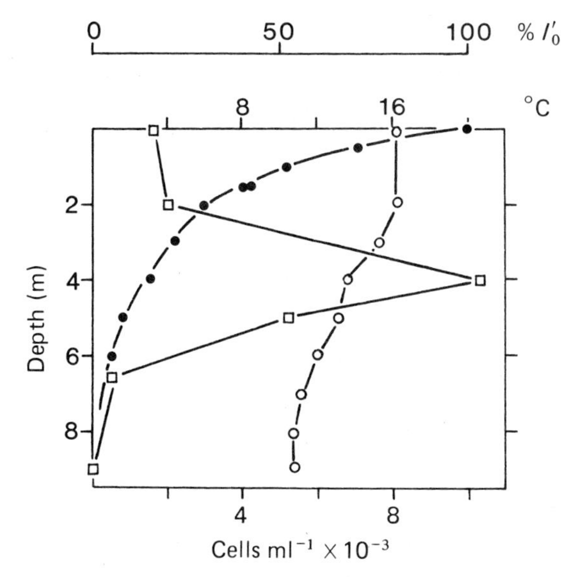

Figure 34. Depth profiles of *Anabaena circinalis* (□, cells ml^{-1}) in relation to temperature (○, °C) and relative light penetration (●, % I'_0) in Crose Mere on 28 May 1975. (Redrawn from part of Figure 1 of Reynolds, 1978*a*.)

(25 July, 5 September). This process is shown in greater detail in Figure 33. Large *Oscillatoria* populations were simultaneously present in two Blelham enclosures (A, B) in June, 1979, that in A being the greater of the two and centred on the 2–3 m depth range; the smaller population in B was stratified at about 4–6 m depth. An early morning breeze on 21 June, associated with the passage of a cold front, rapidly mixed both enclosures to a depth of about 2.5 m. This was sufficient to redisperse the A population within the top two metres while the B population was left relatively intact.

Instances of stratification by other species of cyanobacteria have been observed recently in tropical and subtropical lakes, involving *Lyngbya* cf. *limnetica*, in Lagoa Carioca, a small, sheltered lake in the Rio Doce Valley in Brasil (Reynolds, Tundisi & Hino, 1983*a*), and *Dactylococcopsis* sp. in Solar Lake, Sinai (Padan & Cohen, 1982). Several bloom-forming species of *Anabaena*, *Aphanizomenon* and *Microcystis* (all of which form 'large' colonial units that move proportionately faster than *Oscillatoria* trichomes for the same density difference and are therefore less able to maintain discrete layers with respect to depth) are known to stratify coarsely for brief periods of suitable photic and thermal characteristics (Reynolds, 1978*a*; see also Figures 34, 35) or, more typically, to perform diel migrations in response to diel cycles of night and day (Sirenko *et al.*, 1968; Vanderhoef

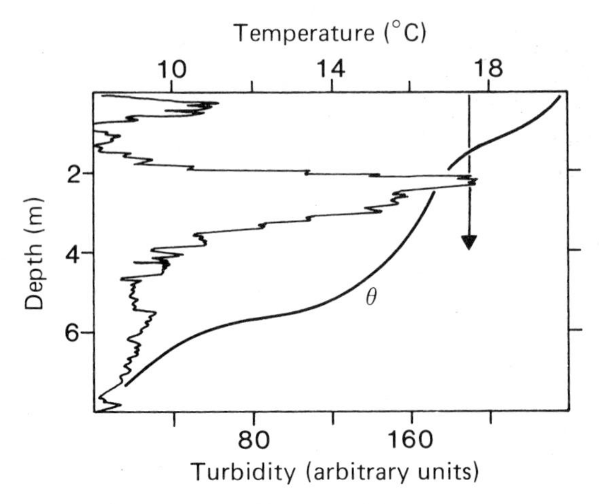

Figure 35. Profile of turbidity owing to *Microcystis* colonies stratified in the water column of Blelham Enclosure B on 12 July, 1978, as detected by horizontal-beam transmissometry, in relation to temperature (θ) and Secchi-disc extinction depth (vertical arrow). (Traced from the same original print-out reproduced in part of Figure 47 of Reynolds *et al.*, 1981.)

et al., 1975; Konopka *et al.*, 1978). The apparent rapidity of these migratory movements presumably derives largely from the greater velocities of the individual units, or aggregations thereof, permitted by relatively small adjustments in buoyancy (i.e. Stokes equation determines that v' is large because r is large). In other words, the buoyancy control mechanism has freed these organisms from the constraints of size and form resistance that influence the anatomy of most planktonic species (see Chapters 1 and 2).

The advantages of 'streamlining' and rapid adjustment of buoyancy are epitomized in shallow, unstable tropical lakes which may be mixed to a depth of several metres once or twice per day, the mixing interludes being separated by intermediate phases of intense thermal stratification. Moreover, the potentially damaging irradiances experienced near the surface must be successfully avoided. Not surprisingly, such lakes tend to be dominated by 'large' cyanobacteria, the outstanding example being the *Microcystis* populations in Lake George, Uganda, studied by Ganf (e.g. 1974*a*), whose data are stylized in Figure 36. Ganf's work emphasized the formation of strong near-surface density gradients ($\leqslant 2.74$ kg m^{-4}; $Ri > 1$) during the typically windless morning period and their subsequent disruption by wind-driven mixing during the afternoons. In the morning *Microcystis* colonies were observed to lose buoyancy (Ganf estimated a maximum sinking rate of *ca.* 1 mm s^{-1}) which led to a reduction of about

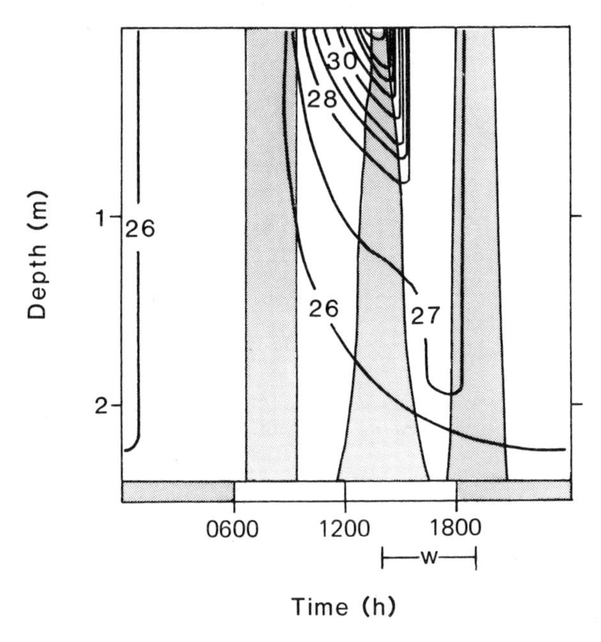

Figure 36. Stylized vertical distributions (as cylindrical curves) of *Microcystis aeruginosa* colonies through the water column of Lake George, Uganda, at three times during the day (08 h, 14 h and 19 h), in relation to the diel cycle of thermal stratification (represented by isotherms). w marks period of wind activity. (Stylized from several sets of data included in Ganf, 1974*a*.)

50% in colony concentration in the top 0.2–0.3 m of the water column. Afternoon winds redistributed the population over the top two metres; some recovery in buoyancy was also observed. The changes in buoyancy are consistent with those expected from the turgor-control mechanism (Dinsdale & Walsby, 1972; see also Reynolds, 1975*a*). The selective value of this response is enhanced by the necessary ability to avoid the high-surface irradiance. It can be surmised that few other planktonic species are capable of adapting to such sharp extremes of mixing and insolation, so it need be no surprise that *Microcystis* dominates the plankton of Lake George so effectively.

3.3 Horizontal distribution

Despite early implicit assumptions to the contrary, patchiness in the horizontal distribution of phytoplankton occurs frequently. As in the case of vertical distribution patterns, the element of scale and its perception is crucial to the interpretation of horizontal distribution and its ecological impact. It is necessary to relate the sizes and longevities of patches both

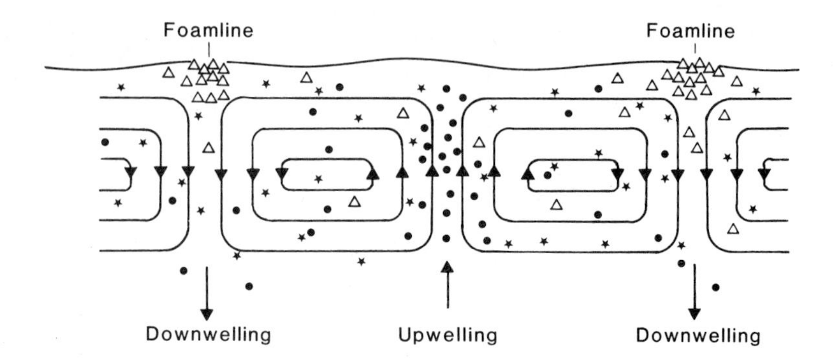

Figure 37. Schematic section across the horizontal direction of flow of Langmuir rotations (cf. Figure 10) to show the likely distributions of strongly negatively buoyant (●), strongly positively buoyant (△), and well-entrained isopycnic (*) organisms. (Based on Figure 7 of Smayda, 1970, and Figure 8b of George, 1981*c*.)

to the processes which contribute to their formation and survival and the generation times of phytoplankton.

3.3.1 *Small-scale patchiness*

At the finest readily-measured scales (mm, s) discontinuities are inevitable, being at the level of presence or absence of a particle in a given parcel of water. Equally, the situation may alter at a given point when one parcel is displaced by the next. The trajectories of the various individual parcels determine the pattern of distribution at higher scales. Thus, if the parcels are part of a more widespread and identifiable flow pattern (e.g. convection cells, Langmuir rotations or open turbulence), patchiness at the smaller scale may appear either integrated within or further differentiated between larger and more stable patches.

The entrainment characteristics of the particles present will be relevant to this scaling, as a relatively well-documented example illustrates. Smayda (1970) predicted the distributions of algal particles, categorized according to their intrinsic settling behaviour, against the known flow characteristics of Langmuir rotations. Later work (e.g. George & Edwards, 1973; Harris & Lott, 1973) has substantiated Smayda's view that downward-settling or -swimming organisms will segregate selectively into the upwellings, buoyant particles will collect at the convergences above the downwellings (usually marked by buoyant foam streaks – the familiar windrows) while well-entrained particles will be distributed randomly; Figure 37 gives a schematic representation of these trends. The scheme is determined by differential buoyant behaviour of the species present and holds as long as specific behaviour remains constant (see also Evans & Taylor, 1980). Since the horizontal width, vertical height and mean current speed of the rotations are all correlatives of wind speed, such patterns may develop within minutes of the generation of the circulation and, potentially, survive for the duration of constant wind action. However, because winds fluctuate in intensity and direction, new patterns may become superimposed upon, and may obliterate, the earlier patterns. The rotations are, moreover, three-dimensional corkscrew motions, progressing broadly downwind (see Figure 10). In lakes, those motions are sooner or later interrupted by shallows, margins or islands and the circulation gives way to new constraints (see later). In other words, any individual organism is unlikely to adopt a permanent position relative to the flow but to be transported elsewhere. Horizontal motions of this scale are unlikely to confine organisms to the extent that the discontinuous distributions are compounded by growth (production of new individuals). While the imposed conditions may well affect the physiological performance of the organisms present, the scale of the fluctuations is accommodated within the existing range of the general environmental characteristics each experiences during its separate life-span and so will not be expected to contribute directly to detectable local differences in growth rate. This is not to deny that small-scale variations are important, for, as will be seen, they contribute to the understanding of larger-scale patchiness in lakes; neither are they irrelevant to the design of sampling programmes that aim to monitor changes in specific populations (see, for instance, Cassie, 1968; Nasev, Nasev & Guiard, 1978; Venrick, 1978; Moll & Rohlf, 1981). Nevertheless, there is good evidence that small-scale patchiness collapses into higher-level responses of more or less confined gross circulations of lakes and estuaries when the characteristic motion is driven by winds exceeding 3–5 m s^{-1} (George & Edwards, 1976; George & Heaney, 1978; Sandusky & Horne, 1978; Therriault *et al.*, 1978; Therriault & Platt, 1981).

3.3.2 *Large-scale patchiness*

Maintenance of large-scale (km, d) patchiness must depend partly upon the positive 'feedback' provided by net population change, whether this is strictly reproductive, enhanced or otherwise by point sources of nutrient enrichment, or whether it is due to locally different removal rates through, for instance, local aggregations of herbivorous animals. It is clear that a patch will be maintained so long as the dynamics of the population change sustaining it exceed the rate at which the patch is diffusively eroded at its edges. Several authors have considered this relationship (e.g. Skellam, 1951; Kierstead & Slobodkin, 1953; Joseph & Sendner, 1958; Ozmidov, 1958; Okubo, 1971, 1978). The so-called 'KISS' model (of Kierstead &

Slobodkin, 1953, and Skellam, 1951) predicts the critical patch size below which reproductive effects are overcome by diffusion. Specifically, Kierstead & Slobodkin (1953) argued that the critical radius of a patch (r_c) is given by:

$$r_c = 2.4048\ (D/k) \tag{19}$$

where D is the horizontal diffusivity and k is the net rate of increase (or decrease) in the population. If values of k appropriate to algae (in the order of 0.1 to 1.0 doublings per day) and to wind-driven diffusivities ($D \sim 5 \times 10^3$ to 2×10^6 cm^2 s^{-1}, from Okubo, 1971) are interpolated, the derived critical patch radii fall between 500 m and 100 km. These agree with general observations on large-scale patches in the open ocean (e.g. Steele, 1976; Okubo, 1978); even under the most favourable conditions of low diffusivity and high growth, patches smaller than one kilometre across are likely to be rapidly eroded. It may be readily appreciated that such formulations for long-term patch stability are applicable to the open systems occurring in the oceans and only the largest lakes, where horizontal diffusion is the dominant dissipating force.

3.3.3 *Advective patchiness*

In smaller, closed basins, horizontal diffusivity at the surface must be compensated by deeper (or, possibly, lateral) return flows, the flow being constrained by the margins of the basins. Here, the mean ('advective') return flow assumes a relatively greater significance: diffusion is simply halted when the patch expands to the greatest dimensions of the basin. Thus, when a planktonic alga is transported in a surface flow whose mean velocity is 0.03 m s^{-1}, it may travel some 2.6 km per generation ($k = 1$ d^{-1}). If the length of the basin is < 2.6 km, it is likely that the alga will already be travelling at depth in the opposite direction before it divides and, potentially, that the daughter cells will be added on to the surface flow at the origin, perhaps on the second or third day: the large-scale patch is retained and comes to occupy the entire basin (i.e. it is indiscernible as a patch). This 'conveyor-belt' motion (cf. George & Edwards, 1976) will tend to integrate the horizontal distribution of populations in smaller lakes, even overcoming local point-sources of nutrient-rich inflows (as George, 1981*b*, has shown). Patchiness can develop only on finer scales, that is, of smaller critical size and of shorter duration than predicted by the 'KISS' model. In instances of advective flow, the model derived by Joseph & Sendner (1958) gives a more useful prediction of patch-size:

$$r_c = 3.67\ u_s/k \tag{20}$$

where u_s is the mean advective velocity. If $k = 1$ division per day and $u_s = 0.005$ m s^{-1}, $r_c = 1.6$ km; if $u_s = 0.02$ m s^{-1}, $r_c = 6.3$ km.

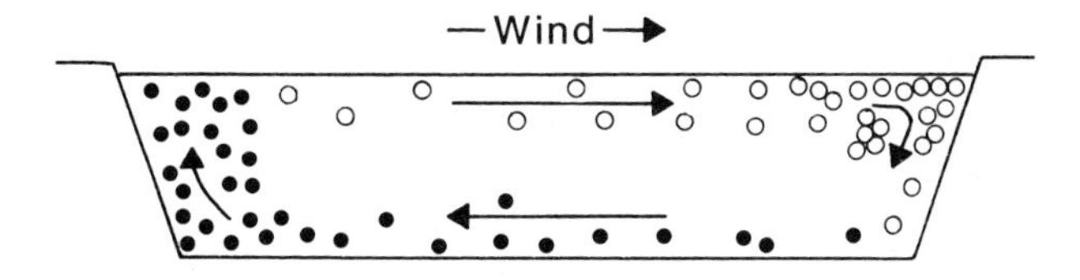

Figure 38. Whole-lake 'conveyor-belt' model of patchiness of strongly negatively buoyant (●) and strongly positively buoyant (○) organisms. (Redrawn from Figure 1 of George & Edwards, 1976.)

However, it must be recognized that the patchiness may be further enhanced as a direct consequence of particle behaviour within the 'conveyor belt' flow, which is analogous to behavioural segregation in Langmuir rotations (Figure 37). Although most planktonic 'algae' show little directed horizontal movement of their own, many species potentially regulate their position in the vertical plane (see §3.2). Organisms that are able to migrate to a preferred depth in the light gradient do not necessarily modify their behaviour even if they are contemporaneously transported laterally by advective flow. Towards the edge of the basin the same directed responses will result in their rapid removal from the regions of downwelling (i.e. the downward movement is enhanced by the dominant flow, until organisms are swept away in the deep-return current) and to their retention in the corresponding upwelling on the opposite side of the lake (where downward movements go against the current). For buoyant organisms, the opposite will be true. Patches of the former will develop upwind; those of the latter, downwind (Figure 38). Distributions of this type have been observed for zooplankton species (Colebrook, 1960; George & Edwards, 1976), *Ceratium* (Heaney, 1976; George & Heaney, 1978) and *Microcystis* (George & Edwards, 1976). Two examples are represented in Figure 39: both were based upon the concentrations of organisms in a series of spot-samples collected at different horizontal stations in rapid succession (i.e. as near to being simultaneous as practicable).

George & Edwards (1976) were able to relate the incidence of advective patchiness in the small Eglwys Nynydd reservoir to wind speed and direction, by plotting the index of mean crowding ($\overset{*}{x}/\bar{x}$: see §3.2) in populations of buoyant cyanobacteria against 'accumulated wind effect' (which allows for the exponential decay of momentum when the wind drops). Their data are re-drawn here as Figure 40, with the horizontal axis replotted as equivalent steady wind speed. Horizontal patchiness is greatest when winds are light but weakens as wind speed approaches 3.5 m s^{-1}, disappearing at $W > 5$ m s^{-1}. The work on *Ceratium* in Esthwaite

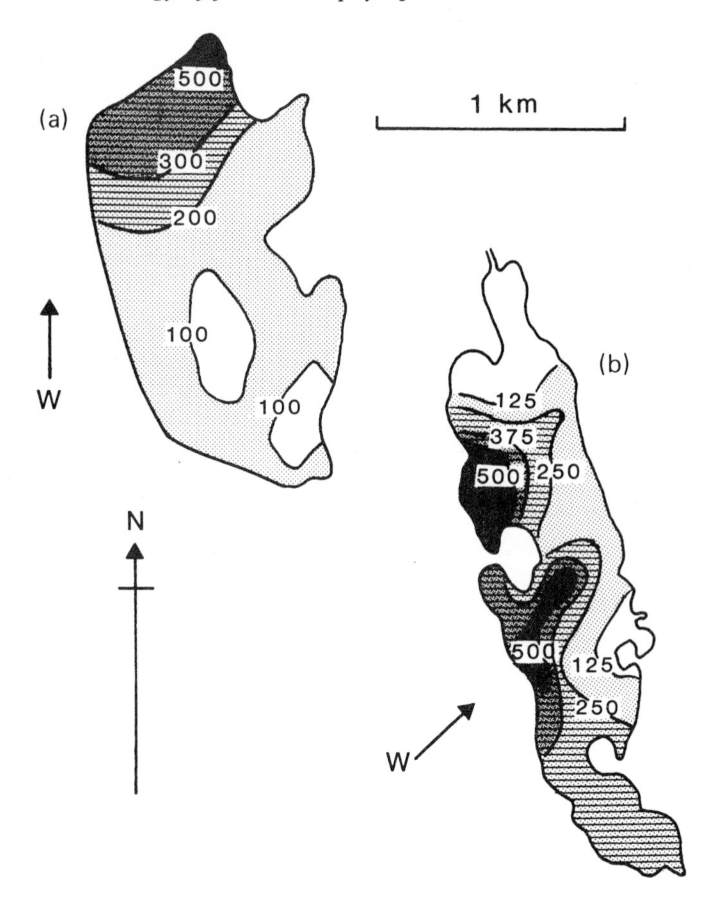

Figure 39. Advective horizontal patchiness of planktonic organisms in small lakes, relative to wind direction: (a) positively buoyant *Microcystis aeruginosa* (isopleths in μg chlorophyll *a* l^{-1}) in Eglwys Nynydd reservoir and (b) surface-avoiding *Ceratium* (cells ml^{-1}) in Esthwaite Water. ((a) is redrawn from Figure 12b of George & Edwards (1976); (b) is redrawn from Figure 4 of Heaney, 1976.)

Water (Heaney, 1976; George & Heaney, 1978; Heaney & Talling, 1980*a*,*b*) consistently points to horizontal patchiness occurring only at wind speeds < 4.0 m s^{-1}.

These considerations show clearly that the incidence of advective patchiness in small basins is consequent upon the interaction between behaviour in the vertical and the effect of the stable, advective flow patterns generated by light winds and return currents. As was shown in §3.2, the horizontal velocity of flow and its gradient with depth directly influence the ability of the organism to regulate its vertical position. The conditions that allow 'algae' to accumulate in the vertical are precisely those under

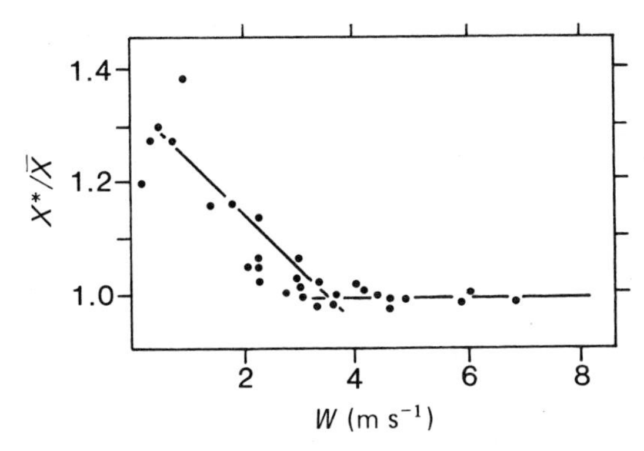

Figure 40. The relationship between horizontal patchiness of *Microcystis* in the Eglwys Nynydd reservoir and wind speed (W). (Redrawn from Figure 14 of George & Edwards, 1976.)

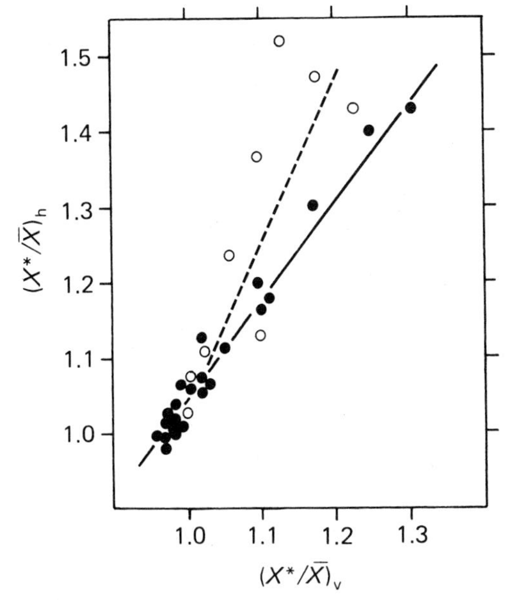

Figure 41. The relationship between horizonal $(\overset{*}{x}/\bar{x})_h$ and vertical patchiness $(\overset{*}{x}/\bar{x})_v$ of *Microcystis* (●, Figures 28, 40) and of *Daphnia hyalina* (○) populations, studied by George & Edwards (1976). (Redrawn from Figures 15 and 18 of George & Edwards, 1976.)

which self-regulating populations accumulate up- or down-wind. If turbulence dominates the surface flow, the vertical and horizontal components are simultaneously suppressed. Apart from being subject to similar critical wind speeds ($W = 3$–5 m s^{-1}), the extent of horizontal patchiness is directly correlated with the extent of vertical patchiness (Figure 41). The incidence of horizontal patchiness is thus predictable in terms of basin size, the vertical behaviour of the dominant species and the immediate history of wind action on that lake.

To summarize this consideration of horizontal plankton patchiness and its causes, it can be stated that the contributory mechanisms have still to be fully resolved. Nevertheless, it is apparent that the responses of plankton vary according to the scale and type of the imposed flow pattern: large-scale patchiness (km, d) reflects the responses of populations (compounded by growth and attrition) to diffusivity. At progressively smaller scales, patchiness is a behavioural response to flow patterns (advective, convective) which alter over smaller intervals of space and time (m, h).

3.4 Temporal variations in abundance and composition of phytoplankton

The discussion in this chapter has so far centred on spatial and temporal variations occurring in populations *already in existence* and the immediate prospects for their further development. This line of reasoning may be briefly developed as a basis for later chapters in which the effects of major fluctuations in the physical, chemical and biotic characteristics of the ambient medium upon the wax and wane of specific populations and the consequent, distinctive progressions in the structure of the planktonic communities are considered.

Seasonal changes in the abundance and composition of phytoplankton may be observed to occur on temporal scales measured in weeks or months, during which mean specific population densities may increase or decrease through 6–9 orders of magnitude (say from 10^{-3} to 10^{6} cells ml^{-1}), or the equivalent of 20 to 30 cell divisions. The time scale is such that the amplitude range of environmental oscillations is itself likely to alter significantly. In temperate lakes, for instance, the environmental variables that are subject to such profound changes include day-length, total irradiance and its attenuation within the underwater light field, temperature, thermal structure and nutrient availability, as well as the abundance of herbivorous animals. The ecological responses of the phytoplankton to these changes are generally well-defined, although factor interaction often prevents a clear appraisal of cause and effect and may even lead to misinterpretation. Correct interpretation is essential if year-to-year variations and long-term responses of phytoplankton composition to climatological change and eutrophication effects are to be adequately understood and predicted.

There has been a large number of studies made and descriptions given of phytoplankton periodicity, though few have seriously attempted to account for the patterns observed. Several classical studies on individual lakes or groups of temperate lakes (e.g. Birge & Juday, 1922; Ruttner, 1930, 1938; Pearsall, 1932; Grim, 1939; Findenegg, 1943) established the clear year-to-year similarities in changes in abundance and composition and, to a lesser extent, similarities from lake-to-lake. The vast body of data which has been added to the descriptive literature serves to emphasize that patterns of seasonal change are broadly repeated in lakes of widely dispersed geographical locations but sharing similar morphometric, climatological and chemical properties. It may be presumed that similar factors critically influence the cycles observed in each of them.

Hutchinson (1967) has made one of the few attempts to categorize the seasonal sequences into a general descriptive framework for temperate lakes. In a more recent paper, Reynolds (1980*a*) described similar progressions in terms of species assemblages and attempted to relate community changes, rather than fluctuations in specific populations, to environmental characteristics which were reproduced in a series of British lakes. A selection of these generalized sequences, examplified by published data for given lakes, is summarized below. The categories must not be regarded as being complete, neither do they necessarily accommodate any given sequence exactly. They are presented essentially to illustrate that such generalizations can be formulated and to give context to the evaluation of environmental impact upon phytoplankton abundance and composition developed in the following chapters.

3.4.1 *Seasonal variations in the total quantity of phytoplankton*

It is convenient first to consider the variations in abundance of the total phytoplankton present in freshwater systems. The basis for equating the biomass of different species may be dry-weight (preferably ash-free), carbon or chlorophyll *a* content or total cell volume. Chlorophyll content (or approximations thereof, according to relationships with original data, developed in Chapter 1) has been selected for the six sequences presented here (Figure 42). The simplest case (Figure 42a) is of an increase in phytoplankton biomass through spring and early summer to a maximum in July and a subsequent decrease to a winter minimum, exemplified by Millstätersee, Austria, during 1935 (Findenegg, 1943). Single summer peaks were also observed in other Carinthian Lakes (Findenegg, 1943), Bodensee (Grim, 1939) and the less-productive lakes of the English Lake District (Pearsall, 1932: Wast Water, Ennerdale Water). In all these examples, phytoplankton abundance is subject to a lasting and overriding control by chronic nutrient deficiency (see also

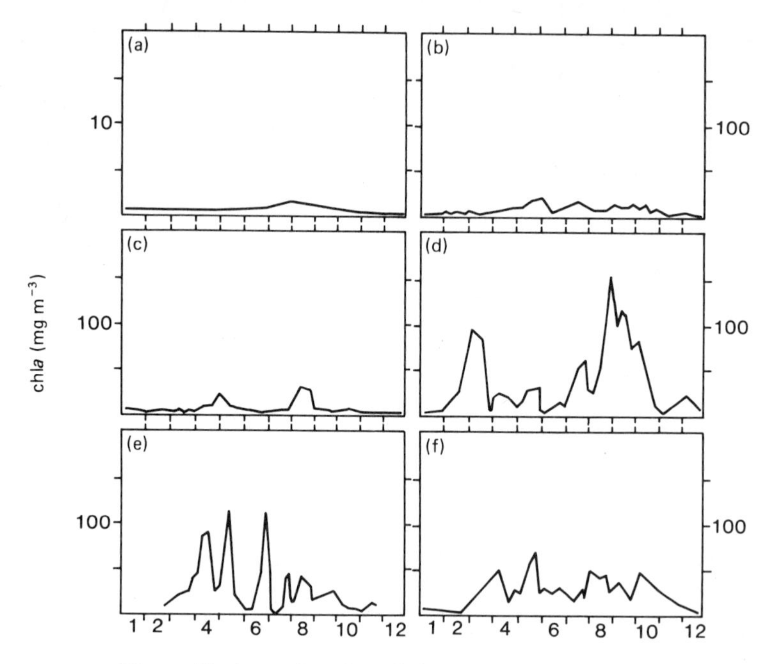

Figure 42. Annual cycles of phytoplankton biomass, measured (or approximated from volume measurements) as chlorophyll *a* concentration, in temperate lakes: (a) Millstättersee, 1935; (b) Windermere, North Basin, 1978; (c) S. Erken, 1957; (d) Crose Mere, 1973; (e) Blelham Enclosure A, 1978; (f) Hamilton Harbour, 1976. (Figure reproduced from Reynolds, 1980*b*.) Note that the vertical scale of (a) is 10 times greater than that of (b)–(f).

Chapter 5), the peak biomass occurring when the physical conditions most favour it (i.e. when the water temperature reaches its maximum and the stratification is most strongly developed).

The second example (Figure 42b) illustrates the biomass fluctuation in the north basin of Windermere, England, 1978, where low nutrient concentrations (especially of phosphorus) also determine a low average phytoplankton crop, though less severely so than in the first example. The sequence is typical of many larger, unproductive temperate lakes in which the 'mid-summer' maximum is reached earlier (in late spring in this example) and extended through an irregular succession of smaller peaks, finally disappearing in the autumn. Presumably, chemical and physical conditions determine the earlier attainment of the maximal biomass, while later changes may be more directly dependent upon the external supply of nutrients and upon internal recycling mechanisms. The third example (Figure 42c: Erken Sjön, Sweden, 1957; based on Nauwerck, 1963) shows this trend more clearly, wherein the earlier maximum occurs during spring (shortly after the lake becomes ice-free) while a second peak, associated with the early stages of autumnal destratification, exceeds any other during the year. This 'diacmic' pattern is probably typical of many 'mesotrophic', shallow (< 30 m) temperate lakes (Hutchinson, 1967). It is supposed that the two peaks reflect the coincidence of physically-suitable growth conditions with relatively high concentrations of limiting nutrients, shortages of which prevent the attainment of large biomasses during the midsummer period.

This chemical uncoupling from the physical control is accentuated in the fourth example (Figure 42d) which depicts the sequence in a shallow (< 10 m), productive lake, Crose Mere (England), in 1973. Here, the diacmic sequence is preserved but intermediate fluctuations, due to different dominant species, each able to elaborate a substantial biomass from the nutrients available, characterize the summer period. Similarly more complex cycles are to be found among other productive temperate lakes (for examples, see Hutchinson, 1967; Reynolds, 1978*b*).

If nutrients are never limiting, either because they are always present in large quantities or they are frequently or continuously added, there would seem to be no reason why intermediate peaks should not be as large or larger than those occurring in spring and autumn, or that they should not coalesce into a single 'monacmic' span covering most of the year, with physical limitations becoming dominant. Some small rich 'farm ponds' have been shown to approach the latter condition (e.g. Folly Pool, near Shrewsbury, England: Reynolds, 1973*c*) but there is little information available on which to base a general statement. Uhlmann (1971) showed that artificial sewage ponds could maintain continuously high standing crops provided that the rate of hydraulic throughput did not exceed the rate of increase in biomass.

However, large temperate lakes appear not to behave in this way. The experiments in the large limnetic enclosures in Blelham Tarn, described by Lund & Reynolds (1982), were successful in controlling nutrient loading and reproducing quantitative sequences closely resembling three of the first four examples (Figure 42b, c, d). When nutrients were replenished weekly, the average standing crops were certainly higher, but the annual cycle was always characterized by abrupt fluctuations in total phytoplankton abundance (Figure 42e shows the sequence in Enclosure A, 1978). These changes are influenced by several factors, the most important apparently being those involving the basic biologies of the participating dominants (when they grow and how they disappear) and the ways in which these interact with alterations in column stability (Reynolds, 1980*a*,*c*; see also Chapter 8). The latter explanation has already been invoked to explain the natural sequence exemplified in Hamilton Harbour, Ontario (Haffner, Harris & Jarai, 1980), redrawn here as Figure 42f. In the most nutrient-rich lakes, then, physical constraints appear to exert the major influence on phytoplankton abundance, although they are not the same ones influencing the unproductive lake sequences (Figure 42*a*).

Relatively much less information is available for lakes nearer to the equator though it seems likely that similar controls operate. In less-productive lakes, nutrient poverty restricts the production of large crops, though these have a longer suitable growing season and populations may be maintained for relatively longer (e.g. Melack, 1979; Hecky & Kling, 1981). More productive lakes either maintain large crops of persistent, self-regulating populations of cyanobacteria, presenting an overall impression of general ecological stability (Lake George, Uganda, provides an outstanding example: Ganf & Viner, 1973), or they reproduce 'temperate' sequences over truncated time-spans, which may be more or less recapitulated two or more times per year: this deduction is possible from a detailed work on Lake Lanao, Philippines, carried out by Lewis (e.g. 1978*a*, *b*). Many more investigations on tropical lake sequences are required before the validity of these tentative generalizations can be tested.

3.4.2 *Temporal variations in the composition of phytoplankton communities*

Hutchinson's (1967) review of qualitative variations in the phytoplankton established that broadly similar cycles of specific abundance were evident among a wide range of temperate lakes. He cited several examples of spring and early summer populations dominated by diatoms: *Cyclotella comensis* in Findenegg's (1943) Carinthian Lakes, *Cyclotella comta* and related species in Bodensee (Grim, 1939), the Lunzer Untersee (Ruttner, 1938) and the less productive lakes of the English Lake District (Pearsall, 1932), *Asterionella formosa* and *Tabellaria fenestrata* (together with *Melosira italica*: see Lund, 1954, and Lund, in Macan, 1970) in the richer examples, and *Asterionella* with *Stephanodiscus* spp. in more productive lakes in North America (quoting Birge & Juday, 1922; Hutchinson, 1944) and Europe (see also Reynolds, 1978*b*; Ridley, 1970; Coveney *et al.*, 1977).

During the summer months, diatoms were frequently replaced, at least partially, by other dominant species: generally the more productive the lake, the more complete is the eclipse. Thus, while *Cyclotella* spp. remained dominant in Millstätersee (indeed they were achieving their maximum biomass), an increasing fraction of the total biomass became attributable to *Gomphosphaeria*; in the neighbouring Klopeinersee, dinoflagellates (*Peridinium*, *Ceratium*) became increasingly conspicuous (Findenegg, 1943).

In moderately productive lakes, like Windermere (Pearsall, 1932; Lund in Macan, 1970; Reynolds, 1980*a*), Grasmere (also in the English Lake District: Reynolds 1980*a*), Schleinsee (Germany: Vetter, 1937) and Linsley Pond (Massachussets: Hutchinson, 1944), the diatoms all but disappeared from the plankton (or from the epilimnia at least: Ruttner described *Asterionella* populations surviving in the hypolimnion of Lunzer Untersee), to be replaced by communities dominated by chrysophytes (*Dinobryon*, *Mallomonas*, *Uroglena* and/or chlorophyte genera such as *Sphaerocystis*, *Gemellicystis*, *Coenococcus*, *Oocystis*, *Didymogenes* and *Dictyosphaerium*) and, later on, by *Gomphosphaeria*- or *Ceratium*-dominated maxima. In many of these lakes, diatoms (chiefly *Asterionella*, *Fragilaria* and *Tabellaria* spp.) and, in some of them, desmids (*Cosmarium*, *Staurastrum*, *Staurodesmus*, *Xanthidium* spp.) contributed to the biomass in late summer.

In productive lakes of Europe, such as Crose Mere and Rostherne Mere (Reynolds, 1978*b*, 1980*a*), Erken (Rodhe, Vollenweider & Nauwerck, 1958), and in North America, especially within the mid-western areas of glacial deposition in Canada and in the Wisconsin-Minnesota region (e.g. McCombie, 1953; Brook, 1971; Lin, 1972; Kling, 1975), the disappearance of the vernal diatom populations is progressively followed by maxima attributable to colonial Volvocales (*Eudorina*, *Pandorina*, *Volvox* spp.), filamentous cyanobacteria (*Anabaena*, *Aphanizomenon*) and, ultimately, by large standing populations of *Ceratium* and/or *Microcystis*. Even here, diatoms are not excluded, there being occasional summer pulses dominated by *Asterionella*, *Fragilaria* or *Melosira granulata*.

A fourth category of generalized cycle might be added, although it is not clearly defined. Many small, highly-enriched water bodies (and even some sizeable lakes receiving high nutrient discharges, e.g. Hamilton Harbour, L. Onondaga, New York State – see Haffner *et al.*, 1980; Sze & Kingsbury, 1972; Sze, 1980) are typically dominated by mixtures of smaller diatom spp. (e.g. *Nitzschia*, *Stephanodiscus hantzschii*, *Diatoma*), Chlorococcales (*Scenedesmus*, *Ankistrodesmus*, *Monoraphidium*, *Crucigenea*, *Tetraëdron*, *Oocystis*, *Coelastrum*, *Pediastrum*) and cyanobacteria (*Aphanocapsa*, *Aphanothece*, *Synechococcus* etc); see also Pavoni (1963).

Though concerned with a relatively small selection of British lakes, Reynolds' (1980*a*) attempt to group together species which collectively made up the 'assemblages' and were, potentially, alternative dominants, provides further evidence of the 'reproducibility' of plankton cycles in given lakes. Several of these sequences have been imitated in the Blelham Tubes (Reynolds, 1980*a*; Lund & Reynolds, 1982). The same approach has been followed in the production of Figure 43, which attempts to summarize the major kinds of seasonal progressions. The original alphanumeric

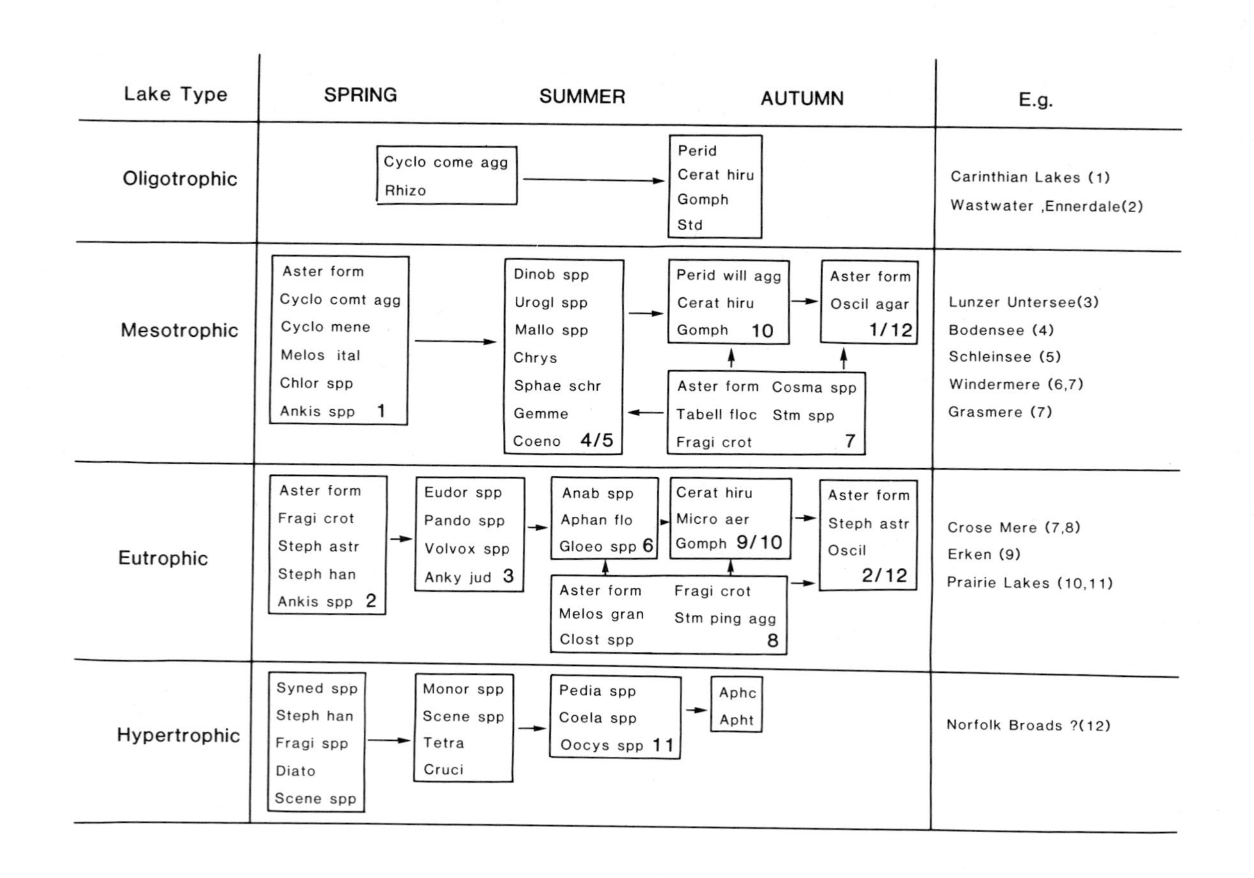
Lake Type
SPRING
SUMMER
AUTUMN
E.g.
Oligotrophic
Cyclo come agg
Rhizo
Perid
Cerat hiru
Gomph
Std
Carinthian Lakes (1)
Wastwater ,Ennerdale(2)
Mesotrophic
Aster form
Cyclo comt agg
Cyclo mene
Melos ital
Chlor spp
Ankis spp 1
Dinob spp
Urogl spp
Mallo spp
Chrys
Sphae schr
Gemme
Coeno 4/5
Perid will agg
Cerat hiru
Gomph 10
Aster form
Oscil agar
1/12
Aster form
Cosma spp
Tabell floc
Stm spp
Fragi crot
7
Lunzer Untersee(3)
Bodensee (4)
Schleinsee (5)
Windermere (6,7)
Grasmere (7)
Eutrophic
Aster form
Fragi crot
Steph astr
Steph han
Ankis spp 2
Eudor spp
Pando spp
Volvox spp
Anky jud 3
Anab spp
Aphan flo
Gloeo spp 6
Cerat hiru
Micro aer
Gomph 9/10
Aster form
Steph astr
Oscil
2/12
Aster form
Fragi crot
Melos gran
Stm ping agg
Clost spp
8
Crose Mere (7,8)
Erken (9)
Prairie Lakes (10,11)
Hypertrophic
Syned spp
Steph han
Fragi spp
Diato
Scene spp
Monor spp
Scene spp
Tetra
Cruci
Pedia spp
Coela spp
Oocys spp 11
Aphc
Apht
Norfolk Broads ?(12)

group-labels of Reynolds (1980*a*) are included for reference, where appropriate, but the system has been expanded to include the wider range of available data. The assemblages include small species which tended to be overlooked in the early surveys (Hutchinson, 1967).

No attempt is made here to explain these sequences; this is left until Chapter 8, after other aspects of anabolism, growth and attrition, and specific differences therein, have been considered. In the intermediate chapters, wherein these topics are reviewed, several underlying questions are, however, kept to the fore: why do diatoms (or any other group) not remain continuously dominant throughout the year? Why are they replaced? What factors determine that some species are better adapted to

Figure 43. Generalized successional pathways in lakes of different trophic status: the boxes represent assemblages of species (whose names are listed below) one or more of which may be abundant; bold numbers refer to assemblage-labels of Reynolds (1980*a*). Examples of each sequence are referenced 1–12. (1), Findenegg (1943); (2), Pearsall (1932); (3), Ruttner (1930); (4), Grim (1939); (5), Vetter (1937); (6), Lund, in Macan (1970); (7), Reynolds (1980*a*); (8), Reynolds (1978*b*); (9), Nauwerck (1963); (10), Lin (1972); (11), Kling (1975); (12), Leah *et al.* (1980). Abbreviations to species: Ankis, *Ankistrodesmus* spp.; Anky, *Ankyra judayi*; Anab, *Anabaena* spp.; Aphan flo, *Aphanizomenon flos-aquae*; Aphc, *Aphanocapsa*; Apht, *Aphanothece*; Aster form, *Asterionella formosa*; Cerat hir, *Ceratium hirundinella*; Chlor, *Chlorella* spp.; Chrys, *Chrysochromulina*; Clost spp includes *Closterium aciculare*, *C. acutum*, *C. tortum*; Coela, *Coelastrum* spp.; Coeno, *Coenococcus*; Cosma, *Cosmarium* spp. (*C. abbreviatum*, *C. contractum*, *C. depressum*); Cruci, *Crucigenea* spp.; Cyclo come agg, *Cyclotella comensis* aggregate; Cyclo comt agg, *C. comta* aggregate; Cyclo mene, *Cyclotella meneghiniana*; Diato, *Diatoma*; Dinob spp, *Dinobryon* (chiefly *divergens*); Eudor spp, *Eudorina* spp.; Fragi spp, *Fragilaria* spp.; *Fragilaria crotonensis* (Fragi crot) is distinguished; Gemme, *Gemellicystis* cf. *neglecta*; Gloeo, *Gloeotrichia*; Gomph, *Gomphosphaeria* spp., including forms ascribable to *Coelosphaerium*; Mallo, *Mallomonas* (e.g. *M. caudata*); Melos gran, *Melosira granulata*; Melos ital, *M. italica*; Micro aer, *Microcystis aeruginosa*; Monor spp, *Monoraphidium* spp.; Oocys spp, *Oocystis* spp. (e.g. *O. borgei*); Oscil, *Oscillatoria* spp., usually ascribable to *O. agardhii* (Oscil agar); Pando spp, *Pandorina* spp. (e.g. *P. morum*); Pedia spp, *Pediastrum* spp. (*P. boryanum*, *P. duplex*, *P. tetras*); Perid, *Peridinium* spp.; the *P. willei* aggregate (Perid will agg) is distinguished; Rhizo, *Rhizosolenia*; Scene, *Scenedesmus* spp.; Sphae schr, *Sphaerocystis schroeteri*; Std, *Staurodesmus*; Steph astr, *Stephanodiscus astraea* (= *S. rotula*); Steph han, *Stephanodiscus hantzschii*; Stm spp, *Staurastrum* spp.; the *S. pingue* aggregate (Stm ping agg) is distinguished; Syned spp, *Synedra* spp.; Tabell floc, *Tabellaria flocculosa*; Tetra, *Tetrastrum*; Urogl spp, *Uroglena* spp.; Volvox spp, *Volvox* spp. (Reproduced, with permission, from Reynolds, 1982.)

living during one part of the year when others are not? Do they involve specific photic or nutrient preferences, or are they biotically determined? In short, what are the mechanisms of seasonal change?

3.5 Intra-assemblage structure and competition

There is, however, one essential point that emerges from the discussion so far, which demands the immediate attention of the general ecologist, and that requires some initial consideration here. It will have been recognized already that assemblages comprise several species which are potentially alternative dominants or simultaneous codominants. Two common assumptions, that the limnetic epilimnion is homogeneous, as nearly as any habitat can be expected to be, and that the species inhabiting it compete directly for the same material resources, should determine that only one of very few species inhabit the plankton simultaneously. In reality, tens of species will often be encountered in the examination of a small volume (< 10 ml) of natural lake water. This apparent 'Paradox of the Plankton' (Hutchinson, 1961) offends the competitive exclusion principle, formulated by Hardin (1960; though still frequently referred to as Gause's Hypothesis[1]) which states that complete competitors cannot coexist. Several attempts have been made to explain its apparent inapplicability to phytoplankton, the most widely accepted apparently being that of Richerson, Armstrong & Goldman (1970). They developed Hutchinson's belief that the main answer to the plankton paradox was due to a violation in the assumption that species are at equilibrium. Owing to frequent variability in the habitat, conditions are neither constant nor ideal for any one competing species and the competitive advantage swings from one species to another before the latter has had the opportunity to replace the former. Incidentally, this view implies that the various species are not adapted to perform identical functions. From an analysis of a large number of epilimnetic samples from Castle Lake, California, Richerson *et al.* (1970) demonstrated a highly patchy distribution for many of the species present. Therefore, a mixed epilimnion could not be regarded as homogeneous; the authors argued that the observed variations were less successive than contemporaneous. There were thus always in existence temporary *niches* in which growth conditions differed from those elsewhere. Such niches were, however, frequently obliterated and reconstituted at random. The small size of the niches permitted high between-niche diversity, directly favouring the maintenance of several species.

[1] I can find no evidence that Gause formally stated this principle, although much the same reasoning had been previously expressed by Monard, the French freshwater ecologist, in 1920, and, perhaps, by Grinnell, an American Biologist, as early as 1904.

The concept of niche diversification is most clearly appreciated in stably-stratified layers where localized differences in growth conditions may be more readily related to gradients in light penetration and nutrient concentrations. The niches are tangible, inherently more stable in space and time. Indeed, several authors have related the increased diversity in the total community of stratified water columns to discontinuous vertical distributions of the component species (Hutchinson, 1967; Moss, 1967; Margalef, 1968; Reynolds 1976*a*). Under such conditions, the various species present exhibit not only distinctive micro-habitat (niche) preferences but also specific differences in their means of self-regulation (see also §3.2).

In the mixed environment, diversification of niches may involve additional criteria which are essentially physiological or behavioural. An important development of this concept was made by Petersen (1975) whose model of nutrient-limited growth, based on Michaelis-Menten kinetics (see Chapter 5), was offered as an explanation of Hutchinson's plankton paradox. Coexistence might be attributable to species simultaneously experiencing different specific limiting controls; that is they were not in direct mutual competition. The distinction between Petersen's view and that of Richerson *et al.* is a narrow one; the importance of Petersen's consideration lies in its corollary – that the fewer limiting factors that are operating, then the fewer is the number of available 'physiological niches' and, by deduction, the fewer is the number of species in simultaneous existence. It does seem generally true that in lakes maintaining high availabilities of nutrients, species diversity is usually lower (Margalef 1958, 1964; Reynolds, 1978*c*). A simple illustration of this is provided in Figure 44, which plots annual courses in Margalef's (1958) index of diversity, d_s (according to equation 21

$$d_s = (s-1)/\log_e N \tag{21}$$

where $s =$ the number of species and $N =$ the number of individuals) through the course of the year for three temperate lake systems, Windermere (in which nutrients, especially phosphorus, are chronically limiting), Crose Mere (in which they are rarely so), and in one of the Blelham enclosures (in which they are assumed never to have been so). In this series, species diversity drops from the range 1.78 to 6.44 to 0.24 to 1.75 in the experimental enclosure; the sudden increase in September was directly attributable to collapse of part of the enclosure wall and the intake of Tarn water into the system.

Annual fluctuations in each plot show certain other similarities which are characteristic of seasonal progressions: in each case d_s is lowest during the vernal circulation, when diatoms are dominant, and increases as the water columns become stratified. Detail variations within this framework

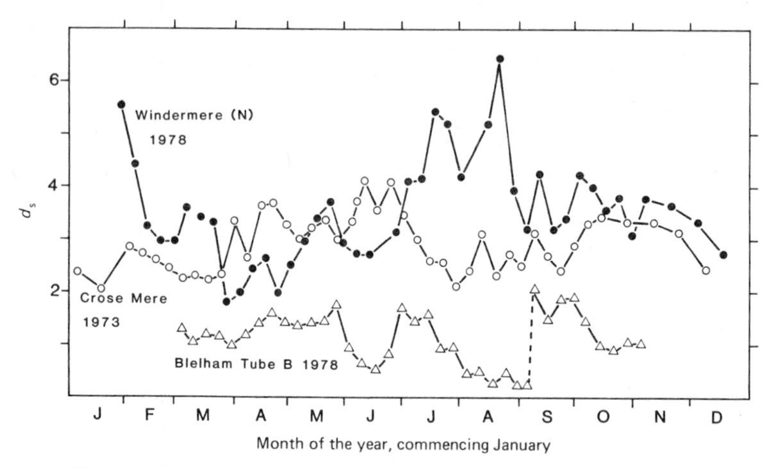

Figure 44. Plots of Margalef's species diversity index, d_s, calculated for the phytoplankton community of a nutrient-deficient lake (Windermere, North Basin), a nutrient-rich one (Crose Mere) and an experimental enclosure fertilized with nutrients each week (Blelham Tube B). (Original, calculated from previously published data in Reynolds, 1976*a*, 1980*a*, Reynolds & Wiseman, 1982.)

may be correlated with major changes in dominance, the peaks corresponding with the replacement of one assemblage by the next, the troughs with more stable community structure (cf. Reynolds, 1980*a*). Since many of the more abrupt changes in composition are themselves consequential upon sharp changes in structural organization of the community, the plots help to substantiate the major conclusion, which is that the diversity of plankton communities is essentially related to the number and longevity of simultaneously coexistent exploitable niches, in the broadest sense.

4

Photosynthetic activity of phytoplankton

'Make hay while the sun shines'
Traditional

4.1 General features of planktonic photosynthesis

It was stated in Chapter 1 that the paramount requirement of planktonic autotrophs either to prolong suspension in, or to gain frequent access to, the upper illuminated zone of the water column was directly consequential upon the requirement for light energy, in order to drive photosynthetic carbon fixation and anabolic growth. In essence, the latter requirement is no different from that of any other chlorophyll-containing plant inhabiting the earth's surface. There are considerable differences, however, in the photosynthetic responses of planktonic 'algae' intermittently transported within and beyond an underwater light field that is itself characterized by relatively higher coefficients of light absorbance and scattering than occur in the atmosphere and has a different spectral composition. These differences are reviewed in the present chapter.

No attempt is made here to describe in detail either the structure or the biochemical functioning of the photosynthetic apparatus; several recent reviews are available and should be consulted (e.g. Halldal, 1970; Whittingham, 1974; Gibbs *et al.*, 1976; see also the individual contributions in Stewart, 1974). It is necessary, however, to summarize those features which are relevant to the physiology and ecology of photosynthetic organisms in the plankton. Much of this chapter is based on Harris' (1978) excellent review of planktonic photosynthesis, to which the reader is also referred.

The photosynthetic process is usefully summarized by the well-known equation describing the fixation of carbon dioxide and the manufacture of organic carbon skeletons for cellular metabolism and growth.

$$6CO_2 + 6H_2O \xrightarrow{\text{light}} C_6H_{12}O_6 + 6O_2 \qquad (22)$$

Photosynthesis can be viewed as a redox system, involving several component processes. Electrons are stripped from a donor substance (H_2O

in this case, but it may be substituted by H_2S or other reduced organic carbon compound) and are passed through a light-driven transport system to generate high-energy phosphate bonds (phosphorylation) which, in turn, are used to fix carbon (carboxylation) in the Calvin Cycle.

The photosynthetic electron transport system in 'algae' conforms to the Hill & Bendall (1960) model of two separate photochemical systems, worked out for higher plants: one (photochemical system II, or PSII) is associated with the non-cyclical transport of electrons derived from water molecules and the concomitant release of oxygen molecules; the other (PSI) transports the electrons which are eventually involved in the reduction of nicotinamide adenine dinucleotide phosphate (NADP to $NADPH_2$). The two light reactions are linked by a cytochrome-carrier system, through which the high-energy phosphate compound (adenosine triphosphate, or ATP) is generated (non-cyclic phosphorylation), augmenting the ATP produced in the cyclical return flow of PSI (cyclic phosphorylation).

These light-reactions are located in specialized membranes organelles, the thylakoids. In the prokaryotic cyanobacteria, the thylakoids are distributed in the cytoplasm; in eukaryotes, the thylakoids are enclosed in membrane-bound plastids. Two types of particle occur in close association with the thylakoid surface, corresponding with the two photochemical systems.

Also closely associated with the thylakoid surfaces are the photosynthetic pigments – chlorophylls and accessory pigments – where they act as 'traps' for light energy or quanta. Maximal absorption by chlorophyll *a* occurs in two wavelength bands, peaking at about 430 and 660 nm. The accessory pigments have maximal absorption at different wavelengths: chlorophyll *b*, which occurs in the chlorophyta, peaks at about 450 and 645 nm; β-carotene, a major pigment in most freshwater algae, peaks in the range 450–470 nm; xanthophylls, which are also widely distributed among the algal groups, peak between 480 and 540 nm; phycobilins, e.g. phycoerythrins (absorbing at 540–560 nm) and phycocyanins (610–630 nm) are present in Rhodophyta, cryptomonads and in cyanobacteria. The accessory pigments thus spread the adsorption spectrum of the trap; the energy they absorb is readily transferred to the chlorophyll trap. The Hill & Bendall model requires that there are two sets of quantum traps (one each for PSI and PSII); each comprises about 300 chlorophyll molecules.

Most of the absorbed quantum energy reaching PSII is consumed in raising the electrons of the donor molecule to a higher energy level and transferring them to the acceptor molecule; the remainder 'spills over' to

PSI or is emitted as chlorophyll fluorescence. PSII fluorescence peaks at about 685 nm; the weaker fluorescence from PSI peaks between 700 and 720 nm, emphasizing the functional distinction between the two photochemical systems. PSII fluorescence can be conveniently and continuously measured *in vivo* in a fluorometer; Lorenzen's (1966) original technique has been widely used as an index of phytoplankton biomass (for critical literature reviews, see Harris, 1978; Yentsch, 1980). Because the fluorescence yield of chlorophyll is not constant, but is subject to physiological control, this relationship has to be interpreted with caution (cf. Heaney, 1978).

The reduced NADP and the ATP produced by non-cyclic phosphorylation are used to fix carbon dioxide. The initial fixation product of the Calvin Cycle is 3-P-glyceric acid (3-PGA); one molecule each of carbon dioxide, water and ribulose-1,5 diphosphate (RuDP) yield two molecules of 3-PGA. ATP and $NADPH_2$ react with 3-PGA to give 3-phosphoglyceraldehyde (3-P-GAP), which now incorporates the high-energy phosphate from the ATP. The 3-P-GAP is then metabolized through a series of sugar-phosphate intermediates, culminating in the production of hexose and, ultimately, polysaccharides which can be stored in the plastid or in the cytoplasm of the cell. (The Calvin Cycle is closed by steps leading to the regeneration of RuDP.)

The cycle is not immutable, however. One physiologically important compound which may be produced is glycolate. Glycolate is formed in several reactions, including the breakdown of the RuDP-carboxylase enzyme at high partial pressures of O_2 and from the oxidation of sugar monophosphates. In phytoplankton, glycolate may be either excreted into the water (e.g. Fogg, 1971) or metabolized in the cytoplasm, releasing CO_2 (photorespiration).

Release of CO_2, as a consequence of light-photosynthesis, is contrary to the expected flux of CO_2 uptake and fixation. Indeed, some of the CO_2 produced may be photosynthetically 're-fixed'; CO_2 loss from the cell is therefore rarely a direct measurement of photorespiration; in any case, living cells will continuously produce CO_2 as a respiratory product of carbohydrate metabolism. These processes have a considerable bearing upon the interpretation of measurements of phytoplankton photosynthesis and its application to population dynamics (see Harris 1980*a,b* and Chapter 6). For full reviews of carbon metabolism and respiration in 'algae', see Lloyd (1974), Raven (1974) and Harris (1978).

The biochemical aspects of algal photosynthesis that are directly interrelated with the physiological ecology of phytoplankton may thus be summarized: photosynthetic rate and, indirectly, the supply of the organic carbon skeletons essential for growth will be dependent upon the quality, intensity and subsurface attenuation of the solar radiation, upon the supply of carbon dioxide, the efficiency of the chlorophyll light-trap, and the effect of temperature; the subsequent fixation, storage and metabolism of carbon compounds will depend upon the physiological condition of the algal cells. Moreover, the photosynthetic responses of 'algae' to fluctuations in environmental conditions may potentially influence species composition and change in planktonic communities. These interrelations are explored in the following sections.

4.2 Photosynthetic behaviour in isolated samples of natural phytoplankton

Photosynthetic rate in phytoplankton is traditionally measured either from the uptake of dissolved ^{14}C-labelled carbon dioxide into algal particles or from the net change in the dissolved oxygen concentration of the suspending medium (as detected by Winkler's titration or by the use of suitable oxygen electrodes). Suspensions of natural or cultured 'algae' are exposed to the natural underwater light field or to artificial laboratory light sources in small closed glass vessels. The oxygen method requires the use of parallel darkened controls (wherein no photosynthesis can occur) in order to correct the net change in oxygen concentration for respiratory consumption. Neither method is without its faults. It is incorrect to assume that ^{14}C incorporated in 'algal cells' represents gross photosynthesis, for some of that taken up may be subsequently recycled into the medium; indeed, it tends towards a measure of net photosynthesis (Dring & Jewson, 1979). It is also wrong to assume that 'respirational' losses are identical in light and darkened bottles (see §4.1 above). In both types of measurement, it also needs to be recognized that the external concentration of carbon dioxide is altered by photosynthetic uptake and that, as a result, the enclosed micro-environment becomes significantly altered with respect to the prototype it is designed to imitate. These shortcomings are to some extent overcome if the shortest practicable exposure times are adopted. A more serious criticism of bottle experiments as a quantitative index of photosynthetic production is that they remove from the phytoplankton cells the dimension of vertical movement within the light field that most cells would experience in the natural environment. This point assumes considerable relevance when measured photosynthetic production is translated into potential growth.

Measurements of photosynthesis in closed bottles exposed at selected depths in the water column nevertheless yield consistent generalized results, which provide a convenient starting point for an analysis of photosynthetic behaviour. Figure 45 depicts a typical depth profile of

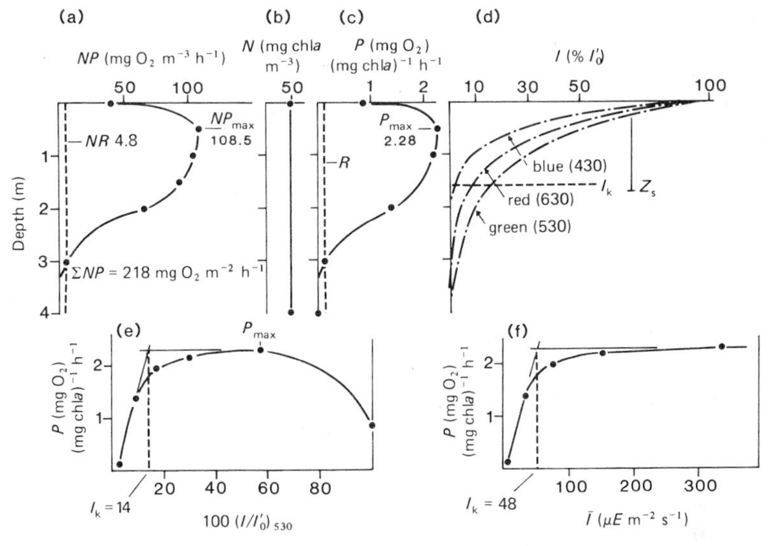

Figure 45. Specimen depth distributions of (a) total gross photosynthetic rate (*NP*) and total gross respiration rates (*NR*); (b) the photosynthetic population, *N*, in terms of chlorophyll; (c) specific photosynthetic rate, $P(= NP/N)$, and respiration rate, $R(= NR/N)$; and (d) the underwater irradiance (*I*), in each of three spectral blocks (expressed as a percentage of I_0', the irradiance in the corresponding spectral block obtaining immediately beneath the water surface), in a mixed temperate lake in late winter. *P* is replotted against *I*, either (e) as a percentage of I_0' in the green spectral block (peak: 530 nm) or (f) as the reworked estimate of absolute irradiance intensity (in $\mu E\ m^{-2}\ s^{-1}$). For further explanation, see text. (Original constructions, based on experimental data obtained at Crose Mere, 25 February, 1971, and presented (in part) in Reynolds, 1978*b*.)

photosynthetic exposures in relation to light penetration during the vernal (isothermally-mixed) phase of phytoplankton growth in Crose Mere. The exposures were of unconcentrated phytoplankton collected from the depth at which they were to be exposed (*in situ* exposure method); in the experiment concerned, the composition and total chlorophyll content of the samples were almost identical. The experiment was carried out in the middle part of the day under cloudless sky; $\theta \sim 5$°C. This discontinuous pattern of photosynthesis is directly related to light availability. Three components are generally recognizable: the lowermost part of the curve is directly correlated with the irradiance level (I), where the rate of photosynthesis may be said to be *light-limited.* Above this, the increment in photosynthesis rate is uncoupled from light, reaching a maximum (P_{max}) at some vertical distance below the surface. At this point, the photosynthetic rate is said to be *light-saturated* (i.e. no further increase in light will enhance

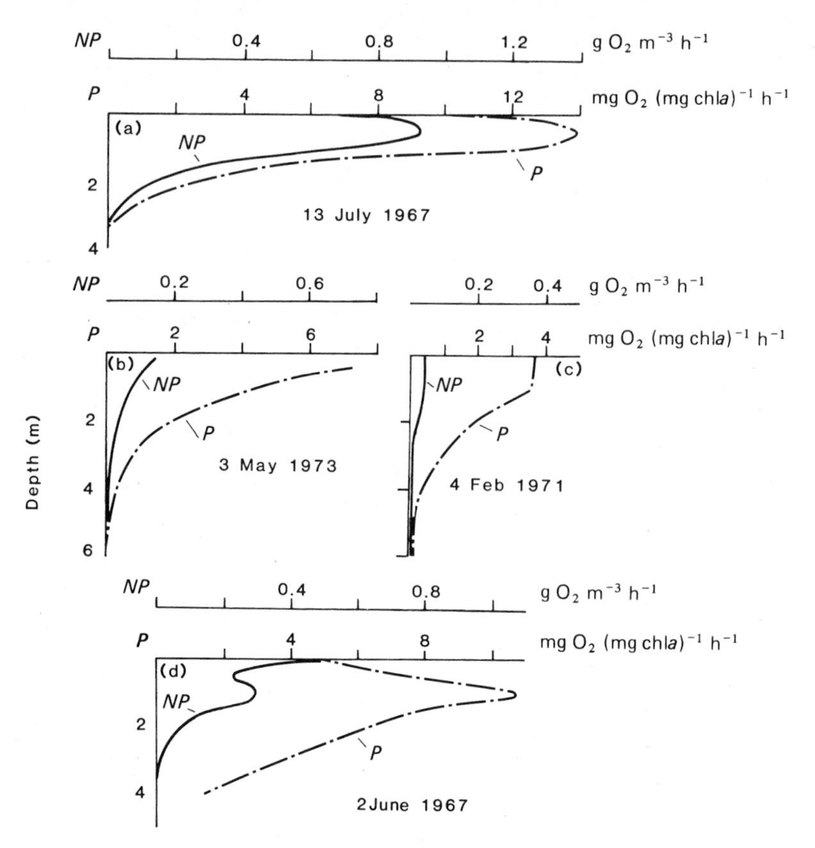

Figure 46. Variations in the basic form of depth profiles of gross (*NP* v. *z*) and specific (*P* v. *z*) photosynthetic rate. (Redrawn from figures in (a,c,d) Reynolds, 1978*b* and (b) Reynolds, 1975*a*.)

photosynthetic rate). The irradiance at the depth at which light saturation is detected (where the *P*-curve straightens and begins to turn away from the *I*-curve of light penetration) is separately identified as I_k. Nearer the surface, photosynthetic rate is apparently depressed, ostensibly because it is *light-inhibited.*

These same three components can be recognized in other profiles represented in Figure 46, when due allowance is made for special environmental factors obtaining. The first example (Figure 46a) depicts an instance of photosynthesis in turbid waters (the turbidity was caused largely by a high population of 'algae'). The rapid extinction of irradiance with depth (see also §4.3) is the cause of the vertical truncation of the *NP* and *P* profiles which nevertheless retain the three components of limitation, saturation and inhibition. Under dull conditions, or even under cloudless conditions shortly after dawn and at dusk, the surface irradiance may be too low to inhibit photosynthesis and may barely saturate it (e.g. Figure 46b). Photosynthetic rate may sometimes be limited by factors other than light: Figure 46c shows the profile of the response of a low-biomass community at low water temperature, in which the elements of limitation and photo-inhibition are separated by a wide spread of maximal light-saturated, temperature-limited measurements. Community photosynthesis (*NP*) responses are also sensitive to discontinuous vertical distributions of the phytoplankton: the complex *NP*-profile in Figure 46d was derived from an experiment carried out during an *Anabaena* bloom, when much of the population was located near the surface. Expressed as photosynthetic rate per unit of population (in this case, as per unit of chlorophyll), a more typical profile results.

The three characteristic components to the relationship between P and I are best treated separately; this is done in the following subsections.

4.2.1 *Light-limited photosynthesis*

The curve of P (see Figure 45c) closely parallels that of I_{530} below a depth of about 1.8 m. It may be noted that shorter wavelengths of photosynthetically active radiation (*PhAR*), (generally within the range 360–700 nm) are rapidly extinguished in this example, the red and green wavelength blocks penetrating to relatively greater depths and, by implication, sustaining light-limited photosynthesis. Regressing the curve of P against the curve of I in the wavelength block ('green': 475–600 nm, peaking at about 530 nm of minimal extinction), it may be shown (Figure 45e) that the initial slope of P is about 0.16 mg O_2 (mg chl$a$)$^{-1}$ h^{-1} per 1% of I'_0 (the incoming irradiance obtaining immediately beneath the surface). A better measure of the photosynthetically active energy at a given depth can be calculated as $I_z = (I_{430} + I_{530} + I_{630})/3$; I'_0 was approximated as $\sim 800\ \mu$E m^{-2} s^{-1}. The slope of P v. I (Figure 45f) then resolves as 0.048 mg O_2 (mg chl$a$)$^{-1}$ h^{-1} (μE m^{-2} s^{-1} *PhAR*)$^{-1}$ or, assuming equimolar exchanges between CO_2 and O_2, as 0.018 mg C fixed (mg chl$a$)$^{-1}$ h^{-1} (μE m^{-2} s^{-1} *PhAR*)$^{-1}$, that is, $\sim$ 5.0 mg C (mg chl$a$)$^{-1}$ E^{-1} m^2.

This value is a measure of photosynthetic efficiency at low irradiance. It falls within the range of observed photosynthetic efficiencies available in the literature (e.g. Jørgensen, 1964; Platt & Jassby, 1976; see also Harris, 1978) which, when suitable adjustments to the units of measurement are made, fall within the range 2–37, peaking between 6 and 18 mg C (mg chl$a$)$^{-1}$ E^{-1} m^2.

The light-limited rate of photosynthesis in most populations, expressed per unit of chlorophyll a, apparently varies little; this conservatism is attributable to the structure of the photosynthesis apparatus. Considerably greater variations result, however, when the slope of P v. I is expressed per cell, per unit volume or per unit dry weight. In the example represented in Figure 45, the population was dominated by *Asterionella* cells, containing *ca.* 1.7 mg chl*a* (10^9 cells)$^{-1}$; the photosynthetic efficiency approximates to 8.4 mg C (10^9 cells)$^{-1}$ E^{-1} m^2 if the population was monospecific. A week earlier the chlorophyll content of *Asterionella* had been only 1.3 mg chl*a* (10^9 cells)$^{-1}$ and a month previously, before net increase had commenced it had been 2.3 mg chl*a* (10^9 cells)$^{-1}$. Had the same photosynthetic efficiency, in terms of chlorophyll, applied to the earlier data, the efficiencies per cell would have fluctuated between 6.4 and 11.4 mg C (10^9 cells)$^{-1}$ E^{-1} m^2. By varying their pigment content, cells are able to regulate their photosynthetic efficiency, although the change is not rapid, being necessarily spread over one or more generation times (cf. Harris, 1978). Cells grown under continuously low irradiances have relatively higher photosynthetic efficiencies than those grown under high light, primarily because of their higher relative pigment content and, hence, improved capacity to absorb light (Steeman-Nielsen, Hansen & Jørgensen, 1962; Steeman-Nielsen & Jørgensen, 1968). Such cells are said to be adapted to low light intensity.

4.2.2 *Light-saturated photosynthesis*

As irradiance is increased, the curves of P v. I flatten out to a maximum (P_{max}), in this instance, at 55% I_0' in the green wavelength block or about 335 $\mu E\ m^{-2}\ s^{-1}$. It should be noted at once, however, that the value of P_{max} is largely independent of the intensity and spectral composition of the photosynthetically-active radiation. Higher published field values of maximum photosynthetic rate (up to 31 mg O_2 mg chl*a*$^{-1}$ h^{-1}) measured in closed bottles and averaged out over exposure periods amounting to several hours, all come from warm tropical lakes in Africa (Talling, 1965; Talling, Wood, Prosser & Baxter, 1973; Ganf, 1975); in temperate lakes, values rarely exceed the equivalent of 20 mg O_2 mg chl*a*$^{-1}$ h^{-1} observed by Bindloss (1974). The ultimate control is physiological because, when no external factors are limiting, the C-fixation rate is determined by the flux-capacity of the molecular photosynthetic pathway (for a full discussion, see Harris, 1978). Within this ultimate limit, the supply of carbon dioxide can be critical (see §4.2.5) and, as is implied by the data above, P_{max} is, to some extent, temperature dependent (see also §4.2.4). The onset of photo-inhibition also occurs at variable intensities (but generally $< 800\ \mu E\ m^{-1}\ s^{-1}$ *PhAR*).

The onset of light-saturated photosynthetic rate, in terms of light

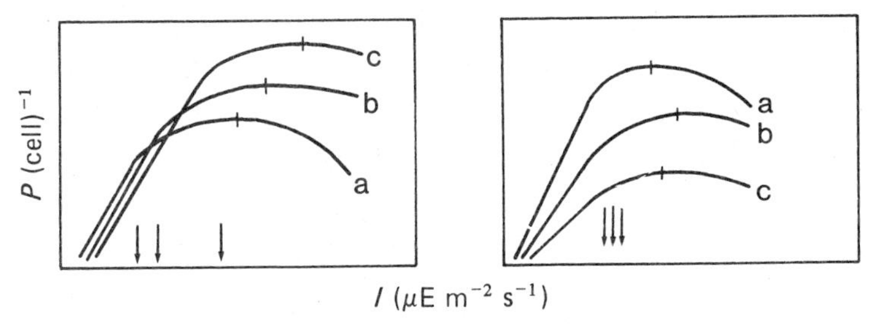

Figure 47. Hypothetical P v. I curves to show different seasonal photosynthetic behaviours and relationships between I_k (arrowed) and P_{max} (vertical tags): curves a,b,c are in order of increasing temperature; (left) epilimnetic populations receiving 'constant' photic conditions; (right) deep-mixed populations receiving fluctuating irradiance levels. (Based on Figures 9 and 10 of Harris, 1978.)

intensity, is rarely abrupt. I_k is the intensity at which the extrapolated light-limited portion of the P v. I curve predicts the observed P_{max} will occur. In our example (Figure 45f), the increment of P v. I begins to decrease at an estimated 36 μE m^{-2} s^{-1}, but the I_k value is about 48 μE m^{-2} s^{-1} *PhAR*. Again, the literature indicates I_k values in the range 20–300 μE m^{-2}s^{-1}, with the mode probably occurring between 60 and 100 μE m^{-2} s^{-1} *PhAR* (Harris, 1978). By virtue of its calculation, the photosynthetic rate at I_k is superficially related to P_{max} and is, therefore, subject to similar regulatory mechanisms. For a given light-limited photosynthetic efficiency, some seasonal variation in I_k is to be expected. Harris (1973) presented data for *Oocystis*-dominated plankton populations in Lake Ontario: in separate experiments, carried out at temperatures between 8 and 21°C, light-limited P v. I curves were almost identical, but P_{max} was about 2.7 times higher at 21°C than at 8°C; I_k was a direct function of temperature.

However, interspecific differences were marked. In populations dominated by diatoms, Harris (1973) observed that light-limited photosynthetic efficiencies (expressed per cell) were not only conspicuously variable but were also directly correlated with P_{max}. In ease case, I_k fell within the same limited range, that is, it was approximately constant (see Figure 47). Both the highest photosynthetic efficiencies and P_{max} values occurred in populations from cold, deep mixed populations (that is, those which were, to some extent, adapted to low average light intensities). In populations yielding lower photosynthetic efficiencies, some reduction in the pigment content and even shrinkage of the plastids themselves was clearly evident. These differences suggest that the recent 'light histories' of cells (whether they are consequential upon deep-mixing condition experienced by the diatoms or related to more stable ambient photic conditions experienced by the epilimnetic populations of green algae in the stratified lake) play a major part in determining their photosynthetic capacities and the onset of light saturation of photosynthesis. Clearly these possibilities (summarized in Figure 47) should influence the interpretation of photosynthetic productivity profiles and seasonal changes in species composition.

4.2.3 *Photo-inhibition*

Marked reduction in the relative photosynthetic rate of phytoplankton incubated in bottles near the surface is frequently observed, though the effect is generally less noticeable in early morning and late afternoon profiles and in those carried out under dull skies. As Harris' (1978) review shows, near-surface photosynthetic rates (P_0) are, really or apparently, depressed in relation to P_{max}, when surface irradiance intensity exceeds 200–800 μE m^{-2} s^{-1} *PhAR*, and the inhibition may become total (i.e. $P_0/P_{max} \rightarrow 0$), when $I_0 > 1400$ μE m^{-2} s^{-1} *PhAR*. Photo-inhibition is clearly a function of incident irradiance, coupled with the vertical extinction rate against depth (see §4.3, below).

The precise causes of photo-inhibition are not clear and, indeed, are evidently not the same in every case. One factor is ultraviolet radiation (which is rapidly attenuated with depth) which, besides having a directly inhibitory effect upon the photochemical systems, may also damage the organic structures in the cell, including the thylakoids and chlorophyll. The plastids may also experience shrinkage. High oxygen concentrations may accelerate these processes, which are perhaps most spectacularly manifest in the photo-oxidative lysis of surface blooms of cyanobacteria (e.g. Abeliovich & Shilo, 1972). 'Algae' may tolerate sublethal irradiances for short periods but reduction in carbon fixation rates generally occurs within a matter of minutes. Recovery, however, often involving renewed synthesis of chlorophyll, the re-expansion of shrunken chloroplasts, replacement of thylakoids and the reactivation of the photosystems, is typically slow.

Changes in the cellular rates of exchange between carbon dioxide and oxygen (upon which much of the evidence of photo-inhibition rests) may be symptomatic of the activation of alternative metabolic pathways of fixed carbon. The net rate of carbon fixation can be controlled by direct extracellular release of organic carbon compounds (including sugars and glycolate) or by active photorespiration, resulting in the production of carbon dioxide (see §4.1 above). Recognition of the importance of photorespiration was slow in coming, largely because experimentation has been conducted under conditions (initially relatively high carbon dioxide and low oxygen concentrations) which are not conducive to photorespiration, nor to its detection; moreover, extracellular release of glycolate and other organic compounds was already well-known (e.g. Fogg, 1966; Watt, 1966) before its significance was fully appreciated.

Glycolate excretion occurs when cell growth rates are limited by nutrient limitation and carbon is photosynthetically fixed in excess of immediate requirements, or when the cells are senescent (e.g. Berman & Holm-Hansen, 1974; Berman, 1976*a*; Sharp, 1977): elimination of Calvin-cycle intermediates may be seen to be one option for dealing with 'unwanted' carbon. By analogy, excess carbon fixed at high photosynthetic rates sustained by high light intensities, would follow similar metabolic pathways but would be manifest as photo-inhibition. This is thought to be less likely to occur in healthy cells (Sharp, 1977), except when placed under conditions of physiological stress (high light, low CO_2, low nutrients).

Alternatively, glycolate is oxidized (photorespired) and eliminated as carbon dioxide. The relevant compounds and enzymatic pathways do not occur universally among the 'algae' groups, and current knowledge is still confused. Harris & Piccinin (1977) determined photosynthetic rates (from oxygen production, monitored with electrodes) in suspensions of *Oocystis*, exposed to high light intensities (1300 μE m^{-2} s^{-1}) and temperatures (> 20°C) for varying lengths of time. Their results suggested that an initially high photosynthetic rate was maintained for ten minutes or so but declined rapidly after longer exposures, during which $^{14}CO_2$ was released and O_2 evolution was reduced. Subsequent studies (summarized in Harris, 1978) on other 'algae' showed variable evidence of photorespiratory activity but that different mechanisms, symptomatic of photo-inhibition, exist. *Ceratium* showed no evidence of photorespiration, the photosynthetic rate declining over three hours or so. In suspensions of natural *Asterionella*- or *Fragilaria*-dominated populations, the net rate of P was initially enhanced by irradiance levels in excess of I_k, but again, no evidence of photorespiration was observed. These differences may well be genuine, reflecting the different phylogenetic extractions of the dominant species observed.

Nevertheless, differential adaptations may be advantageous depending upon recent and current hydrographic conditions. 'Shade-algae', drawn from depth and characterized by a high pigment content and high photosynthetic efficiency, are liable to rapid photo-inhibition at high light intensities; 'algae' grown in well-insolated epilimnia of stratified lakes have lowered photosynthetic efficiencies but are less susceptible to photo-inhibition. Certainly, populations taken from differing environments respond to exposure to alternately rising and falling light intensities, in the

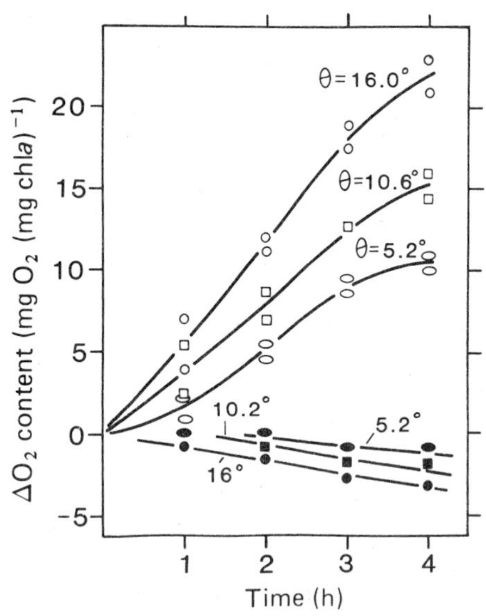

Figure 48. The course of gross photosynthesis in laboratory exposures of natural plankton suspensions ($\equiv 12.6 \pm 0.5$ μg chl *a* l^{-1}), dominated by *Asterionella*, to three temperatures (θ, °C) but constant illumination ($\sim$ 126 μE m^{-2} h^{-1} *PhAR*), as detected by changes in oxygen concentration; solid figures show corresponding changes in darkened bottles ('respiration'). (Original based on author's unpublished data.)

predicted manner: deep-mixed vernal populations are photo-inhibited by irradiances > 300 μE m^{-1} s^{-1} within 5–10 min and subsequent recovery is prolonged; epilimnetic cells from stable epilimnia photo-inhibit only after several hours, while their photosynthetic rate is more directly related to light intensity (Harris, 1973; Harris & Piccinin, 1977). These findings are extremely important to the interpretation of experimental photosynthetic measurements, to the calculation of photosynthetic productivity integrals in open waters and to the understanding of phytoplankton ecology generally, as will be borne out in later chapters.

4.2.4 *The effects of temperature*

'Algal' photosynthetic rates are influenced by several factors other than the quality, intensity and duration of irradiance. These include temperature and the carbon supply, which are now considered separately.

Many cellular processes are temperature dependent, their rates accelerating with increasing temperature, to maximal values occurring between 25° and 40°C. In terms of real temperatures, rates typically increase according to non-linear, exponential functions. Such progressions are generally described by Q_{10} values, i.e. the factor by which the rate increases

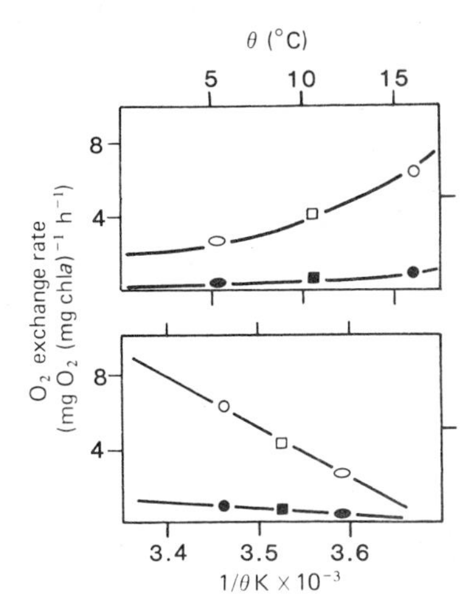

Figure 49. The slopes of the mean photosynthetic and 'respiration' rates, from the curves fitted to the 1–3-h (incl.) portions of Figure 48, plotted against (above) customary temperature and (below) $1/\theta$K. (Original.)

per 10° (Celsius) rise in temperature. Alternatively, these curves can be normalized on Arrhenius plots against reciprocal expressions of absolute temperature ($1/\theta$K).

Time courses of photosynthesis of natural *Asterionella*-dominated phytoplankton suspensions subject to constant laboratory illumination rated at 126 μE m^{-2} s^{-1}, at three separate temperatures, are represented in Figure 48. Regressions are fitted to the 1–3 h (incl.) portions of these curves, the slopes of which correspond to the mean gross photosynthetic rate. In Figure 49, the same rates are plotted against customary temperature (θ, °C) and against $1/\theta$K. The Q_{10} extrapolated from the linking curves is approximately 2.18 between 5° and 15°C and is similar to published values for natural phytoplankton photosynthesis which, between 2° and 25°C, generally lie within the range 2.0–2.3 (e.g. Clendenning, Brown & Eyster, 1956; Talling, 1957*a*; Jewson, 1976).

4.2.5 *The effect of carbon supply*

By taking up dissolved carbon dioxide from the water at a faster rate than fresh supplies invade from the atmosphere, photosynthetic activity, especially in dense populations, can cause significant carbon depletion. Moreover, as the depletion progresses, there is an upward drift in pH as the equilibrium between dissolved CO_2, bicarbonate (HCO_3^-) and carbonate ions (CO_3^{2-}) moves in favour of the latter. Potentially, photosynthetic rate may be limited by the carbon depletion *per se*, the shift in the dominant carbon species present and by the high pH. The effects of these limitations are not easy to separate and their ecological significance, either singly or in concert, is still unclear (Talling, 1976*a*).

It would appear that most algal cells take up CO_2, which diffuses readily across all membranes (e.g. Steeman-Nielsen & Jensen, 1958; Felföldy, 1962), when freely available ($> 0.2\ \mu$mol l^{-1}, pH $\leqslant 8.6$; Lund, 1970). At lower external concentrations, 'algae' may either be limited or they use $(HCO_3)^-$ as the carbon source. Of the former, in some *Chlorella* strains which seem to be restricted to CO_2 as carbon source (Felföldy, 1960) photosynthetic rate at pH > 8.8 is directly limited by the rate of $(HCO_3)^-$ uptake and its intracellular dissociation (e.g. Gavis & Ferguson, 1975; see also Lehman, 1978). Turbulent mixing and movement of the cell relative to the medium may enhance diffusion transport, by breaking down cell boundary layers and maintaining favourable diffusion gradients. By itself, this process will not achieve high productivities at high external pH. More rapid and sustained uptake of $(HCO_3)^-$ ions depends upon active transport mechanisms. Yet photosynthetic rates can be sustained at pH 9–10: indeed, they may be maximal within this range; some reduction occurs at pH > 10, due, presumably, to the effects of high pH *per se*.

The causes of these differences are unclear but they may be interrelated with the carboxylating enzyme systems of the cells (see Harris, 1978), and they may be, to some extent, adaptive (Felföldy, 1960). Moss (1973*a*) concluded, from a series of experiments and a consideration of the literature, that the relative inability to grow at high pH was characteristic of 'oligotrophic' (in this sense, soft-water 'algae': mainly desmids) species. Talling's (1976*a*) more detailed experiments on the capacity of freshwater phytoplankton to remove dissolved inorganic carbon from the water established a series (*Melosira italica* → *Asterionella formosa* → *Fragilaria crotonensis* → *Ceratium hirundinella*/*Microcystis aeruginosa*) of increasing capacity for CO_2-depletion and increasing capability of staging large population maxima under alkaline, CO_2-depleted conditions. His conclusion (Talling, 1976*a*), which differs slightly from that of Moss (1973*a*), is that the CO_2-system in natural waters plays 'a large part in determining the qualitative composition as well as the photosynthetic activity of the freshwater phytoplankton'. This conclusion applies equally to enriched soft-water lakes, whose plankton may be qualitatively similar to those of calcareous lakes, both rich and poor in plant nutrients (see Chapter 6). The conclusion is also relevant to the problem of eutrophication, where argument has centred on whether or not carbon is a major limiting factor (e.g. Kuentzel, 1969; Schindler, 1971). In theory, atmospheric CO_2 invasion is usually sufficient to meet the potential yield determined by limiting nutrients but even rare instances of limitation could have an overriding effect on species composition (Lehman, Botkin & Likens, 1975*a*).

4.2.6 *Respiration*

Compared to estimates of photosynthetic production, there is surprisingly little information on respiratory oxygen uptake. The experimental data included in Figure 48 suggest that the rates of oxygen consumption in darkened controls (0.32 to 0.81 mg O_2 (mg chl*a*)$^{-1}$ h^{-1}) were about one-seventh to one-eighth of the rates of photosynthetic oxygen production at 126 μE m^{-2} s^{-1} *PhAR* at the corresponding temperatures. In the field experiment (Figure 45), the rates were uniform at all depths (equivalent to ~ 0.1 mg O_2 (mg chl*a*)$^{-1}$ h^{-1} or about 1/22 of P_{max}). The limited information available suggests that specific rates of respiration (*R*) typically lie within the range 0.1–2.0 mg O_2 (mg chl*a*)$^{-1}$ h^{-1}. As fractions of photosynthetic rate, these data probably span the typical variation of *R* in natural waters, most falling between 0.04 and 0.1 P_{max} (e.g. Talling, 1957*b*, 1971; Ryther & Guillard, 1962; Steel, 1972; Jewson, 1976; Harris & Piccinin, 1977).

The Q_{10} of the *R*-values in Figure 49 is approximately 2.3, i.e. about the same as for photosynthesis, when related to chlorophyll content: respiration rates apparently show little temperature adaptation. However, there is evidence (reviewed in Harris, 1978) that adjustments in cellular chlorophyll and protein content do yield relatively larger fluctuations in cell respiration rates and that some respiratory adaptation to varying temperatures is possible, particularly among diatom populations.

4.3 Photosynthetic behaviour in non-isolated natural communities

Allusion to the problems associated with direct extrapolation of bottle experiments to describe photosynthetic behaviour of natural phytoplankton has already been made. Prevention of vertical migration and turbulent mixing in an *in situ* bottle chain may give false impressions of photo-inhibition and photorespiration, particularly if experiments are carried out during the sunniest part of the day. Because 'algae' are able to respond rapidly to variations in the irradiance received under natural, mixed conditions, bottle techniques can underestimate integral photosynthetic productivity of free-floating populations: incubation within bottles at a fixed depth or irradiance does not match the response time of cells (Harris, 1980*a*). At least two experimental designs have attempted to overcome this problem: artificial circulation of cells across light gradients, both within enclosed loops (Jewson & Wood, 1975) and in mechanically circulating bottles (Marra, 1978*a,b*), have yielded increased production estimates. Part of this effect may be attributable to the inevitable agitation of the enclosed cells which prevents the build-up of 'diffusion shells' around the organisms: simply rotating bottles *in situ* can result in significantly greater gaseous exchange (Kowalczewski & Lack, 1971). This observation does not invalidate the results of the circulation experiments but casts further doubt on static bottles. The major uncertainty remains the significance of photo-inhibition, if it occurs at all, in cells rapidly transported into and out of depths exposed to saturating irradiances.

Another approach is to measure changes in chemical composition of water resulting from plant and animal metabolism therein: large diurnal changes in the concentrations of oxygen and carbon dioxide and, in soft waters at least, in pH, can be used to estimate primary productivity. The major difficulty here arises in the quantification of inputs and outputs of the gases from the unenclosed system (for instance, through advection and, particularly, exchanges with the atmosphere) and of other sources of consumption or generation of the same gases elsewhere in the same system (community respiration involving micro-organisms and animals) both in the water and in the bottom sediments. Techniques are available for the estimation of these factors, and for their correction for temperature and concentration effects (see e.g. Westlake, 1974; Owens, 1974; Talling, 1974; Kelly, Church & Hornberger, 1974) but their applications to natural communities have had variable success (see, for instance, Kelly, Hornberger & Cosby, 1974; Lingeman & Vermij, 1980). Further development of this aspect of phytoplankton productivity is required.

At the present time, we are bound to rely on model solutions and extrapolations based upon experiments with closed microsystems. These will continue to assume that carbon metabolism is regulated by the responses of intracellular photosynthetic apparatus and enzyme systems to environmental factors; at the same time, they must take account of the behaviour of intact cells in natural water columns. Thus, photosynthetic physiology and productivity will be influenced by sinking and buoyant movements (Chapter 2) and the effect of turbulence and stratification thereon (Chapter 3) or, in short, to the extent and frequency of fluctuations in the light régime. These processes are time-dependent. Fluctuations in fluorescence, chlorophyll content, chloroplast distribution, photo-inhibition and photorespiration are also time-dependent, but the frequency scales (minutes to days) are generally greater than the high-frequency environmental effects (second, minutes; cf. §3.1). The question resolves into one of phytoplankton residence times at different depths relative to the time taken for cells to adapt. Under mixed conditions cells move up and down in the water column over time periods which are sufficiently short for inhibition at the surface (even among diatoms which photo-inhibit relatively rapidly) not only to be largely avoided but to contribute to the maintenance of high photosynthetic rates (Harris, 1978; see also Harris & Piccinin, 1977; Marra, 1978*b*). In quiescent columns, swimming and sinking can contribute to the avoidance of photo-inhibiting irradiances (> 200–$600\ \mu E\ m^{-2}\ s^{-1}$ *PhAR*) and, in some instances, to maintenance of station at more favourable irradiance levels. Physiological (e.g. photo-respiration, extracellular excretion) and morphological (e.g. plastid shrinkage) effects seem to be largely confined to cells unable to escape surface irradiance levels in the above ways (for instance, diatoms 'stranded' near the surface when the water column stratifies or buoyant cyanobacteria lodged at the surface during bloom formation).

Considerable insight into the photosynthetic potential of natural populations in natural water is gained through the linking of quantitative estimates of photosynthesis to the components of the general photosynthetic environment, or 'light climate' (cf. Talling, 1971).

4.3.1 *The underwater light climate.*

Three interacting components dominate the underwater light climate to which phytoplankton is exposed: (1) the intensity and duration of surface-incident radiation, (2) its attenuation with depth, and (3) the vertical extent of column mixing.

4.3.1.1 *Surface-incident radiation.* It has already been argued that at very low irradiance intensities, when the immediate sub-surface irradiance level (I_0') is less than 100–200 $\mu E\ m^{-2}\ s^{-1}$ *PhAR*, photosynthesis will be continuously and directly regulated by light. In many instances, however, I_0' will fall within the range 200 $\mu E\ m^{-2}\ s^{-1}$ (under overcast skies) to 1600 $\mu E\ m^{-2}\ s^{-1}$ (under full tropical midday sun) during at least part of the day. If photo-inhibition is assumed not to occur, there will be an upper portion of the water column in which the available light ($I_0' \rightarrow I_k$) supports a light-saturated photosynthetic rate (NP_{max}); below the depth at which I_k occurs, photosynthesis is light limited. Talling (1957*a*) has shown that the area enclosed by the rectangle $NP_{max} \times z_{0.5I_k}$ is approximately equal to that described by the plot NP v. depth (z), the instantaneous integral of community photosynthesis ($= \Sigma NP$; see Figure 45). The depth, $z_{0.5I_k}$, is related to I_0' through the vertical extinction coefficient, ε, as follows:

$$z_{0.5}I_k = \ln(I_0'/0.5I_k)\cdot\varepsilon^{-1} \tag{23}$$

Talling's solution to ΣNP may be written thus:

$$\Sigma NP = NP_{max}\cdot\ln\,(I_0'/0.5\ I_k)\cdot\varepsilon^{-1} \tag{24}$$

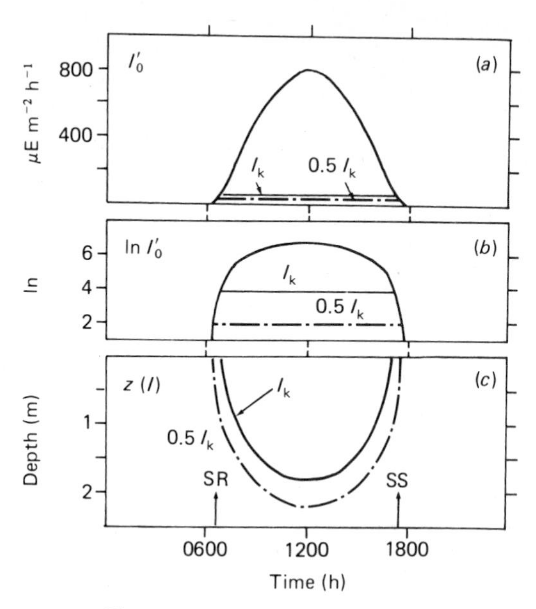

Figure 50. (a) Hypothetical plot of the immediate subsurface irradiance intensity (I'_0) on the day of the experiment, 25 February 1971, assuming I'_0 at midday $\equiv$ 800 μE m^{-2} h^{-1} and the sky to be cloudless throughout; (b) the same on semilogarithmic plot. I_k and 0.5 I_k are represented on both plots; the depths at which these intensities (z_{I_k}, $z_{0.5I_k}$) obtained through the day are plotted on to (c). (Original, following constructions based on Vollenweider, 1965, and Talling, 1971.)

This solution can apply only with $I'_0 > I_k$, although, in reality, the condition will be operative over most (two-thirds to five-sixths) of the solar day. Outside this period, or on very dull days, more complex derivations of $\Sigma\, NP$ will apply (see, for instance, Vollenweider, 1965; Patten, 1968).

The daily course of solar radiation intensity varies essentially as a sine function (an example is plotted in Figure 50a). The same curve is reduced to a logarithmic plot and compared with representations of I_k and $0.5I_k$ in Figure 50b, and with the diurnal variations in the depths at which I_k and $0.5I_k$ occur in Figure 50c. Total daily photosynthesis ($\Sigma\Sigma\, NP$) can be approximated from sequential profiles of ΣNP through the day (e.g. Figure 51) or, assuming that I_k, the population (N) its photosynthetic capacity (P_{max}) and the vertical extinction coefficient (ε) remain approximately constant throughout the day, from an integral derived from observations of these quantities made at around midday.

Integral solutions depend upon accurate empirical evaluation of the diurnal course of I'_0 and in its relationship with I_k. Mathematical integration of $\Sigma\, I'_0$ is not simple, though acceptable approximations can be derived planimetrically. Vollenweider (e.g. 1965) has shown that the daily total

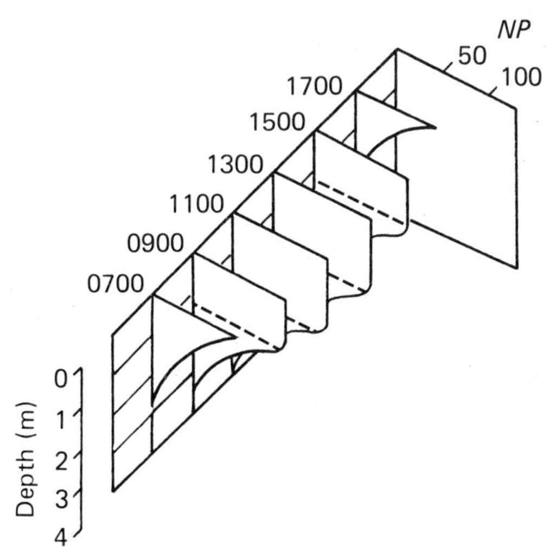

Figure 51. ΣNP profiles at selected times through the day (0700, 0900 h etc.) predicted from Equation 24, with interpolated values of I'_0 (from Figure 50) and I_k (from Figure 45).

irradiance experienced immediately beneath the surface may be approximated as $\Delta\,[(0.63 \text{ to } 0.77)\, I'_{0\,max}]$, where $\Delta =$ the day length in h. Allowing for the period of the day when $I'_0 < I_k$, the daily photosynthetic integral approximates to:

$$\Sigma\Sigma\, NP = [0.67 \text{ to } 0.83]\, NP_{max} \cdot \Delta \cdot \ln\left([0.63 \text{ to } 0.77]\, I'_{0\,max}/0.51 I_k\right)\varepsilon^{-1} \quad (25)$$

Talling's (1957*a*) solution to the integration problem was to treat the absolute daily light integral as equivalent to a mean intensity (I'_0) operating over the entire day (Δ). I'_0 was converted into intensity measured in *light divisions* (LD, $= \ln(I'_0/0.51 I_k)/\ln 2$), having the dimensions of time, and into the required integral by multiplying by the day-length in hours Thus:

$$LDH = \Delta\,[\ln(I'_0/0.5 I_k)/\ln 2] \quad (26)$$

Whence,

$$\Sigma\Sigma\, NP = NP_{max} \cdot \ln 2 \cdot \varepsilon^{-1} \cdot LDH \quad (27)$$

These various solutions are applied to the experimental example (Figure 45) in §4.3.1.4, below, employing evaluations of the vertical extinction coefficient (ε).

4.3.1.2 *Underwater light attenuation.* Owing to the non-linear (hyperbolic) attenuation of light and to alteration of its spectral composition with depth, there is no straightforward evaluation of the vertical extinction

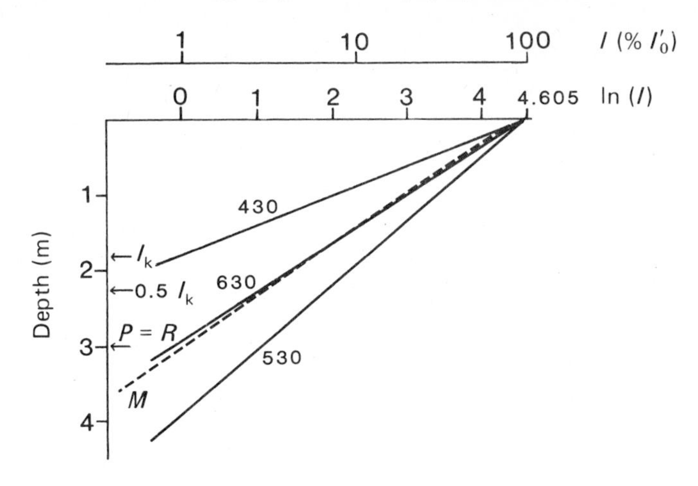

Figure 52. Normalized plots of I, in each of three spectral blocks, peaking at 430, 530 and 630 nm, and the calculated mean of all three (M) against depth. (Redrawn from data in Figure 45d.)

Table 13. *Estimates of* ln (I) *at various depths in Crose Mere on 25 February 1971 and the calculated vertical extinction coefficients for each of three spectral wavelength blocks* (*peaking at* 430, 530 *and* 630 nm *respectively*) *and for the mean of all three; the depth of 1% penetration of* I_0' ($\ln(I) = 0$) *is also tabulated*

nm	430	530	630	mean
ln ($I_{0.5}$)	3.332	4.007	3.795	3.750
ln ($I_{1.5}$)	0.718	2.833	2.197	2.234
ε(0.5–1.5)	2.614	1.174	1.598	1.516
depth where ln (I) = 0, (m)	1.76	3.92	2.88	3.04

coefficient (ε). The plots of light penetration in each of three spectral blocks, shown in Figure 45d, are redrawn in normalized form in Figure 52, together with the (non-linear) function $\ln[(I_{430}+I_{530}+I_{630})/3]$ corresponding to the penetration of *PhAR*. Various derivatives from the plots, set out in Table 13, include the vertical extinction coefficients (ε, m^{-1}) for each spectral block and for the mean (supposed) *PhAR* (ε, m^{-1}). In the example, the greatest relative penetration (and least relative extinction) occurs in the ('green') spectral block peaking at 530 nm and it is predominantly light in this block that sustains potosynthesis between two and four metres depth (Figure 45a, c). Models of planktonic photosynthesis (§4.3.1.1) should clearly relate to this *minimum extinction coefficient* (ε_{min}; in this instance, $\varepsilon_{min} = 1.17$ ln units m^{-1}). Relative to the derived ε value,

$\varepsilon_{min} = 0.77\,\varepsilon$, i.e. $\varepsilon = 1.29\,\varepsilon_{min}$. The combined experience of many workers from many such comparisons consistently points a a factor of between 1.25 and 1.35; Talling (e.g. 1957*a*, 1971) has adopted the value 1.33 as being typical. ε is put to 1.33 ε_{min} in the solutions to the model equations (25 and 27, above) compared in §4.3.1.4.

The attenuation of light is only partly due to absorption by the water itself which, depending upon its dissolved organic ('Gelbstoff') content, may be biased towards the blue end of the spectrum. Frequently, however, absorption is dominated by particulate matter in suspension in the water, which may often be largely 'algal' in origin. In other words, 'algae' contribute to their own 'light climate' and, as will be shown later, may actually become *self-shaded.* The algal contribution to the total extinction (ε_a) is gauged from:

$$\varepsilon_a = \varepsilon_t - (\varepsilon_w + \varepsilon_p) \tag{28}$$

where ε_w and ε_p are the extinction coefficients attributable to the water and to inert suspended particulate matter respectively; ε_t is the observed (total) extinction coefficient. The spectral distribution and absorption coefficients of given waters will normally be constant (e.g. Figure 53). Typical values vary within the range 0.02–0.2 ln units m^{-1} in the green wavelength block, but may be considerably higher in waters with a high Gelbstoff content or ones 'stained' with peat or other organic catchment-derived leachates (e.g. Kirk, 1976). Silts and suspended detritus can obviously vary conspicuously in shallow lakes or ones receiving relatively large fluvial discharges: settlement contributes to increased transparency; floods and re-suspension through wind- and wave-action increase the turbidity. Reynolds (1979*b*) showed that the mean non-algal attenuation in surface waters of Crose Mere (ε_w 0.08 to 0.11 m^{-1}) is equivalent to 0.28 when the lake is stratified but $\sim$ 0.52 ln units m^{-1} during the winter months: the increase was attributed to prolonged suspension of inert detrital particles.

The remainder of ε_t is due to 'algae'. Investigations have been made into the scattering, reflection and absorption characteristics of marine (e.g. Kiefer, Olson & Wilson, 1979; for a review, see Yentsch, 1980) and freshwater phytoplankton (e.g. Kirk, 1974, 1975*a*). In addition, there have been many comparisons between ε_t (generally as ε_{min}; the relationship with ε across the *PhAR* spectrum is rather more complex) and chlorophyll content (e.g. Talling, 1960; Bindloss, 1974; Ganf, 1974*b*; Berman, 1976*b*; Jewson, 1977; Dubinsky & Berman, 1979; Reynolds, 1979*b*). An example based on data for Crose Mere (Reynolds, 1979*b*) is shown in Figure 53. The slope of the regression ε_{min} v. chlorophyll *a* content is a measure of mean ε_a expressed per unit of chlorophyll, or ε_s; i.e.

$$\varepsilon_a = N\varepsilon_s. \tag{29}$$

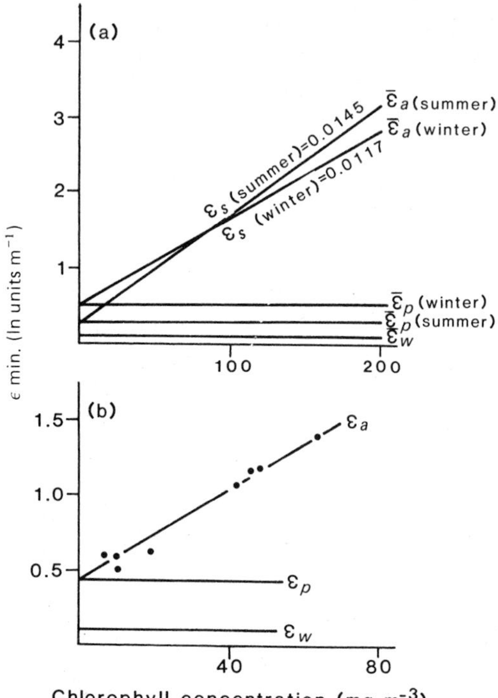

Figure 53. Vertical light extinction coefficients plotted against chlorophyll *a* concentration in Crose Mere; (a) the intercepts of regression equations solved for Crose Mere compared with absorbance (ε_w) of the filtered water (Reynolds, 1979*b*), the difference being equivalent to the background due to particulate non-algal material in suspension (ε_p); (b) values pertaining to January–March, 1971, plotted on an (expanded) section of Figure 53(a). (Original.)

The values of ε_s indicated are typical of the range [0.006–0.02 ln units (mg chl*a*)$^{-1}$ m^2] encountered among natural populations. To judge from published derivations of ε_s from (virtually) monospecific populations, the variability of the range is specific. In particular, ε_s varies with the size (volume) of the algal units (Figure 54): ε_s decreases as unit volume increases above ~ 250 μm^3. This is to be expected since the chlorophyll diffused among a large number of small particles will interfere relatively more with the passage of light than the same amount of chlorophyll concentrated into a smaller number of larger units. This has sometimes been referred to as the 'sieve effect' (Talling, e.g. 1971). In a different context, Edmondson (1980) analogized the variation in light scattering properties of large v. small cells to the difference in the appearance of a stick of

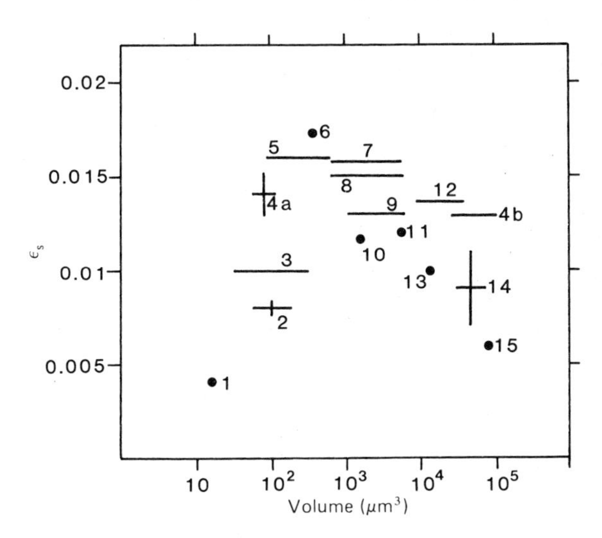

Figure 54. Measurements of the minimum specific extinction coefficient (ε_s) attributable to chlorophyll in suspensions of various algae, plotted against cell volume. From the literature: 1, *Synechococcus* (Talling, in Harris, 1978); 2, Loch Leven nanoplankton (Bindloss, 1976); 3, *Oscillatoria redekei* (Jewson, 1977); 4a, *Microcystis aeruginosa* cells (Harris, 1978); 4b, *Microcystis aeruginosa* colonies (Reynolds, unpublished); 5, cyanobacteria generally (Ganf, 1974*b*); 6, *Scenedesmus* (Harris, 1978); 7, *Asterionella formosa* (Reynolds, 1978*b*); 8, *A. formosa* (Talling, 1960); 9, *Anabaena circinalis* (Reynolds, 1975*a*); 10, Hamilton Harbour netplankton (dominated by colonial chlorophytes: Harris, 1978); 11, *Melosira* (Jewson, 1977); 12, *Stephanodiscus astraea* (Reynolds, 1978*b*); 13, *Oscillatoria agardhii* (Jewson, 1977); 14, *Ceratium* (Harris *et al.*, 1979; Reynolds, unpublished); 15, *Peridinium* (Berman, 1976*b*). (Modified after Harris, 1978.)

blackboard chalk in a beaker of water from that of the same contents after the chalk is crushed into a powder. That ε_s should apparently decrease when mean cell size drops below 250 μm^3 is therefore surprising: variation in ε_s is evidently more complex than has been commonly assumed. For instance, ε_s will be depressed by increases in specific chlorophyll content of (e.g.) low-light adapted cells and increased as a consequence of possessing relatively large quantities of accessory pigments: ε_s is higher in cyanobacteria containing phycobilins than in green algae wherein chlorophylls make up the main pigment complement (Kirk, 1975*b*).

The variability in ε_s has considerable impact upon the 'algal' biomass, in terms of cell volume or dry weight, that can be supported, primarily through its contribution to the light climate. The evaluation of ε in equations (25) and (27) as applied to the example experiment (Figure 45), approximates to

$$\begin{aligned}\varepsilon &= 1.33\,(\varepsilon_{w\,530}+\varepsilon_{p\,530}+N\varepsilon_{s\,530})\\ &= 1.33\,(0.095+0.327+47.6\times 0.016)\\ &= 1.33\times 1.174\\ &= 1.561 \text{ ln units m}^{-1}\end{aligned}$$

There are further, potentially useful, characterizations of the photosynthetic 'light climate' that can be deduced from the vertical extinction coefficient (and from ε_{min}, in particular). The light-limited gross photosynthetic rate (in Figure 45) falls to zero at about 3.2 m, where the *PhAR* remaining (M in Figure 52) is almost exactly 0.8% I'_0 (about 6 μE m^{-2} s^{-1}). This point (*the euphotic depth*, or z_{eu}) marks the lower limit of the water layer (euphotic zone) in which net photosynthesis is possible. The exact quantum value (in terms of I'_0) will vary as I'_0 varies and in relation to all the same factors regulating P, but it generally occurs in the range 0.5–3% I'_0. The definition of the euphotic depth $z_{eu} = z(I = 1\%\ I'_0\ PhAR)$, is thus a valid approximation of the photosynthetic limit in a majority of cases. Apart from direct graphical deductions, z_{eu} can be approximated as $\ln I'_0/\varepsilon$ (provided I'_0 is expressed as 100%, i.e. $\ln I'_0 \simeq 4.605$). In terms of ε_{min} the fraction becomes $4.605/1.33\ \varepsilon_{min}$, i.e. $3.46/\varepsilon_{min}$. It will be evident from Figure 52 that, in the experiment concerned, the fraction of the irradiance in the wavelength band of minimum extinction remaining at 3.2 m was ~ 2.3% $I'_{0(530)}$ (0.84 ln units), whence z_{eu} may be restated as $(4.605–0.84)/\varepsilon_{min}$, i.e. ~ $3.76/\varepsilon_{min}$. Most published determinations of the numerator in this relationship, derived from work in a wide range of lakes of diverse geographical distribution (e.g. Talling, 1960, 1965, 1971; Talling *et al.*, 1973; Ganf, 1974*b*; Bindloss, 1976; Jewson, 1977; Reynolds, 1979*b*), fall within the range 3.2–4.2; Talling's (e.g. 1971) average 3.7 holds for most generalizations.

A further application is the prediction of the phytoplankton population in the euphotic zone (N, in terms of chlorophyll *a* beneath unit area of lake surface. If extinction is due entirely to chlorophyll,

$$3.7/\varepsilon_{min} = 3.7/\varepsilon_a = 3.7/(N\varepsilon_s) = z_{eu}$$

Then:

$$N_{eu} = 3.7/\varepsilon_s \qquad (30)$$

Accordingly, maximum euphotic contents of 185–620 mg m^{-2} may be predicted.

Although digressing from the central theme, it is convenient to mention here the use of Secchi-disk readings as an auxillary measure to characterize light attenuation in natural waters. The Secchi disk is a weighted circular plate, usually 30 cm in diameter, painted white or with alternate black-and-white quadrants, which is lowered into the water, in order to locate the depth (z_s) at which it just disappears from sight. Its general use characterizing the optical properties of lakes is not in dispute. In the absence of more detailed photometric information, broad deductions concerning the photosynthetic environment may be possible but they should be applied with considerable caution (Vollenweider, 1974; Edmondson, 1980). The major limitation is that the Secchi disk is relatively more sensitive to scatter and less so to penetrating irradiance than selenium photometer cells, so that it is not measuring the same effect. Moreover, the depth of disappearance varies with the observer. Several authors, however, have attempted to establish relationships between z_s and z_{eu} or ε_{min} (see e.g. Vollenweider, 1960; Stewart, 1976). For a given lake, the product ($\varepsilon_{min} \times z_s$) should be approximately constant; typically, that constant falls within the range 1.4–3.0, with a mean of about 2.2 (Vollenweider, 1974). Then the light intensity remaining at the Secchi depth will range between 5 and 24% I_0 with a mean of about 15%, whence $z_{eu} \sim (1.2 \text{ to } 2.7)z_s$, with a mean of about 1.7 z_s. In the experiment described in Figure 45, $z_s = 1.7$ m, $z_{eu} = 1.76\ z_s$ and $z_s \varepsilon_{min} = 2.0$.

4.3.1.3 *Column mixing*. Downward extension of the mixed layer (z_m) beyond the euphotic depth (i.e. $z_m > z_{eu}$) increases the proportion of aphotic- to euphotic-zone volume and, hence, the relative time spent in darkness by 'algal' cells. While gross photosynthesis is not necessarily affected (except by light-adaptive responses of the 'algae' to increase their photosynthetic efficiency), the I income is 'diluted' across a larger proportion of the suspended population and respirational losses will be increased in relation to photosynthesis (Talling, 1971).

There are several ways of representing the relationship. The simplest of these is through the direct use of the ratio of mean mixed depth (z_m) and the mean euphotic depth (z_{eu}). However, it must necessarily take into account the bathymetry of the basin, so that the depth of either zone is expressed as its volume over lake area: thus, from the data of Reynolds (1975*b*), represented in Figure 55,the mean winter mixed depth of Crose Mere (730 000 m^3/152 000 m^2) is 4.8 m and the mean depth of the euphotic zone on the occasion of the experiment (Figure 45) was (375 000 m^3/152 000 m^2), 2.47 m, yielding a ratio z_m/z_{eu} of 1.94. The less is the mixed depth (e.g. when the lake is stratified) or the greater is the clarity (e.g. at low biomass) then the more the ratio will tend to 1 and may well fall below it. The effective limit, however, is unity, since the model cannot permit

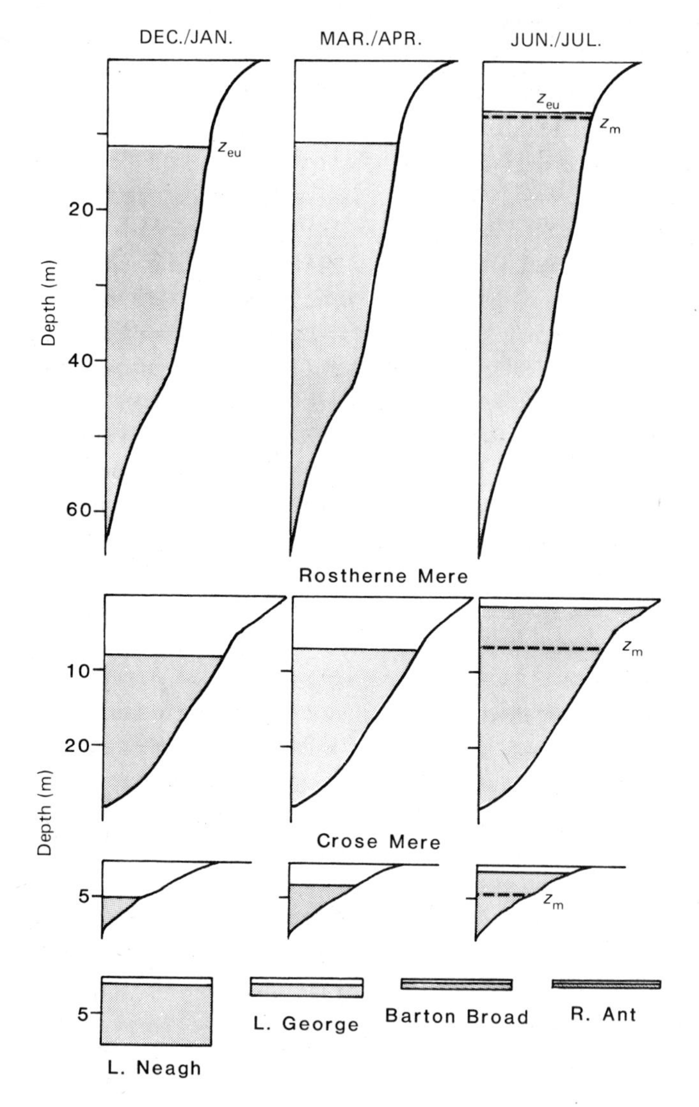

Figure 55. The seasonally-changing relationship between euphotic depth (z_{eu}) and mixed depth (z_m) in lakes of differing morphometries (Windermere North Basin, Rostherne Mere and Crose Mere) and their typical relative volumes in some other habitually mixed environments. Data for Windermere based on Figure 6 of Talling (1971); for Rostherne and Crose Mere on various bathymetric data summarized in Reynolds (1976*b*); for L. Neagh from Jewson (1977); for L. George, Ganf & Viner (1973); for Barton Broad from Osborne & Moss (1977); for R. Ant, Dr J. Hilton (pers. comm.).

phytoplankton to effectively increase its exposure to light for a proportion of day length exceeding one!

An alternative is derived from Talling's (e.g. 1957*a*, 1971) expression of natural depths (z) in terms of units on an *optical depth* scale, each equivalent to one halving of I, so that z is restated as $z\varepsilon_{min}/\ln 2$. From the profile of ε_{min} in Figure 52, natural depth must be multiplied by 1.7 to give the corresponding optical depth; 1 O.D. unit corresponds to 0.59 m, the depth increment giving one halving of I_{530}. The optical depth of the mixed layer is then equivalent to ($z_m\varepsilon_{min}/\ln 2$), i.e. 8.16 in the example given.

Yet another approach is to directly compare $\Sigma\Sigma\, NP$ with the community respiration rate, which is assumed to apply throughout the mixed layer and throughout 24 h; i.e. $\Sigma\Sigma\, NR = 24zNR$. Community respiration is difficult to measure accurately in the field, since oxygen uptake will be enhanced by bacterial and zooplankton respiration and by biochemical oxidation of organic material. So far as the phytoplankton is concerned, NR must be verified in prepared suspensions under controlled conditions (see Figure 48); as shown in §4.2.6, NR may be assumed to fall within the range 0.04 to 0.14 NP_{max}.

Situations conducive to high ratios of mixed to euphotic depth (and, by implication, of column respiration to daily photosynthetic rate) occur in deep lakes during isothermal mixing, in nutrient-rich lakes supporting dense phytoplankton biomasses, and in waters containing large quantities of suspended silt or other non- 'algal' material (e.g. turbid rivers). The examples represented in Figure 55 show how the mixing component interacts with light penetration in the seasonal regulation of the light climate.

4.3.1.4 *Interactions: photosynthetic models.* Pooling together these various components of the light climate, daily photosynthetic integrals can be solved for values interpolated from the example experiment in Crose Mere (Figure 45). Restated, the Vollenweider approximation (equation 25) is:

$$\Sigma\Sigma\, NP = \frac{[0.67 \text{ to } 0.83]\, NP_{max} \cdot \Delta \cdot \ln\left([0.63 \text{ to } 0.77]\, I'_{0\,(max)}/0.5 I_k\right)}{1.33\,(\varepsilon_w + \varepsilon_p + N\varepsilon_s)}$$

Interpolating $N = 47.6$ mg chl*a* m^{-3}, $P_{max} = 2.28$ mg O_2 (mg chl*a*)$^{-1}$ h^{-1}, $\Delta = 10.8$ h, I'_0 (max) $= 800$ and $0.5I_k = 24$ μE m^{-2} s^{-1} *PhAR*, $e_w =$ 0.095, $\varepsilon_p = 0.327$ and $\varepsilon_s = 0.0158$ ln units m^{-2}, $\Sigma\Sigma\, NP$ resolves within the range 1531–2020 mg O_2 m^{-2} d^{-1}.

The Talling solution (equation 27) predicts:

$$\Sigma\Sigma\, NP = \frac{NP_{max} \cdot \ln 2 \cdot \Delta \cdot [\ln(I'_0/0.5 I_k)/\ln 2]}{1.33\,(\varepsilon_w + \varepsilon_p + N\varepsilon_s)}$$

Putting $I'_0 = 406\ \mu E\ m^{-2}\ s^{-1} (\equiv 15.8\ E\ m^{-2}\ d^{-1}/10.8 \times 3600)$ and all other components as above, $\Sigma\Sigma NP$ is resolved at 2119 mg O_2 m^{-2} d^{-1}. For comparison, the integral sum of the profiles represented in Figure 51 is approximately 2057 mg O_2 m^{-2} h^{-1}.

From these estimates of gross photosynthetic production in a small, shallow lake, for one 24-h period at the end of February, the respiration of the suspended community, throughout 24 h, must be deducted.

$$24\ z\ NR = 24 \cdot 4.8\ \text{m} \cdot 47.6\ \text{mg chl}a\ \text{m}^{-3} \cdot 0.101\ \text{mg O}_2\ (\text{mg chl}a)^{-1}$$
$$\simeq 554\ \text{mg O}_2\ \text{m}^{-2}\ \text{d}^{-1}$$

Thus, net photosynthetic production $= \Sigma\Sigma\ NP - 24zNR$
$$\simeq 2057 - 554$$
$$= 1503\ \text{mg O}_2\ \text{m}^{-2}\ \text{d}^{-1}$$

Relative to one unit of the population, expressed in terms of its chlorophyll *a* content (47.6 mg chl*a* $m^{-3} \times 4.8$ m = 228 mg chl*a* m^{-2}), the net production is equivalent to 6.59 mg O_2 (mg chl*a*)$^{-1}$ d^{-1}.

It is possible to convert both the production and the biomass to common units, e.g. of carbon, if it is assumed that 32 mg $O_2 \equiv 12$ mg C and 1 mg chl*a* $\equiv$ 50 mg organic carbon. Thus, the net productivity is:

$$(6.59 \times 12/32)\ \text{mg C per 50 mg C biomass}$$
$$= 0.0494\ \text{mg C (mg C biomass)}^{-1}\ \text{d}^{-1}.$$

This means that the population may be expected to increase its biomass by not more than 4.94% each day.

Where suitable data exist, gross photosynthetic integrals can be derived and applied to trace seasonal fluctuations in net production. Each derivation will be peculiar to a particular lake on a particular occasion. Knowledge is insufficiently developed to permit more than rough approximations to be extended to all situations, though these may sometimes be useful. The models can be used in a generalized way, however, to predict certain features of the production ecology of phytoplankton. For instance, it is clear that gross production ($\Sigma\Sigma\ NP$) is likely to be enhanced by increased temperature and freely-available CO_2 (through their effect on P); I'_0 and I_k are important, for the greater is the difference in their values, so the greater is the value of the light function $\ln (I'_0/I_k)$; ε_{min} and Δ have a relatively greater quantitative effect on $\Sigma\Sigma\ NP$ but the derivative is particularly sensitive to N. Thus, we should expect *production* to be highest among large populations at summer water temperatures. Examples of high photosynthetic production (6–18 g O_2 m^{-3} h^{-1}) come from warmer regions, e.g: Lake George, Uganda, (Ganf, 1975), Lac Tchad, Chad (Lévêque *et al.*, 1972) and Red Rock Tarn, Australia (Hammer, Walker & Williams, 1973). On the other hand, in high-latitude, nutrient-deficient lakes (supporting low biomasses) such as Char Lake, Canada (74°N),

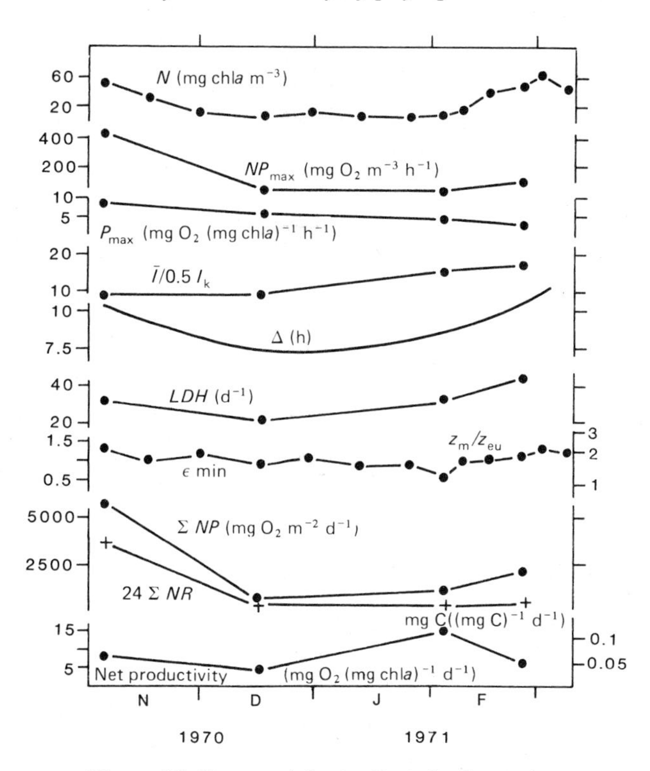

Figure 56. Temporal fluctuations in the various components affecting the calculation of net photosynthetic productivity (as mg O_2 (mg chl*a*)$^{-1}$ d^{-1} or mg C (mg C)$^{-1}$ d^{-1}) in Crose Mere through the winter of 1970/1. (Original, from data in Reynolds, 1978*b*, and developed in Chapter 4.)

maximal production rates of < 10 mg O_2 m^{-3} h^{-1} have been noted (Kalff & Welch, 1974). Net production or *productivity* (i.e. production per unit biomass), however, is sensitive to high N, largely through its effect upon ε and (hence) optical depth and through large respiratory losses. Productivity will therefore be greatest under the same conditions favouring high P but when biomass (N) and attenuation of irradiance are low. Both production and productivity are depressed by increased mixing. At low z_m, production can theoretically lead to an indefinite increase in the productive biomass but this is confined to a progressively shallower layer beneath the surface (see §4.2).

It may be presumed that high net productivity in mixed layers will lead to some increase in biomass and, in turn, to greater potential production and reduced productivity. Indeed, the relationship has been used to monitor the development of specific populations or populations dominated by specific organisms: Figure 56 presents the various components measured in, or deduced from, four experiments through the winter of 1970/1, in

Crose Mere. The modelled net productivity was highest in the early February experiment, even though higher standing crops had been observed at higher water temperatures during the earlier part of the winter. Reynolds (1978*b*) used these data to support the contention that the evident limitation upon winter *Asterionella* increase (and the vernal 'outburst' in particular) was directly consequential upon the light climate in that lake. Substantial net increase in N (see Figure 56) began at the end of January, after having fallen to a seasonal minimum. The interaction of increasing day length (> 8.7 h d^{-1}) with relatively low ε_{min} (> 0.8 ln units m^{-1}) and, hence, low z_m/z_{eu} (~ 1.43) appears to have been critical. This principle is reviewed in the next section.

4.3.2 *Photosynthetic regulation of the phytoplankton community*

Photosynthetic models can be used to define upper limits on the standing biomass of phytoplankton under given limnological conditions. The simplest deduction is that the population that is present in the euphotic zone, defined as $3.7/\varepsilon_{min}$, cannot exceed $3.7/\varepsilon_s$ mg chl*a* m^{-2}. Depending upon the specific ε_s characteristic of the dominant species, the upper limit for the *euphotic content* is between 185 [$\varepsilon_s = 0.02$] and 620 ($\varepsilon_s = 0.006$) mg chl*a* m^{-2}. This limit does not necessarily mean that, once achieved or approached, further net increase is impossible. Provided always that $z_m < z_{eu}$, there remains scope for further population increase: the attenuation increases, the euphotic depth is decreased and the euphotic content remains stable. The productive population is merely confined to a progressively narrower band near the surface.

Sooner or later, however, the condition breaks down, i.e. that $z_m > z_{eu}$. A more complex relationship is then required to describe the biomass limitation. Talling (e.g. 1957*a*, 1971) and Steel (1972, 1973) have addressed this problem, each proposing model equations to predict maximum standing crops under given limnological conditions. The models are, in concept, very similar and are formulated on the identical premise that limitation occurs when the 'column compensation point' is reached, that is, when $\Sigma\Sigma\, NP = 24\Sigma\, NR$. The method selected to estimate $\Sigma\Sigma\, NP$ is less important than the subsequent derivation: here I shall use the Talling derivation (equation 27).

At column compensation point,

$$\Sigma\Sigma\, NP/24\Sigma\, NR = 1,$$

Or, in full

$$\frac{NP_{max}\cdot \ln 2\cdot \Delta\cdot \{\ln(\bar{I}_0'/0.5I_k)/\ln 2\}}{1.33\,(\varepsilon_w+\varepsilon_p+\varepsilon_s N)\cdot 24\,\bar{z}_m NR} = 1 \tag{31}$$

Table 14. *Model calculations of the hypothetical maximum biomass* (N, *in mg chlorophyll* a m^{-3}), *that can be supported in Crose Mere* (*mean depth* 4.8 m) *at the winter solstice, the equinox* (*column fully mixed*) *and at the summer solstice* (*mean depth of epilimnion*, 3.1 m), *assuming different values of* P_{max}/R and ε_s; $(\varepsilon_w+\varepsilon_p)$ *is put equal to* 0.52 ln *units* d^{-1} (*mixed column*) *and* 0.28 ln units d^{-1} (*stratified*); *LDH calculated as* 3.97Δ

	Δ(h)	*LDH*	Z_m (m)	ε_s(ln units (mg chl*a* m^{-2})$^{-1}$)	N when P_{max}/R = 7	N when P_{max}/R = 25	Observed maximum
Winter solstice	7	27.8	4.8	0.013	28	202	20
				0.016	23	164	
Equinox	12	47.6	4.8	0.013	76	374	196
				0.016	62	304	
Summer solstice	17	67.5	3.1	0.011	233	1050	382
				0.016	189	721	

The term N may be cancelled from the numerator and the denominator, and P_{max}/R may be replaced by a factor of between 1/0.04 to 1/0.14 (see §2.6). Thus,

$$\frac{(7 \text{ to } 25)\ln 2\cdot\Delta\cdot\{\ln(\bar{I}'_0/0.5I_k)/\ln 2\}}{1.33\,(\varepsilon_w+\varepsilon_p+\varepsilon_s N)\cdot 24\,\bar{z}_m}=1$$

Whence,

$$N=\frac{1}{\varepsilon_s}\times\frac{(7 \text{ to } 25)\ln 2\cdot\Delta\cdot\{\ln(\bar{I}'_0/0.5I_k)/\ln 2\}}{1.33\times 24\,\bar{z}_m}-(\varepsilon_w+\varepsilon_p) \tag{32}$$

If precise data on I'_0 or I_k are unavailable, an estimated value for the ratio ($I'_0/0.5\,I_k$) may be substituted. The natural logarithm of this quantity generally falls in the range 2.0–3.5, with a maximum frequency between 2.5 and 3.0 (cf. Vollenweider, 1974). Substitution of a factor 2.75 (or $LD = 3.97$, $LDH = 3.97\Delta$) will not result in any great error in the calculation.

In Table 14, calculations of the maximum biomass in Crose Mere are presented for different combinations of P_{max}/R and ε_s, as they might apply at the winter solstice and the equinox (assuming isothermal circulation) and at the summer solstice (water column stratified; epilimnetic depth set notionally at 4 m; $z_m = 3.1$ m). The derived quantities are compared with highest actual estimates of chlorophyll concentration made in the lake at approximately corresponding times of the year. December-January concentrations generally range between 5 and 20 mg chl*a* m^{-3} in this lake; the population that attained a concentration of 196 mg m^{-3} in

March 1968 was dominated by *Stephanodiscus astraea* while the *Ceratium* maxima of August 1968 and August 1973 both accounted for > 377 mg chl*a* m^{-3} (averaged over the mean epilimnetic depth: local concentrations were much higher: see Figure 31a and Reynolds, 1973*d*, 1976*a*). No population has been observed to exceed the maximum concentration hypothesized, but it would be surprising if the assumed optimum conditions were met day-after-day during the course of population increase. Moreover, the model takes no account of loss processes, which can significntly offset the gains from production (see Chapter 7). Nevertheless, the limit calculated with $P_{max} = 7\ R$ is surpassed nearly every spring and summer (see Reynolds, 1973*d*); available data suggest that a P/R ratio of 10–15 gives a more accurate prediction of maximum biomass as chlorophyll.

A more generalized model of the dependence of net production of biomass in terms of the light climate in natural water columns, based on that of Talling (e.g. 1957*a*, 1971), may be developed from equation (31). Restating, we can say

$$\frac{NP_{max} \cdot \ln 2[LDH]}{1.33\ \varepsilon_{min} \cdot 24\ \bar{z}_m\ NR} = 1$$

Whence,

$$\frac{LDH}{\varepsilon_{min} \cdot \bar{z}_m (\ln 2)^{-1}} = \frac{1.33 \times 24 \times R}{P_{max}} \qquad (33)$$

The factor group, $\varepsilon_{min} \cdot z_m (\ln 2)^{-1}$ is Talling's definition of optical depth (§4.3.1.3). Thus, the limiting condition of a mixed water column will be reached when LDH/optical depth is equivalent to $32R/P_{max}$. The light-climate of mixed water columns can be represented as a two-dimensional plot of optical depth (or z_m/z_{eu}) v. LDH, as in Figure 57. The gradients marked on the graph correspond to the limits imposed by various evaluations of P/R; the horizontal dashed lines correspond to the optical depths of the three lakes (Windermere, Crose Mere and Rostherne Mere) described in Figure 55, assuming they have zero active chlorophyll content (i.e. $\varepsilon_{min} = \varepsilon_w + \varepsilon_p$ only), and that they are either fully mixed (as in winter and spring) or are typically stratified, as represented in Figure 55. Net increase in biomass (as chlorophyll *a*), resulting in a further increase of optical depth, is sustainable only when LDH exceeds a critical limit determined by P/R ratio, that is, within the triangle described by LDH, optical depth and the P/R ratio. In this way, it is possible to predict that biomass will be supported in Crose Mere provided LDH exceeds 5 ($P/R = 25$) to 16 ($P/R = 7$). These conditions can be satisfied by mid-winter irradiances except, perhaps, on very dull days. At typical winter chlorophyll levels (10–20 mg m^{-3}; optical depth, 6.5–8), however, critical

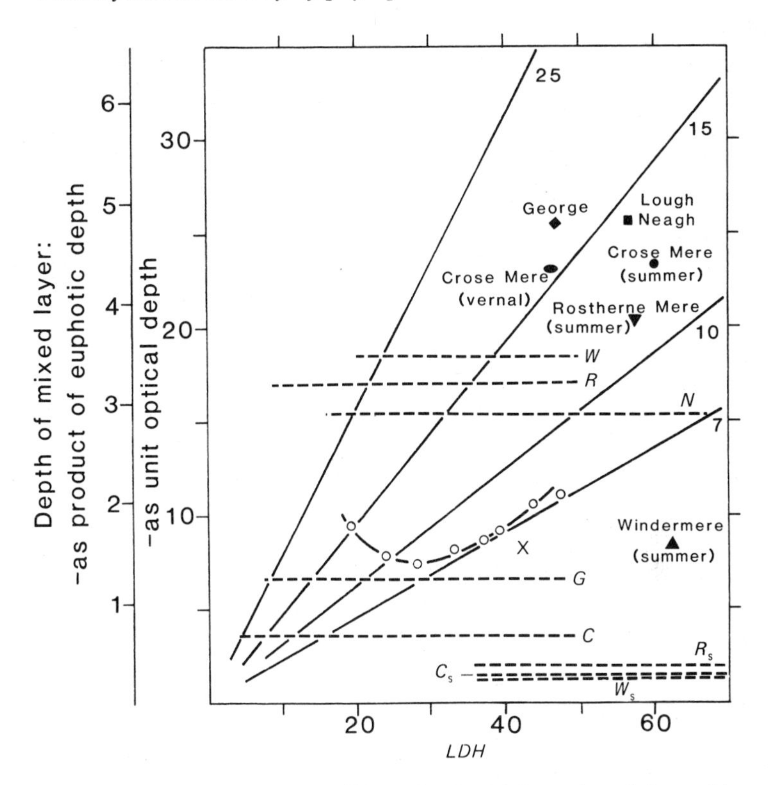

Figure 57. Critical photic conditions (as *LDH*) in various lakes subject to different relative mixing (as a product of euphotic depth or expressed as optical depth). Dashed lines represent the relative depth of the winter mixed layer in a deep relatively unproductive lake (Windermere, W), an optically deep productive one (Rostherne Mere, R) and shallow, productive one (Crose Mere, C), assuming no algae to be present; equivalent values for the summer stratified columns (W_S, R_S, C_S, respectively) and the always-unstratified columns of Lough Neagh (N) and Lake George (G) are also inserted. The plankton biomass that can be supported must lie within the angle subtended by minimum optical depth (dashed) lines and the P/R gradient (those isolines for $P/R = 7$, 10, 15, 25, are inserted). Solid points represent typical maximal levels most of which are most conspicuously grouped around the $P/R = 15$ isoline; Windermere falls well below this line (see text). The series of open circles describe the change in light conditions in Crose Mere between January and March, 1971; the circle marked 'X' represents the occasion of the example experiment on 25 January, 1971. (Original, method of plotting based on Figure 12 of Talling, 1971.)

LDH rises to between 8 and 29. Net increase of the 1971 *Asterionella*-dominated population commenced at approximately 28 *LDH* (corresponding to $\Delta = 8.2$ h d^{-1}: see Fig. 56), with a P/R ratio of ~ 8.5. Subsequent increases in biomass are permitted by increasing *LDH* and, after the onset of stratification, decreasing depth of the mixed layer. The maximum observed mean chlorophyll concentrations in Crose Mere (Table 14) correspond to P/R ratios of ~ 15. The model also enables us to predict the effects of autumnal increases in column mixing and declining *LDH* upon the capacity of the lake to support net phytoplankton growth.

Parallel predictions can be made for Rostherne Mere and Windermere. Both are optically deep water columns in which net increase in chlorophyll concentration is restricted to $LDH > 22$. Indeed, vernal increase is very slow in both lakes until the equinox is passed ($LDH > 33$) but continues to a maximum after the onset of stratification in Windermere (e.g. Lund, 1950), or is truncated altogether in Rostherne Mere (Reynolds, 1978*b,d*). Nevertheless, the latter supports large populations of *Microcystis* and/or *Ceratium* in summer, of the order predicted by a P/R ratio of ~ 12.

The additional information plotted on Figure 57 applies to two further lakes (Lough Neagh, Northern Ireland and Lake George, Uganda) that rarely stratify for more than a few hours on end and which also support large chlorophyll concentrations. In both cases the estimates of optical depth v. *LDH* at the reported biomass maxima are accommodated by the model at a P/R ratio of ~ 15.

The ecological implications of the interactions between incident irradiance, underwater attenuation and column mixing, which together determine the 'light climate' to which phytoplankton is naturally exposed, hinge upon their influence on ambient photosynthetic behaviour, and relative rates of respiration. The foregoing consideration has concentrated on the productive potential of healthy phytoplankton and has established explanations both for how much can be produced and the conditions required. The examples introduced in this chapter have been drawn from lakes which, with the exception of Windermere, are apparently capable of sustaining healthy phytoplankton populations which approach the upper limits that can be predicted from current knowledge of their physiologies. There are, of course, many lakes in which maximal populations fall well short of those limits, often owing to deficiencies in the availability of their chemical requirements. These are considered in the following chapter.

5

Nutrients

'He can take in his beak food enough for a week'
MERRITT, *The Pelican*

5.1 Nutrient requirements of phytoplankton

Of the twenty or so elements which may comprise the tissue of 'algae' (and, indeed, all other plants), either as compounds of varying complexity or as ions maintained within the protoplast, about 11 – C, O, H, N, P, S, K, Mg, Ca, Na and Cl – each typically constitute $\geqslant 0.1\%$ of the ash-free dry-weight and may be referred to as *macronutrients*. The remainder (*micronutrients*), which include Fe, Mn, Cu, Zn, B, Si, Mo, V, and Co, are required to be present in traces (often $\ll 0.1\%$ by weight) but are no less essential to the functioning of the healthy cell. For members of the Chrysophyta and of diatoms, in particular, Si (silicon) is a major component of the cell wall and, hence, occupies the role of a macronutrient.

Before they can be assembled into the structures of the living cell, various nutrients must be obtained from the external medium (by a process which may be simply, if unsatisfactorily, termed '*uptake*'), and almost always in the chemical forms in which they naturally occur. Many are assumed to be present in solution as relatively simple inorganic compounds or ions. Uptake by autotrophic organisms is limited to sources which are both soluble and diffusible, so that they may pass through the semi-permeable membrane (or plasmalemma) into the cell; many soluble complexes and insoluble (particulate) polymers are unavailable. (Heterotrophs are not so restricted, although absorption and final assimilation into the cell still requires that such substances are first broken down by enzymatic digestion.)

Few of even the available, dissolved nutrients are ever present in water in the concentrations that must be maintained within the cell. Thus, passive diffusion rarely meets the uptake requirements of the cell; indeed, the diffusion gradients will often be in the reverse direction. Uptake rather occurs through specific enzyme-transport systems ('ion pumps'), located near the cell surface and which are, thermodynamically, *endothermic* (i.e. consuming energy supplied in respiration or photosynthesis). These mechanisms, even where they are known, will not be considered in detail here.

Of greater ecological importance is that the external concentrations of these nutrients are subject to wide variations in space and time. They are moreover, present in differing amounts relative to the requirements of the active, healthy cell. Thus, growing populations are faced with the prospect that the supplies of one or other of its nutrient sources will 'run out', or become so dilute to be effectively unobtainable. As in the case of the light requirements for photosynthesis, we may recognize that the growth of 'algae' may be limited, saturated or, in some cases, inhibited by the availability of particular nutrients.

That variations in the chemical composition of natural waters might be important in regulating the abundance, composition and the geographical and periodic distribution of phytoplankton was recognized by early ecologists, although the first clear elaboration of hypotheses to explain chemical determination is rightly attributed to Pearsall (e.g. 1930, 1932). Pearsall carried out a series of field studies on the composition of the phytoplankton in relation to dissolved substances in lakes in N.W. England. *Inter alia*, he concluded that: diatoms increased when the water was richest in dissolved silica; that development of chrysophytes (especially *Dinobryon*) was favoured at low silica levels and high ratios of nitrogen to phosphorus; that desmids were associated with waters of low calcium content and a low nitrogen/phosphorus ratio; and that the abundance of cyanobacteria was correlated with the concentration of albuminoid-nitrogen (organic-N).

Pearsall's work stimulated a great deal of research throughout the world on the link between phytoplankton and water chemistry. Whilst few of Pearsall's conclusions remain wholly acceptable in the light of present knowledge, his chemical approach to phytoplankton ecology (and his appreciation of chemical ratios in particular) remains recognizable in many of the current theories concerning nutrient limitation of phytoplankton. Moreover, the subsequent studies have reinforced Pearsall's contention that phosphorus, nitrogen and (in the special case of diatoms) silica are the nutrients whose typically low concentrations in natural waters are most likely to become limiting; some studies have emphasized the importance of certain of the micronutrients ('traces'). Each is considered under separate headings in this chapter, while a further section (§5.6) discusses the interaction between limiting nutrients; limitation by carbon is discussed in the previous chapter (§4.2.5).

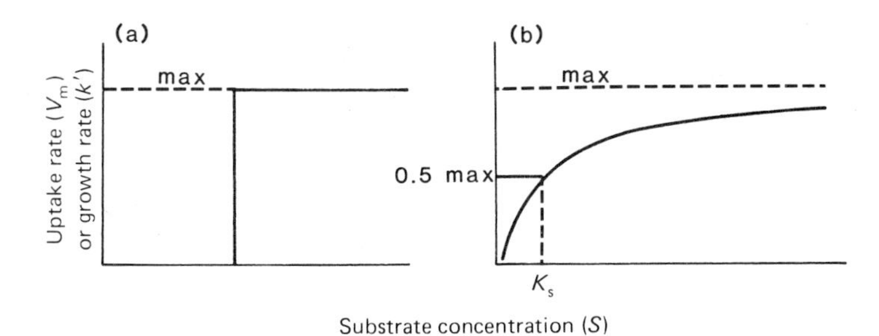

Figure 58. Graphical representation of concepts of 'algal' performance (as growth rate, k', or nutrient uptake rate, V_s) against nutrient substrate concentration (S); (a) according to the Liebig 'model' (but see text); (b) according to Michaelis–Menten kinetics; K_s is the substrate concentration to which the response is 0.5 that of the theoretical maximum. (Original.)

Some initial clarification of what is meant by 'nutrient limitation' may be useful. This may conveniently be done by analogizing the production of 'algal' cells to the assembly of motor cars. So long as the specified selection and number of components are available, it is possible to assemble complete vehicles. As soon as the supply of one of the components (be it wheels or lights or ashtrays) is exhausted, the next motor car cannot be completed. It may arguably function quite well without ashtrays and, for part of the time, without lights, but with the exhaustion of the supply of wheels, further production of useful vehicles ceases: an absolute minimum has been reached. So with 'algae' the elaboration of new biomass cannot reasonably exceed the capacity of the environment to supply the nutrient which is effectively exhausted first: that element is then said to be 'limiting'. This is approximately *Liebig's 'Law of the Minimum'*. Its graphical representation (Figure 58a) assumes that growth proceeds at a rate independent of nutrient concentration until it becomes absolutely limiting.

So far as is known, the absolute minimum cell contents of the macronutrients are, weight for weight, remarkably similar among freshwater species of 'algae' (see Table 15). Apart from the special relationship between diatoms and silica, it is not immediately obvious that the law of the minimum could have a consistent direct impact upon the species composition of natural plankton assemblages: the species harbouring the greatest fraction of the limiting nutrient would be the one with the largest potential biomass. However, there is some evidence from early culture studies (Chu, 1942, 1943; Rodhe, 1948) that species respond differently to different external concentrations of phosphorus: *Dinobryon divergens* and *Uroglena americana* reached their optimum growth rate in media

Table 15. *Minimum and optimum contents of various elements in freshwater 'algae', expressed as percentages of dry-weight or *ash-free dry-weight*

Content	C	N	P	Si
Minimum				
Anabaena flos-aquae	49.7[1]	—	0.40[2]	—
Microcystis aeruginosa	46.5[1]	3.8[3]	0.34[3]	—
Various cyanobacteria	—	4.5[4]	—	—
Asterionella formosa	—	—	—	32[4]
Stephanodiscus hantzschii	—	—	—	20[4]
**Asterionella formosa*	—	3.4[5]	0.03[6]	—
Scenedesmus obliquus	54.6[1]	—	—	—
Scenedesmus quadricauda	—	—	0.59[7]	—
Scenedesmus sp.	—	—	0.10[8]	—
Optimum				
Cyanobacteria	46–49[1]	8–11[9]	0.7–1.1[1],[9]	—
Chrysophytes	—	3.3–5[9]	2.1[9]	—
Chlorophytes	49–56[1]	6.6–1.9[9]	1.2–2.9[1],[9]	—
In General (tentative)[10]	51–56	8.0–10.4	0.8–1.45	—

References; (1), Anon. (1968); (2), Healey (1973); (3), Gerloff & Skoog (1954); (4), Lund (1965); (5), Lund (1950); (6), Mackereth (1953); (7), Nalewajko & Lean (1978); (8), approximated from data of Rhee (1973); (9), Strickland (1965); (10), original, based on various sources. —, No information.

containing $< 20\ \mu g\ P\ l^{-1}$ while *Scenedesmus* and *Ankistrodesmus* showed optima at considerably higher levels.

One way of delaying the onset of limitation by external nutrients is to maintain internal reserves, accumulated in excess of immediate demands while still freely available. There is good evidence for '*luxury uptake*' of this kind among 'algal' species, but storage capacity of suitable condensates in small unicells is restricted. Mackereth's (1953) analyses of the phosphorus content of *Asterionella* cells suggest a capacity for storage equivalent to 24 times the absolute cell minimum, i.e. enough to sustain four full divisions of a cell growing in phosphorus-free medium. Storage of nitrogen and some other nutrients is probably less efficient than this. It is, however, important to realize that a low external concentration of a dissolved nutrient does not necessarily mean that that nutrient is limiting the cells already present.

There is an alternative, more dynamic view, of nutrient limitation. To return to the motor car analogy, the relationship between component stocks and finished vehicles is in reality more complex than a straight arithmetic function. We may assume that the component manufacturers are supplying their goods continuously, though at different rates. While the components are all being manufactured in excess of demand, none is limiting; assembly rate can be maintained to the capacity of the factory. If the delivery of just one of the components (say, wheels) falls below this level, then the rate of assembly drops, or becomes limited by the rate of supply of that component; the greater is the short-fall, the greater is the drop in production. The situation may be eased by reserve holdings of wheels or (without any allusion to motor manufacturers' standards) if wheels are 'recycled' from the breaker's yard, though ultimately, the dependence upon the supply is direct.

The plot of assembly rate (= growth) against supply rate of the limiting component is a hyperbolic curve (cf. Figure 58b). That a similar, positive, hyperbolic relationship exists between the rate of uptake of a limiting nutrient by phytoplankton cells and the concentration of that nutrient was originally demonstrated by Dugdale's (1967) application of the Monod equation describing Michaelis-Menten enzyme kinetics:

$$V_s = V_{s\,max} S/(K_s+S) \tag{34}$$

where V_s is the rate of uptake, $V_{s\,max}$ is the maximum rate of uptake and S is the nutrient concentration; K_s (the half saturation coefficient) represents the nutrient concentration at which the rate of uptake, V_s, is one half of the maximum rate (i.e. $V_{s\,max}/2$). K_s is considered to be species-specific constant and has been recognized to be a measure of the relative ability of a species to use low levels of nutrients. When ambient concentrations are low, low-K_s species theoretically have an advantage over species with higher K_s. Comparative growth rates (in terms of cell doublings per day, or the rate of specific increase in cell biomass per unit time, k') have been assumed (and in many cases shown) to conform with the Monod equation (though the actual value of K_s may be different); thus,

$$k' = k'_{max}\, S/(K_s+S) \tag{35}$$

Droop's (1973) formulation of the relationship between growth rate and the nutrients actually available to the cell (including both external and internal concentrations) represents a further sophistication to the Monod model:

$$k' = \frac{k'_{max}(q-q_0)}{K_s+q-q_0} \tag{36}$$

where q, the 'cell quota', is the amount of limiting nutrient available to each cell and q_0 is the amount when k' is equivalent to zero (i.e. the absolute minimum).

In referring to nutrient requirement and limitation, it is not always possible to say which model is correct. Examples which conform to one or other concept are included in the following discussion. Moreover, limitation is often implied or assumed in comparing the performances of species in natural habitats, without reference to either model. In many such cases, it is appropriate to adopt Gibson's (1971) view of limitation, that is, that a factor is *not* limiting if an increase in that factor produces no significant stimulation in algal growth. This principle is used in the laboratory algal bioassay which measures the growth of algae inoculated into filtered, natural water, otherwise unmodified or 'spiked' with selected nutrients (see also Skulberg, 1964; Maloney, Miller & Blind, 1973).

5.2 Phosphorus

As a component of nucleic acids and of adenosine triphosphate (the basis of enzyme synthesis and intracellular energy transfer systems), phosphorus is essential to the function and growth of all plants. In water, phosphorus usually occurs in the oxidized state, either as inorganic orthophosphate ions (HPO_4^{2-}, $H_2PO_4^-$) or in organic, largely biogenic compounds. Dissolved phosphates are derived from weathering of phosphatic minerals (e.g. apatite) present in catchment soils and are generally present in aqueous concentrations within the range 0.1–1000 μg P l^{-1}; seasonal variations in supply and biological transformations mean that, in any given water, commensurate fluctuations in P concentration are possible. Relatively large amounts of P may be present in particles of varying sizes ranging down to colloidal, and in organic compounds. Partitioning among these various forms, the pathways and rate constants of their transformations and their relationship to inputs, exports and removal to the sediments have been modelled by Lean (1973). Sedimentation of particles, of 'algae' and of the faecal pellets of animals grazing upon them contribute to a flux of phosphorus to the sediments, which thus provide a major reservoir of the element. Events near the lake bottom, especially where this is overlain by an anoxic hypolimnion, lead to further complexing and exchanges of phosphorus with iron, aluminium and organic substances. The rate of 'release' of phosphate, which may be available to 'algae' (Golterman, Bakels & Jakobs-Möglin, 1969), at times exceeds the flux to the sediment.

Dissolved orthophosphate is evidently the major source of phosphorus for phytoplankton and it is taken up rapidly by phosphorus-deficient cells until very low concentrations (< 1 μg P l^{-1}) remain in the water (Rigler, 1966). Direct measurements of phosphate in the water rarely give any accurate measure of the phosphorus available to 'algae'. Moreover, the widely-used molybdenum-blue method fails to distinguish between dissolved and (unavailable) colloidal forms (Rigler, 1968). In modern studies, which acknowledge these difficulties, it is both fashionable and more correct to refer to 'soluble reactive phosphorus' (SRP) and 'total phosphorus' (TP), though these terms in no way resolve the dilemma of deciding exactly what *is* available to algae. A further complication is that many algae can utilize dissolved organic phosphorus sources through the production of alkaline phosphatases. So far as is known (for a review, see Nalewajko & Lean, 1980), phosphatase activity occurs at low external concentrations of available orthophosphates, so that, potentially, low levels of phosphatase production may be an indication of intracellular phosphorus sufficiency.

In view of these considerations, it is impossible to cite phosphorus concentrations that are likely to limit 'algae' in their natural environments without reference to many other variables and to the rates at which phosphorus is supplied to or regenerated within the open water. While SRP levels remain above 5–10 μg P l^{-1}, however the probability is that many phytoplankton species will not be simultaneously limited by phosphorus deficiency.

The kinetics of phosphorus uptake have been the subject of numerous laboratory investigations conducted since the mid-1960s. Much of this work has been reviewed by Nalewajko & Lean (1980): as these authors pointed out, the available data indicate that phosphate uptake by phosphorus-limited cells conforms to the Michaelis-Menten model. However, some aspects of the methodological assumptions have been questioned (Brown, Harris & Koonie, 1978); owing to the mutual interference of biological and chemical processes and phosphate detection by the molybdenum-blue method, reductions in the phosphate concentration of the culture filtrate probably underestimate uptake by the cells present. Methods following the incorporation of labelled phosphorus (^{32}P or ^{33}P) into the particulate (cell) fraction overcome some of these difficulties. Such tracer studies (e.g. Nalewajko & Lean, 1978) have not altered the general conclusion that Michaelis-Menten kinetics apply to phosphorus uptake at concentrations > 10 μg PO_4 P l^{-1} but performance in relation to lower, and perhaps therefore more critical, substrate levels needs further research.

Although half-saturation coefficients and maximal uptakes have been worked out for a number of freshwater species in culture (K_s varying between 11 and 364 μg P l^{-1}; see e.g. Nalewajko & Lean, 1980), no clear, overall picture has yet emerged. Relatively more has been achieved with marine algae: Eppley, Rogers & McCarthy (1969) have concluded, *inter alia*, with respect to nitrogen uptake in 17 phytoplankton species, that K_s values generally increase with cell size and with increasing growth rate and tend to be lower among the species of nutrient-deficient oceans. Analogous characteristics might eventually be shown to apply to phosphorous uptake among freshwater species inhabiting (say) oligotrophic and eutrophic lakes.

The availability of phosphate to cultures limited only by phosphorus is usually manifest in increased growth rate. Under steady-state conditions, the Michaelis-Menten kinetics also describe the relationship between growth rate and substrate concentration, but growth and uptake rates diverge during transient non-steady states (e.g. Gotham & Rhee, 1981), the half-saturation concentrations for growth being generally lower (e.g. Fuhs, Demmerle, Canelli & Miu Chiu, 1972; Droop, 1974; Tilman & Kilham, 1976). When cell substrate is included in the equation, growth rate may be temporarily sustained in excess of uptake rate.

At present, few good data exist on the half-saturation constants of P-limited growth in cultured algae and still less on the behaviour of natural assemblages. Nevertheless, the possibility that interspecific differences in phosphorus limitation of growth rate might influence compositional and dynamic variations in natural waters has been firmly established, especially by the work of Tilman and his colleagues (Titman, 1976; Tilman, 1977; Tilman & Kilham, 1976; Kilham, 1978). Using batch cultures of *Asterionella formosa* and *Cyclotella meneghiniana*, it was found that the species had similar maximum growth rates but that the K_s of *Asterionella* growth was significantly lower (0.02–0.04 μmol l^{-1}, i.e. 0.6–1.2 μg P l^{-1}) than that of the *Cyclotella* ($\sim$ 0.25 μmol, or 8 μg P, l^{-1}). It was shown that, if grown together at low external P concentrations, *Asterionella* usually dominated, as was predicted. The comparisons were extended to examine the effects of low silica concentrations on growth, and later, to semi-continuous cultures of both algae in chemostat cultures in which various ratios of Si/P were maintained. Their results are described below (§5.6), together with other examples of growing selected species against two nutrient gradients.

5.3 Nitrogen

The principal requirement of 'algae' for nitrogen is in the synthesis of amino-acids and proteins, wherein it constitutes about one-eighth to one-sixth by weight; the minimum nitrogen content of cells is about three to four per cent of dry weight. Several commonly-occurring sources of nitrogen are potentially available to 'algae'. These include nitrate, nitrite, and ammonium ions as well as certain dissolved organic nitrogenous compounds, such as urea and free amino-acids and peptides. Certain cyanobacteria are also able to 'fix' atmospheric nitrogen dissolved in the water; nitrogen fixation is discussed separately (§5.3.2).

5.3.1 *Distribution, uptake and assimilation of combined nitrogen*

Inorganic nitrates are extremely soluble and may be abundant (in the order of milligrams per litre) in waters receiving substantial inputs of leachates from agricultural soils, ground water or treated sewage effluent. Without such inputs, maximal concentrations may be one or two orders of magnitude lower (i.e. 10–1000 $\mu g\ N\ l^{-1}$). Freshwaters characteristically deficient in nitrate probably include many tropical continental systems draining forest or well-leached savannah soils (e.g. Prowse & Talling, 1958); low nitrate concentrations are also characteristic of the upper 50–100 m of the world's major oceans. Moreover, seasonal fluctuations in nitrate concentrations are evident in most temperate waters: relatively high winter concentrations become depleted from the upper layers in the summer, as 'algal' uptake exceeds the rate of external supply and thermal stratification restricts internal recycling. In the anoxic hypolimnia of eutrophic lakes, nitrate is reduced to ammonium. Nitrite, which can be produced both chemically and by bacteria either reducing nitrate or oxidizing ammonium, is present at generally low concentrations ($< 60\ \mu g\ N\ l^{-1}$) but may accumulate locally when oxygen tensions are low ($< 1\ mg\ O_2\ l^{-1}$).

Ammonium is also typically present in unpolluted surface waters at concentrations $< 150\ \mu g\ N\ l^{-1}$. Much higher concentrations ($> 1\ mg\ N\ l^{-1}$) can develop in the anoxic hypolimnia of small eutrophic lakes (e.g. Reynolds, 1976*c*, 1979*b*). The principal source of ammonium ions derives from bacterially-mediated degradation of organic matter and from direct animal excretion (McCarthy, 1980); surface concentrations are often temporarily raised following the collapse of 'algal' blooms and during the increased autumnal circulation in stratified lakes. Thus, spatial and temporal variability in ammonium availability potentially occurs on smaller scales than are typical of nitrate. Low concentrations do not always imply ammonium deficiency, because the ion can be continually regenerated in aquatic communities. Equally, direct sedimentation of 'algae' and of the faecal pellets of animals feeding on them transports nitrogen out of the upper water layers of lakes and seas, where nitrogen limitation of phytoplankton remains a possibility. Indeed, nitrogen does appear to be the major nutrient limiting primary production in many tropical freshwater lakes and in the open oceans (cf. Owens & Esaias, 1976).

At times, nitrogen may limit phytoplankton production in temperate eutrophic lakes, especially where phosphate concentrations are relatively high (and N:P ratios are low; see also §5.6) or where low epilimnetic nitrate concentrations obtain in summer (e.g. Gerloff & Skoog, 1957; Lund, 1965, 1970; Edmondson, 1970; Toerien, Hyman & Bruwer, 1975; Reynolds, 1978*b*).

A large literature on 'algal' uptake and metabolism of nitrogen exists and has been the subject of comprehensive reviews (e.g. Morris, 1974; McCarthy, 1980). Phytoplankton actively take up combined nitrogen at low external concentrations (1–2 μg N l^{-1}), through the mediation of enzymes in the plasmalemma. Intracellular assimilation of nitrogen involves a number of reactions, including reductive amination, to form glutamate, and subsequent transaminations to form other amino acids; the substrate for the initial reaction is apparently always ammonium (see Owens & Esaias, 1976, for references). Nitrate and nitrite have to be reduced prior to assimilation, in reactions catalyzed by (respectively) nitrate- and nitrite-reductase. Ammonium is therefore the most energetically favourable source of combined inorganic nitrogen and, not suprisingly, it has been frequently observed (e.g. Strickland, Holm-Hansen, Eppley & Linn, 1969; McCarthy, Taylor & Taft, 1975; Conway, 1977) that ammonium is taken up by marine algae in preference to nitrate when both sources of the element are available, provided NH_4^+-N concentration exceeds 0.5–1 μg-atom l^{-1} (7–14 μg N l^{-1}). The same may be true for freshwater species (Brezonik, 1972; Zevenboom & Mur, 1981); some species (e.g. *Chlamydomonas reinhardii* and *Euglena gracilis*) of hypereutrophic waters may not utilize nitrate at all (Moss, 1973*b*).

Half-saturation concentrations for uptake in N-limited cultures have been determined for a number of marine (Eppley *et al.*, 1969; Caperon & Meyer, 1972; Parsons & Takahashi, 1973) and freshwater species (see Lehman, Botkin & Likens, 1975*b*); generally they are in the range 1–14 μg N l^{-1}, the higher values applying to species of relatively more eutrophic waters, including those species of cyanobacteria for which data are available (e.g. van Liere, Zevenboom & Mur, 1975); the observation has clear ecological repercussions. 'Luxury uptake' (see §5.2) when sources of combined nitrogen are abundant is probable, though the capacity for storing soluble ions, as opposed to assimilates, is restricted and it may be proposed that cellular reserves of nitrogen relative to immediate requirements are potentially exceeded by those of phosphorus.

Under steady-state conditions, half-saturation constants for N-uptake and N-limited growth are probably similar. Moss (1973*b*) found, with few exceptions, that specific growth rates of several laboratory-cultured species of freshwater chlorophytes were similar whether grown with nitrate or ammonium as the nitrogen source; significantly different interspecific growth rates were preserved (see Chapter 6).

5.3.2 *Nitrogen fixation*

Circumstantial evidence that phytoplankton 'algae' were able to assimilate dissolved atmospheric nitrogen (N_2, chemically a relatively inert gas) was first confirmed by Dugdale, Dugdale, Neess & Goering (1959) in a study of *Anabaena* in Sanctuary Lake, Pennsylvania. The biochemistry of nitrogen fixation, based on the nitrogenase enzyme system, has been extensively researched (for reviews, see Stewart, 1973; Hardy, 1977; Granhall, 1978). The introduction of the acetylene-reduction assay for nitrogenase activity (Stewart, Fitzgerald & Burris, 1967) has contributed greatly to the understanding of planktonic nitrogen fixation. It is now clear that the property is exclusive to certain bacteria and cyanobacteria (especially the order Nostocales); the importance of nitrogen fixation in the ecology of cyanobacteria has been stressed many times (e.g. Fogg, Stewart, Fay & Walsby, 1973; Stewart, 1973; Mague, 1977).

Nitrogen fixation in the Nostocales is apparently confined to the heterocysts (Fay, Stewart, Walsby & Fogg, 1968), specialized cells differentiated at intervals along the trichomes or (in *Gloeotrichia*) at their proximal ends, where some 90% of the nitrogenase activity is located (Tel-Or & Stewart, 1977). Heterocysts are readily distinguished under the light microscope by their larger size and thickened walls (for a detailed review of structure and function of heterocysts, see Tyagi, 1975); the latter presumably resist penetration by oxygen, which blocks the nitrogenase enzyme system. There is evidence that the heterocyst: vegetative cell ratio increases in natural populations of *Anabaena* and *Aphanizomenon* as external sources of combined nitrogen fall, below about 300 $\mu g\ N\ l^{-1}$ (Reynolds, 1972; Horne & Goldman, 1972) and, accordingly, the nitrogen-fixing capability is facultative. Nevertheless, strains possessing this capability presumably gain selective advantage over competitors when combined nitrogen limits their growth. Many field measurements of limnetic nitrogen fixation per unit area per unit time have been made (e.g. Brezonik, 1972; Vanderhoef, Huang, Musil & Williams, 1974), and which show that, at times, it can represent the major contribution to the nirogen requirements of the dominant plankton. In some cases, nitrogen fixation by cyanobacteria contributes as much as 50% of the annual nitrogen input into the lake (see McCarthy, 1980).

Nitrogen fixation can occur, or has been induced, in non-heterocystous freshwater cyanobactera (e.g. *Plectonema*: Stewart & Lex, 1970; *Gloeocapsa*: Rippka, Neilson, Kunisawa & Cohen-Bazire, 1971). The maintenance of an O_2-depleted microenvironment, however, remains paramount (Carpenter & Price, 1976). One way of achieving this is through the close adpositioning of trichomes or cells in dense rafts (Sirenko *et al.*, 1968). For many common freshwater genera, however, (e.g. *Microcystis*, *Gomphosphaeria*: Rippka *et al.*, 1971) no such N_2-fixing property has been certainly demonstrated.

Marine nitrogen fixers are represented by non-heterocystous *Oscillatoria* (*Trichodesmium*) spp.; *O. thiebautii* has been shown to derive the bulk of its nitrogen requirements in this way (Carpenter & McCarthy, 1975). Nevertheless, nitrogen fixation is thought to be generally insignificant in the oceans, despite their acknowledged poverty of combined nitrogen (McCarthy, 1980).

5.4 Silicon

Although all phytoplankton have a requirement for the small amounts of silicon involved in protein and carbohydrate synthesis, it is among the chrysophyte genera (and among the diatoms in particular), which obligately strengthen their cell walls with amorphous silica polymers, that the requirement becomes ecologically important. In the diatoms, a pair of siliceous frustules constitute the basic structural unit of the wall; in chrysomonads the silica is incorporated into the delicate scales that clothe the cell; in either case, the completed structure is unique to that species.

Silicon is present in most natural waters as solid or colloidal silicate polymers, derived from catchment soils or ('recycled') from biogenic sources (e.g. dead diatoms). Depolymerization leads directly to the formation of soluble monomeric orthosilicic acid ($SiOH_4$), the 'soluble reactive silicon' (SRS) that is detected colorimetrically by the standard analytical technique (as reduced silicomolybdate: Mullin & Riley, 1955); this is probably the only fraction available to diatoms and other phytoplankton. Modern analytical treatments present data in terms of either $Si(OH)_4$ or SRS concentration; in a large number of earlier, mainly biological texts, equivalent silica (SiO_2) concentrations are cited. Interconversion of units may be derived from $1\ \mu mol\ Si(OH)_4\ l^{-1} \equiv 1\ \mu g\text{-atom Si}\ l^{-1} \equiv 28\ \mu g\ \text{Si (or SRS)}\ l^{-1} \equiv 60\ \mu g\ SiO_2\ l^{-1}$.

SRS concentrations in open waters are typically unsaturated. Maximum concentrations range between 300 μmol (18 mg SiO_2) l^{-1} in some lowland rivers and as little as 2 $\mu mol\ l^{-1}$ in some seas; those in many temperate lakes fall in the order 20–200 μmol (1.2–12 mg SiO_2) l^{-1}. The latter, especially, are subject to marked seasonal variations, increasing during high winter discharges and water circulation and decreasing, sometimes to very low levels ($> 1\ \mu mol\ l^{-1}$), as a direct result of biological uptake. These observations, at least, support the view that silicon availability potentially regulates the growth of diatoms in natural lakes and, as a consequence, the species composition of the plankton.

There has been a number of studies made of morphological and biochemical aspects of silicon assimilation and incorporation by diatoms (for recent reviews see Werner, 1977; Paasche, 1980). The formation of new diatom frustules is restricted to the period just after nuclear division and cell separation and it is apparent, from diatom cultures synchronized to appropriate light-dark alternations, that silicic acid uptake and deposition occurs mainly at this time. Silicic acid is accumulated in the cell by an enzyme-mediated transport system, whose activity is also apparently phased to the division cycle. Each daughter cell eventually receives one of the maternal valves and a new (inner) valve. The bulk of the silicon pool is consumed in the formation of the new valve faces and mantles, which takes some 10–20 min; however, completion of spines and sculpturing and (where applicable) insertion of girdle bands may take several hours. When cultures are grown to the point where external silicic acid is exhausted and new valve formation is prevented, cells having undergone nuclear division and cytoplasmic separation accumulate in the medium.

The evidence suggests that diatoms take up little more silicic acid than is required to complete the next cell division and that 'luxury uptake' and maintenance of reserves of silicon scarcely occur. This view is supported by the relative constancy of diatom silica content (see Table 6 and, e.g., Lund, 1965; Paasche, 1980). Indeed, the combination of precise and accurate determinations of silicon availability and the specific unit requirements of dominant diatom species allow much better assessments of the limits of potential and actual diatom production to be made than is possible from corresponding data for phosphorus or nitrogen. Moreover, growth rate might be expected to remain independent of all but very low external silicon concentrations; equally, diatom species are likely to react more immediately to external silicon deficiencies than to those of (say) phosphorus or nitrogen (see Werner, 1977). In other words, the kinetics of silicon limitation more nearly approach the Liebig model (see §5.1) than those of any other of the major chemical nutrients.

Curves describing short-term uptake versus concentration of $Si(OH)_4$ in diatom cultures do tend to be truncated Michaelis–Menten hyperbolae, which may relate to limitation of the rate of wall formation rather than the capacity of the transport mechanism (Paasche, 1980). Reported half-saturation constants are, however, variable (0.3–5 μmol $Si(OH)_4$ l^{-1} for marine species; 4–8 μmol l^{-1} in freshwater species: see Paasche, *loc. cit.*) emphasizing that interspecific differences in ability to adapt to low external silicon concentrations may exist. Paasche (1980) also pointed to similarities between the maximum rates of specific uptake and growth as an indication of their dependence upon wall formation. Half-saturation constants for silicon-limited growth, even among the freshwater species studied by Tilman & Kilham (1976), remain corresponding low ($< 4\ \mu$mol, 0.24 mg $SiO_2\ l^{-1}$) but interspecific differences are preserved. Many of the studies nevertheless showed that cultured diatoms can be grown to the point where silicic acid in the medium is reduced to the limits of its chemical detection (about 0.2 μmol l^{-1}); the three strains of *Asterionella formosa* studied by Tilman & Kilham (1976) provide a notable exception to this statement, all having failed to grow when the silicic acid concentration fell below 0.4–0.8 μmol (24–48 μg SiO_2) l^{-1}. The latter observation provided evidence that *Asterionella* is not well-adapted to growth at low silicon concentrations, matching Lund's (e.g. 1950) findings on the natural populations of Windermere (see below). However, a subsequent study of *Asterionella* in culture (Hughes & Lund, 1962) indicated that it was capable of using all the available silicon in the medium, provided all other requirements were satisfied.

Lund's classical studies (Lund, 1949, 1950, 1964; Lund *et al.*, 1963) on the seasonal cycle of *Asterionella* in Windermere and other nearby lakes reinforced the link between environmental chemistry and the growth of a phytoplankton 'alga', in a way that was to set the pattern for many subsequent ecological investigations. Detailed records accumulated over many consecutive years repeatedly showed that a net increase in the *Asterionella* population, at a rate equivalent to 1–2 doublings per week, got under way in March and was maintained for 6–8 weeks, until a maximum standing population of some $5–10 \times 10^6$ cells l^{-1} was achieved. Neither the size nor the timing of the maximum was found to be as consistent as its coincidence with the depletion of the dissolved silica concentration to a level of about 0.5 mg SiO_2 ($\sim$ 8 μmol $Si(OH)_4$) l^{-1} from an initial, winter level of between 1.5 and 2.0 mg l^{-1}. Laboratory observations had indicated the silica content of *Asterionella* cells to be typically close to 140 pg SiO_2 cell^{-1} (140 μg, or 2.3 μmol $Si(OH)_4$, per 10^6 cells), so the observed depletion of 1.0–1.5 mg $SiO_2\ l^{-1}$ could be fairly attributed to uptake by the dominant diatom. Although further decrease in dissolved silica was regularly observed and assumed to be indicative of further diatom growth (not necessarily of *Asterionella*, since species of *Cyclotella* and *Melosira* frequently form subdominant vernal populations in the lake), it was not manifest in any net population increase. Indeed, a phase of rapid decline usually followed, much of which could be accounted through sinking out of the upper layers (Lund *et al.*, 1963).

Unequivocal though the evidence for limitation seems, the data introduce us to the problems of interpreting events in natural systems, which are modified by other environmental factors: allowance should be made for the effect of continuous (albeit slow) replenishment of the silica supply from inflows and internal recycling; for losses to grazing, parasitism and sinking during the growth phase; for light and temperature effects; and for the possible effects of limitation by nitrogen, phosphorus (though Lund was unable to correlate the cessation of growth with a fall in their concentrations to any consistent level) or other element. Against this, a further doubling of a population of 3.5×10^6 *Asterionella* cells l^{-1} would theoretically exhaust the dissolved silicon at a concentration of 0.5 mg SiO_2 l^{-1}. The hypothesis that silicon normally limits the growth of *Asterionella* in Windermere is not disproved.

There are many examples given in the literature of silica depletion of natural waters being correlated with planktonic diatom growth, in both relatively silica-deficient (Schelske & Stoermer, 1971; Kilham & Titman, 1976) and silica-rich lakes (Gardiner, 1941; Tessenow, 1966; Reynolds, 1973*a*; Bailey-Watts, 1976; Gibson, 1981), rivers (Swale, 1969), coastal seas (e.g. Pratt, 1965) and open oceans and upwellings (e.g. Ryther *et al.*, 1971; Thomas & Dodson, 1975). In many of the above examples, the concentration of dissolved silica was observed to fall to > 0.03 mg SiO_2 ($\equiv 0.5$ μmol) l^{-1}, effectively restricting any further net increase in the diatom population; where the silica content of individual cells was known, the observed depletion broadly agreed with the amounts required to sustain the observed maximum (see Figure 59). Among the freshwaters considered, the depletion was most extensive in the chemically-rich non-stratifying lakes (Loch Leven: Bailey-Watts, 1976; Lough Neagh: Gibson, 1981) and rivers (Severn: Swale, 1969) which, incidentally, tended to be dominated by centric forms (*Cyclotella*, *Stephanodiscus* and *Melosira*) rather than by *Asterionella*. The relationship is rarely a straightforward one in that supplies (including re-solution from the sediments) continue and cells may be constantly lost from suspension (see Chapter 7). Experiments carried out in closed Lund-Tube systems in Blelham Tarn, Cumbria (described by Lund & Reynolds, 1982; Reynolds & Wiseman, 1982) have nevertheless shown good agreement between the total production of *Asterionella* cells, calculated from the observed uptake of silica and the total cells eventually settling into sediment traps near to the bottom and onto the sediment itself.

Silica limitation is usually followed by a decline in biomass. In the case of Lund's (1950) *Asterionella* a sharp increase in the number of empty cells or ones with disorganized contents indicated that a mass mortality of the cells had occurred. Lund (1950) suggested that while other growth

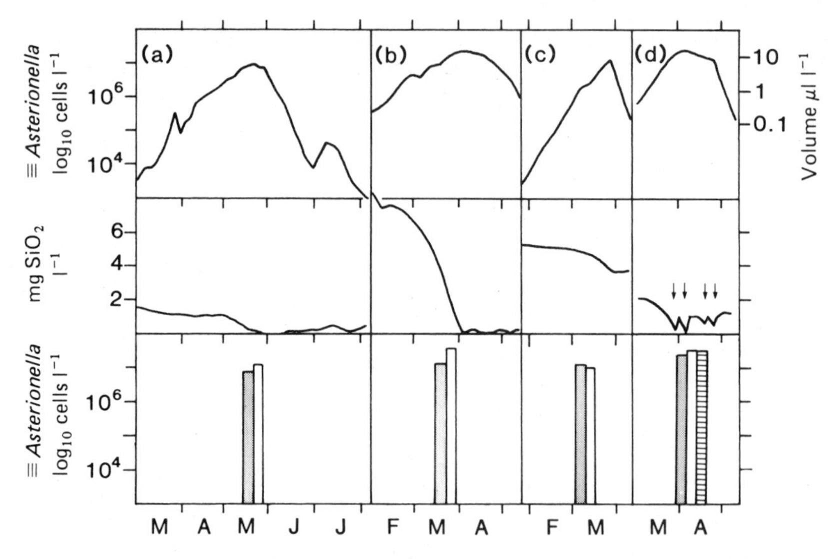

Figure 59. Changes in (top row) the diatom populations, (middle row) silicon concentrations (as SiO_2) and (bottom row) comparisons between the maximum observed standing crops (stippled) and the expected maxima, computed from the silicon uptake in (a) Windermere North Basin, March–July 1946 (Lund, 1950), (b) Lough Neagh, February–April 1978 (Gibson, 1981), (c) Crose Mere, February–March 1967 (Reynolds, 1978*b*), and (d) Blelham Enclosure B, March–April 1978 (Reynolds & Wiseman, 1982); the hatched area in the bottom row indicates the equivalent total sediment trap catch. Diatom populations presented as '*Asterionella* equivalents', assuming one cell ≡ 140 pg SiO_2. (Original.)

conditions remained favourable, the cells were stimulated to maintain growth and nuclear division, but were incapable of obtaining the necessary silicon to complete the formation of daughter cell walls. The experiments of Moed (1973) indicated that silicon-limited *Asterionella* cells do finally die if they are exposed at intervals to illumination > 3 klx (> 50 μE *PhAR* m^{-2} s^{-1}) but that sinking of live cells into deeper (dark) waters provided an escape from the harmful effects of light. Moed (1973) also showed that increased sinking rate was symptomatic of silicon limitation in the laboratory; it is evident from the data of Lund *et al.* (1963) that the rate of sinking loss of live cells from the population increases after the maximum in Windermere. Gibson (1981) came to a similar conclusion in respect of the post-maximal loss of *Stephanodiscus* and *Melosira* cells from suspension in the unstratified water of Lough Neagh: whereas losses of diatoms to the sediments were relatively minor during population increase, the majority (see also Jewson, Rippey & Gilmore, 1981) of cells settle out live and enter the resting state, characterized by dormant, contracted protoplasts (Lund, 1954): there is no mass mortality, though bottom-living cells are liable to consumption by bottom-dwelling invertebrates (Jewson *et al.*, 1981). In both Windermere and Lough Neagh diatoms have the opportunity of surviving the unfavourable conditions (in this case, low silicon concentration) and, potentially, of re-establishing themselves when suitable growth conditions are restored.

Though silicon limitation of diatom crops represents one of the clearest illustrations of a causal relationship between nutrient availability and the ecological response of planktonic populations it must be borne in mind that in many other lakes, vernal diatom maxima are terminated just as rapidly, even though dissolved silica is not limiting (Figure 59(c); see also: Hutchinson, 1944; Reynolds, 1973*a*; Lehman, 1979) and the water can be shown to be chemically capable of supporting diatom growth (Reynolds & Butterwick, 1979). Attendant increases in sinking rate (and in the fractional sinking loss rate from shallower mixed layers) have been demonstrated (Reynolds & Wiseman, 1982; see also §2.6, 3.2, 7.3) that are unrelated to nutrient availability.

5.5 Other nutrients

The collective consignment of 'other nutrients' to a single subsection does not imply that they should be considered less important to phytoplankton. I propose to deal with them in less detail simply because their *ecological* role in regulating species composition and abundance is either relatively minor or it is unclear.

One element that belongs in the second of these categories is calcium. Calcium ions are often (but not always) the most abundant cations in freshwaters, generally ranging in concentration between 0.1, in very 'soft' waters, and $\sim$ 6 milliequivalents l^{-1} (2–120 mg Ca^{2+} l^{-1}) in hard water, 'marl' lakes, saturated with calcium carbonate to the extent that the salt is precipitated from solution. Calcium occupies a central role in the pH-carbon dioxide-bicarbonate system in freshwaters (see Figure 60) and hence, influences the supply of photosynthetically-available carbon and the capacity of the water to buffer fluctuations in pH; both mechanisms play a major part in controlling photosynthetic activity of phytoplankton and, potentially, its species composition (see §4.2.5). It is doubtful whether such variations, which depend on anionic concentrations and transformations, are more than indirectly related to calcium, though it is of interest that Lund (1961) showed that the sparse phytoplankton of Malham Tarn, Yorkshire, a calcareous marl lake otherwise deficient in plant nutrients, was qualitatively similar to that of eutrophic lakes rather than to other nutrient-deficient lakes (cf. Figure 43), even though fluctuation in pH and

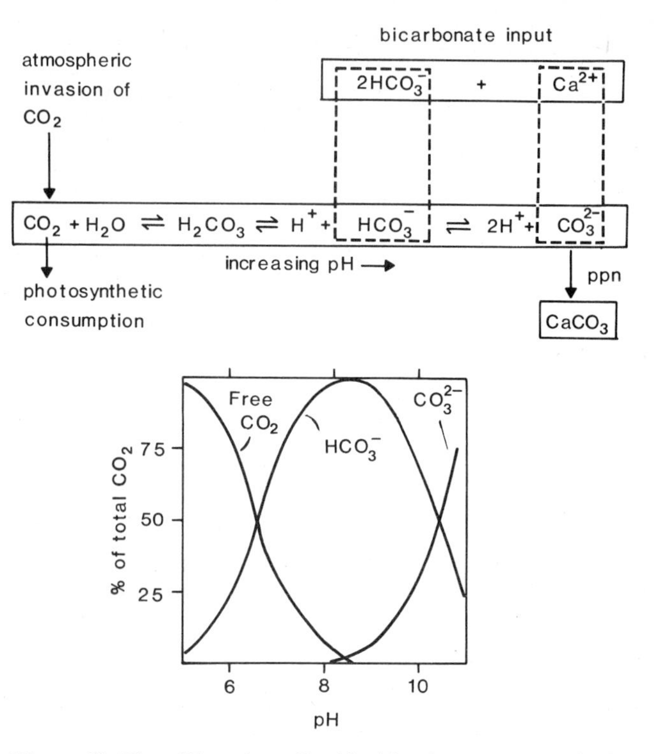

Figure 60. The pH–carbon dioxide–bicarbonate system in freshwaters. The relative quantities of the three components, CO_2, HCO_3^- and CO_3^{2-}, determine the pH of the water, as shown in the inset. Changes in concentration of one component shifts the equilibrium. Photosynthetic withdrawal of CO_2 raises pH to the point where carbonate (CO_3^{2-}) is precipitated as the calcium salt. For a fuller explanation, see Stumm & Morgan (1970).

total carbon dioxide concentrations were relatively minor. Certain algae in the Haptophyceae which deposit coccoliths of calcite crystals near to or at the exterior of the cell presumably have a specific requirement for calcium; however none of the freshwater members of the group produce such elaborate structures as the marine coccolithophorids, so their requirement for calcium must be lower. Freshwater cyanobacteria seem to have an affinity with calcareous waters (e.g. Allen, 1956) that is largely unexplained. Gerloff, Fitzgerald & Skoog (1952) found that 250 $\mu g\ l^{-1}$ of calcium (12.5 $\mu eq\ l^{-1}$) was sufficient to sustain maximal growth of *Microcystis* in culture. Lund (1965) concluded that there seem to be few planktonic 'algae' whose distribution is restricted by the amount of calcium present in natural waters, though he did point to one or two exceptions, citing references to *Phacotus* spp. as indicators of highly calcareous phases in a lake's history.

Magnesium, which forms the other common divalent cation in freshwaters, is an essential component of chlorophyll but there is little evidence of its limiting phytoplankton production in nature (reviewed in Lund, 1965). Similarly, potassium and sodium are rarely considered to have an important influence on 'algal' ecology, because they are usually present in quantities far exceeding 'algal' requirements. Lund (1965) reviewed some reported qualitative effects on algae when potassium salts were substituted for sodium ones, or *vice versa*, but they are difficult to interpret or extrapolate to natural situations.

There is some evidence that the ratio of monovalent (M) to divalent (D) cations may be important in regulating the growth of 'algae' and, because M:D varies conspicuously among natural waters, in determining the species composition of the phytoplankton assemblages they support. Pearsall's (1922) view that diatoms are abundant only in relatively calcareous waters (M:D < 1.5), while many desmids and some chrysophyceans were favoured by higher ratios, was supported by experiments carried out by Provasoli, McLaughlin & Pintner (1954), using species of *Fragilaria* and *Synura*. Talling & Talling (1965) showed that desmids increased in African waters of high alkalinity (up to ~ 2.5 meq l^{-1}) provided that sodium was the dominant cation. The same lakes however also support rich diatom floras. It may be that the majority of freshwater 'algae' can grow over a wide range of M:D (1–14) and that it is rather the effects of high or low calcium (particularly) or, occasionally, sodium (for instance in brackish waters), which are critical.

Apart from bicarbonate, the major anions do not appear to limit algal production in freshwaters. Chloride (with sodium) is high in brackish waters but their distinctive floras owe more to the salinity *per se*. The amounts of sulphate present in natural waters (generally > 0.1 meq l^{-1}, or ~ 4.8 mg SO_4^{2-} l^{-1}) are large in comparison with the sulphur requirements of algae. Sulphide ions, which may be abundant in solution in waters of low redox potential (e.g. anoxic hypolimnia of eutrophic lakes) may be important to those cyanobacteria which can grow in them, both as a source of assimilable sulphur and as an electron donor in the absence of oxygen (Volodin, 1970; Stewart & Pearson, 1970; Oren & Padan, 1978).

The known algal 'micronutrients' (see §5.1) include several metals, the availability of which in natural waters may vary between deficiency and toxic concentrations. Some (B, Va) are required in such small amounts that their specific addition to artificial culture media is often considered unnecessary; their presence as impurities in other chemicals suffice for most practical purposes. However, the requirements for Fe, Mn, Mo, Cu, and Zn are considered sufficient to merit their addition to culture media (Huntsman & Sunda, 1980). Nevertheless, uncertainties concerning their speciation and chemical transformations in natural waters have, until recently, hampered the interpretation of their role in phytoplankton ecology. Thus, although measurements of the total content of each metal may suggest relative superabundance, precipitation and complex formation may render a large fraction unavailable to 'algae'.

It is now clear that the supply of 'heavy' metals to 'algae' is regulated by organic '*chelating agents*'. These have a high affinity for metal ions and so reduce their activity, but nevertheless maintain the metals in solution. Natural chelators are assumed to include humic and fulvic acids ('Gelbstoff'), the chemistries of which are still poorly understood. Their effects are imitated in culture media by the addition of known chelating agents, such as citrate, nitrilotriacetic acid (NTA), trishydroxymethyl-aminomethane (tris, which also serves to buffer pH) and, especially, ethylenediamine tetra-acetic acid (EDTA, usually as the sodium salt), or of humus-extracts of soil. The chelated molecules are presumably too large to be taken up freely by algae. The essential step is the slow dissociation of chelate to release small concentrations of metal ions as they are removed from the medium. Thus, chelates potentially provide exploitable reserves of essential metals and buffer against the toxic effects of excesses.

Iron has long been considered to be the most important trace metal for phytoplankton, principally because the low solubilities of hydrous ferric oxides in aerated neutral waters (between 10^{-8} and 10^{-14} mol l^{-1}) mean that very little iron is available in true solution (except at very low pH) and, without the addition of appropriate chelators, it is difficult to maintain in culture media (e.g. Shapiro, 1967). In the past, the frequently-observed stimulation of growth by the addition of inorganic or chelated iron has been taken as evidence of iron limitation. Applications of the algal bioassay technique for assessing the fertility of filtered natural lake waters (unmodified or spiked with various combinations of nutrients) from the laboratory growth of inoculated test algae, frequently identify iron as a limiting nutrient (e.g. Lund, Jaworski & Butterwick, 1975; Reynolds & Butterwick, 1979); fertility is restored by the addition of iron-EDTA. Reynolds *et al.* (1981) showed that about one third of the total iron (498 μg Fe l^{-1}) in Rostherne Mere water, sampled at a time when the lake supported a dense population of *Microcystis*, was removable with the organisms on a 50-μm filter, and suggested that much of that was already within, or attached to the surface of, *Microcystis* (cf. Elder, 1977). A further half penetrated a 0.45-μm filter, evidently mainly colloidal particles that were unavailable to the assay algae; there was thus no reason to suppose that the natural phytoplankton was iron-limited. Many lakes which support large phytoplankton biomasses have relatively much lower total iron contents than Rostherne Mere (e.g. Lake Victoria, Africa: < 10 μg Fe l^{-1}; Talling & Talling, 1965). Existing experimental evidence for absolute iron limitation of phytoplankton is inconclusive but the possibility that growth rates are dependent upon specific reactions involving chelates and the cells themselves remains.

A further complication is that probably many species of algae liberate organic compounds during growth, which may act as chelators. Certain organisms including some cyanobacteria are known to produce iron-specific 'siderochromes' (Neilands, 1973; Simpson & Neilands, 1976) and this raises the interesting possibility that they could be critical in determining the outcome of interspecific competition for the 'available' iron sources (Murphy, Lean & Nalewajko, 1976).

Information about the other trace metals is also difficult to interpret. Cellular contents of manganese and zinc rank next to those of iron but this may reflect external adsorption on to the cell walls rather than true requirements: Goldman (1964) found that the addition of manganese (as $MnCl_2$) stimulated uptake of ^{14}C by phytoplankton but, in general, the evidence for manganese limitation is inconclusive. The same is true of copper. Excesses of both manganese and copper inhibit many 'algae', and copper sulphate is still widely used as an algicide in the water industry. The toxicity varies interspecifically and with organic content of the water as well as with the concentration of the metal. Additions in the order of micrograms per litre of molybdenum have increased the rate of primary production in several lakewaters (Goldman, 1960, 1964; Dumont, 1972). Cobalt, like molybdenum, is sometimes analytically undetectable; its requirement by 'algae' may be linked with the need for vitamin B_{12}, of which it is a constituent (Lund, 1965).

Because many lakes are able to support much larger crops in response to enrichment of phosphorus (see §5.7) and others can attain light-limited populations when sufficient nitrogen and phosphorus are supplied, the role of trace nutrients in regulating biomass generally seems of subordinate importance. Nevertheless, more subtle controls of growth kinetics and species composition seem probable; a better understanding of these effects may be gained in the light of further research into the complex aquatic chemistries of the nutrients themselves. For more detailed discussion of recent developments in trace metal chemistry and their bearing on phytoplankton, the reader is directed to the review by Huntsman & Sunda (1980).

5.6 Nutrient interactions

It is rarely the case that any given nutrient continuously limits phytoplankton production, although an overall effect on growth rates and biomass of a single limiting nutrient is often clear: Schindler (1977, 1978) has argued that the maximum biomass in many temperate lakes is ultimately limited by the phosphorus supply, much in the way that ocean phytoplankton is limited by nitrogen. Apparent nutrient concentrations, often well below the Michaelis–Menten K_s values for growth observed in continuous cultures, similarly imply frequent P limitation in lakes during the course of the year. Yet even in these waters, low winter temperatures or deep wind mixing (see Chapter 4) may, for example, impose limitations on growth and biomass at lower levels, that is, within the capacity of the nutrient supply. 'Switching' between limiting environmental factors is, potentially, a major feature conditioning growth in natural waters and is itself a powerful selective factor in the ecology of phytoplankton generally, discussion of which is developed in subsequent chapters. The purpose of the present subsection is to identify possible responses of phytoplankton to short-term interactions between different nutrients and to show how these might be conditioned by gross physical characteristics of the medium.

5.6.1 *Responses to interactions of limiting nutrient gradients*

Once the growth kinetics of a given species to a given limiting nutrient have been characterized there are two logical 'next steps'. One is to compare the growth of several species against the same limiting gradient (e.g. Mur, Gons & van Liere, 1978; Gotham & Rhee, 1981). The other is to test the same species against two or more simultaneously limiting gradients (e.g. Rhee, 1978; Rhee & Gotham, 1980). The two approaches are brought together in Tilman's (e.g. 1977) resource-based competition theory which can be tested experimentally and compared with field observations: an outline of the theory is represented in Figure 61. Using batch cultures of *Asterionella formosa* and *Cyclotella meneghiniana*, Tilman & Kilham (1976) established that, though both species exhibited similar nutrient-saturated rates of growth, *Asterionella* had a significantly lower K_s (0.02–0.04 μmol l^{-1}) for P-limited growth than did *Cyclotella* (*ca.* 0.25 μmol l^{-1}; see §5.2) but a significantly higher K_s than *Cyclotella* for Si-limited growth (3.9 against 1.4 μmol l^{-1}). It was correctly predicted that, when growth together under conditions of phosphate limitation for both, *Asterionella* would be dominant but *Cyclotella* would dominate when silicic acid was limiting (Titman, 1976; Tilman, 1977). When both elements are simultaneously limiting, the molecular ratio between them becomes

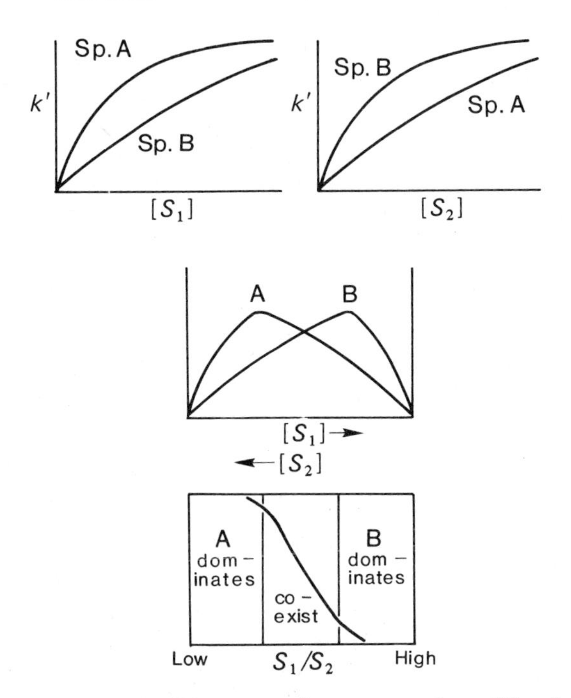

Figure 61. Diagrammatic representation of the elements of Tilman's resource-based competition theory. The top row represents the growth rates of two algae, A,B in relation to increasing concentrations of nutrients S_1 and S_2. When grown together in S_1- or S_2-limited cultures (middle row) species A should dominate when $[S_1]$ is low and species B should dominate when $[S_2]$ is low. When both elements are simultaneously limiting (bottom) the ratio S_1/S_2 becomes critical: A dominates at a low ratio; B dominates at a high one. (Based on figures in Tilman, 1977, Tilman & Kilham, 1976, and Kilham & Kilham, 1980.)

crucial: the ratio of the specific half-saturation constants is nominally between 97 and 195 for *Asterionella* but only 5.6 for *Cyclotella*. The actual ratios of the specific subsistence quotas of these elements were found to be 169 and 15, respectively (Tilman & Kilham, 1976). Thus, *Asterionella* should dominate at high Si:P, *Cyclotella* when the ratio is low. Again, Tilman's (1977) experiments at various Si:P ratios verified the predictions. Moreover, both species could co-exist over that part of the intermediate gradient where *Asterionella* growth was Si-limited while that of *Cyclotella* was simultaneously P-limited. When applied to field observations of the relative abundances of these two species along a natural Si:P gradient in Lake Michigan, the resource-based competition model accounted for some 70% of the total variance (Kilham & Titman, 1976; Tilman, 1977).

Although impressive as an illustration of interaction between species with differing nutrient requirements, the major contribution of this work has been to introduce a powerful new concept to phytoplankton ecology: that interspecific competition along other resource-ratio gradients might play a leading role in influencing the composition and dominance of phytoplankton assemblages and the phasing of observable temporal changes (e.g. Kilham & Kilham, 1980). The concept also permits several species to coexist if they are not limited by the same factors or to the same extent (see also Petersen, 1975). Rhee (1978) suggested that differing optimum N:P ratios of green algae and cyanobacteria influence their mutual competition along N:P gradients and later work (Rhee & Gotham, 1980) showed that differences in specific optimal ratios of cell N and P quotas during the normal growth (when both storage effects and deficiencies should be minimal) of eight cultured species ranged between 7 for the diatom *Melosira binderana*, 9 for *Microcystis* and 20–30 for three species of Chlorococcales. These data may be compared with the Redfield (1958) 'average' atomic ratio of 15–16 (about 7 N:1 P by weight) in healthy, growing cells (see Chapter 1 and §5.1). Significant departures from this ratio have sometimes been taken as evidence of nutrient stress. Rhee's work shows that two-fold variations may be perfectly normal and they are further influenced by storage capabilities and by turnover rates which can alter the ratio over a 24-h light/dark cycle.

The extent to which competitive interactions along natural N:P gradients significantly influence the species composition of phytoplankton communities, as opposed to either nutrient being limiting in its own right, has yet to be fully resolved. Schindler's (e.g. 1977) whole lake fertilizations giving low N:P ratios ($\sim$ 5) promoted the dominance of nitrogen-fixing cyanobacteria, but higher ratios did not. In the (phosphorus-rich) Shropshire meres, seasonally falling nitrogen (reducing N:P) is followed by dominance by *Anabaena* or *Aphanizomenon*, but Chlorococcales dominate those in which the supply of nitrate or ammonium remains relatively high (Reynolds, 1978*b*). These differences may be interpreted in terms of competition for nitrogen along the N:P gradient. However, if the phosphorus supply is critical (N:P high), a quite different algal assemblage, dominated by colonial chlorophytes (e.g. *Sphaerocystis*) or Chrysophyceae (e.g. *Dinobryon*), is more typical (see Figure 43, cf. Reynolds, 1980*a*). Its development is presumably influenced primarily by the outcome of competition for the available phosphorus, independent of the nitrogen concentration. The interpretational difference is not serious, for it is still the interaction between the total amounts of nutrient and their partitioning between the various standing populations and the medium which will bias community structure in a particular direction, and to which specific N:P quotas may testify.

5.6.2 *Physical-chemical interactions*

While the resources for which algae compete extend beyond nutrients (e.g. for light) and that these impinge upon the relative competitive status of the various species present, it must be recalled that lakes, oceans and rivers generally lack the steady-state characteristics of (say) continuous cultures, owing to the (sometimes overriding) effects of water movements. Moreover, horizontal and vertical heterogeneities in the distribution of phytoplankton biomass (see Chapter 3) must lead to local variations in nutrient uptake and subsequent availability.

The water columns of thermally-stratified lakes permit considerable and evolving environmental differentiation and the establishment of striking gradients in resources. Stratification initially develops when the incoming radiation heats the surface water (and so lowers its density) to the point where the available turbulent energy (mostly wind-generated) is no longer sufficient to dissipate the heat throughout the water. A temperature gradient develops which then largely confines further heat income to the upper layers and so accentuates the density difference with the lower layers. The stratification becomes progressively more stable and persists until seasonal cooling, abetted by increased wind action, reverses the process (see also §2.2.4 and §8.1.2).

While stratification persists, other differences between the upper ('epilimnetic') and lower ('hypolimnetic') layers develop. The warmer, well-illuminated epilimnion is more conducive to 'algal' growth than either the cold, darker hypolimnion or the erstwhile fully-mixed column. Phytoplankton growth, however, probably depletes the epilimnetic nutrients at a faster rate than they can be replenished. As cells die, or sediment into the hypolimnion, or are consumed by animals whose faecal pellets sink rapidly, the epilimnion becomes progressively more depleted of nutrients. Hypolimnetic decomposition processes consume oxygen and, in shallow, productive lakes, the hypolimnia often become anoxic. Such conditions (low redox potential) favour the reduction of carbon dioxide to methane, sulphate to sulphide, nitrate to ammonia, and Fe^{3+} to Fe^{2+}, which processes are accentuated by microbial activities. Other nutrients (e.g. phosphate) are released from the sediments into the water. In this way, the lake becomes steadily segregated into a nutrient-limited epilimnion and a light-limited but nutrient-rich hypolimnion.

The vertical distributions of several factors potentially controlling phytoplankton growth in a eutrophic lake in summer are represented in Figure 62. The individual gradients interact to produce a vertical spectrum of 'microhabitats', each offering nearly unique conditions for growth, and

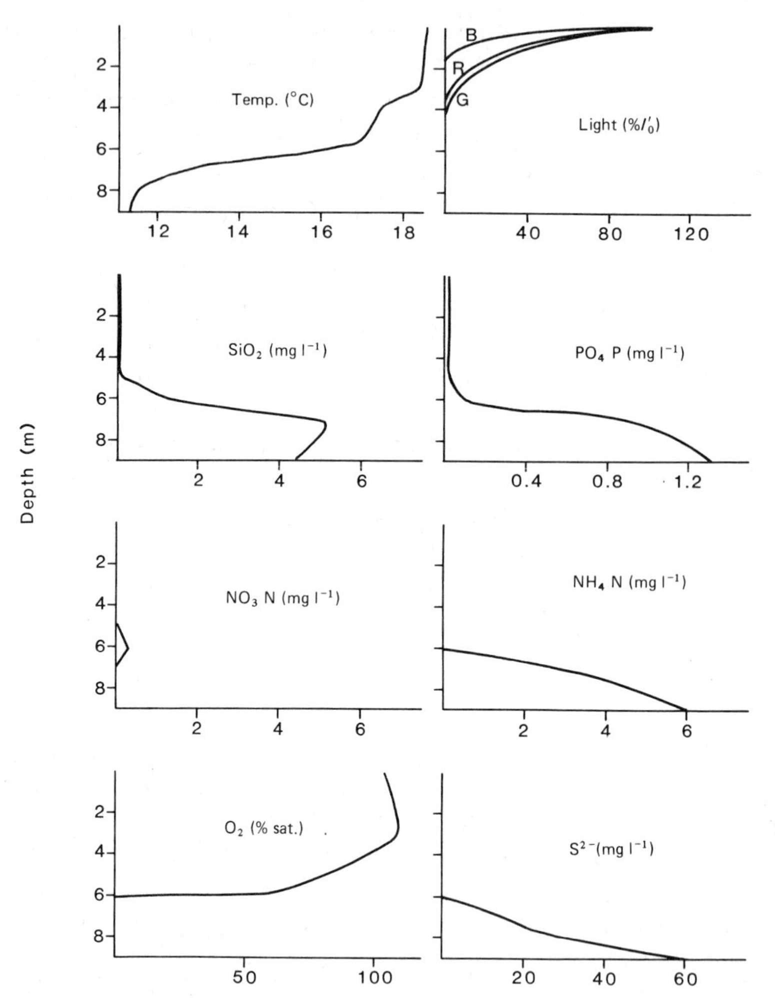

Figure 62. Vertical distributions of various physical and chemical components describing the environment of a stratified eutrophic lake (Crose Mere: 23 July 1974). (From the author's analyses, previously used in Reynolds, 1976*b*, *c*.)

perhaps the conditions favouring the growth of particular species of phytoplankton 'algae' and bacteria. The simultaneous existence of these various microhabitats potentially enables a relatively more diverse flora to be maintained per unit area of lake, not all of which will be directly competing.

In well-mixed environments, cells are likely to experience frequent and rapid changes in the irradiance levels to which they are exposed. Such conditions bias against the attainment of any steady state because they are, as it were, moved frequently and rapidly along gradients in both directions. Nevertheless, during a given period, fluctuations may occur within definable ranges and about some average condition which applies to the population as a whole. These ranges (or '*spectra*') of resources may or may not extend to limiting or inhibiting levels; those which do are more likely to limit the population as a whole. As an illustration, let us envisage a population mixed over several metres; if the euphotic depth is significantly less, then competition for light is critical, selecting for species with low-light tolerances; if phosphorus availability is low, then this too may be just as critical or more so, and the community response may well be different. Equally, low water temperatures may place constraints on 'algae' in another direction, such that neither the light nor nutrients need be limiting. Phytoplankton ecology has not evolved sufficiently that these 'resource spectra' can be described adequately. Yet the communities that do develop show certain common characteristics (Figure 43): these averaged, low-frequency responses must be ultimately related to interactions between physical and chemical factors (Reynolds, 1980*a*).

Harris (1980*b*) has argued that the necessary integration of responses to high-frequency environmental variability takes place at the cellular level (i.e. the variability is perceived *within* the generation time of the cell; cf. §3.1): essentially physiological responses, to fluctuating light – including changes in carbon metabolism, fluorescence (§4.3) – or to fluctuating nutrient availability – such as luxury nutrient uptake (permitting the short-term uncoupling of growth rate from uptake rate) – 'serve to buffer the growth rate of the cell against [high-frequency] fluctuations in the external environment'. The species composition of the phytoplankton community should thus be relatively insensitive to the high-frequency environmental changes themselves but it could be expected to evolve or change in response to the interacting resource spectra, selecting those species best fitted, or adapted, to exploit them. The various nutrients are not the only critical resources, but they often play a leading role in regulating the size and composition of phytoplankton populations, through the various mechanisms discussed in this chapter.

5.7 Eutrophication

Long-term changes in the nutrient supply to lakes represent very large-scale (decades to millenia) shifts in the balance and spectra of resources to which phytoplankton abundance and quality are known to respond. Such changes occur naturally, at different rates, in either direction and have a variety of causes. Relatively dramatic increases in the amounts of nutrients reaching lakes and reservoirs in recent decades, especially in the developed countries, and attendant changes in their biologies have served to focus a great deal of attention to the problems of nutrient enrichment, or *eutrophication.* To review the work done would require an entire volume, one rather larger than this book. Of necessity, this section attempts no more than to reiterate some of the major conclusions, in the context of phytoplankton ecology.

Ecologists use the adjective 'eutrophic' (= 'well-feeding') to describe biological systems into which there is a high input of otherwise limiting nutrients, and which therefore support a high level of organic production. 'Oligotrophic' ('few feeding') describes the opposite, nutrient-deficient condition. The same applies to freshwater systems: eutrophic waters receive relatively high nutrient loadings and can be distinguished from oligotrophic ones by their larger average standing crops of organisms (not just primary producers), and by other metabolic characteristics. The criteria for their separation were established by Thiennemann (1918), although the present terminology owes to Naumann (1919). Perhaps the most important characteristic of a eutrophic lake is that oxygen concentration in the hypolimnion becomes significantly reduced during summer: the greater the epilimnetic production per unit area, then the more intense is the hypolimnetic reduction. Although all the contributory factors were not fully appreciated at the time, it was soon realized that the deep, soft-watered lakes of mountainous areas were generally oligotrophic; on the other hand, eutrophic lakes were frequently shallow, hard-watered and located in lowland districts.

For a long time, it was assumed that oligotrophic lakes were 'primitive' and that they naturally 'evolved' into eutrophic ones. Lakes *do* age, over relatively short periods of geological time, as catchment-derived silts are in-washed and accumulated as lake sediment. Thus, even supposing that no change occurs in the amount of nutrient reaching a basin, its effect becomes more concentrated in the diminishing volume of the lake while the reduction in the hypolimnetic volume accentuates the severity of oxygen depletion during the stratified period. This process has been termed '*natural eutrophication*' (Reynolds, 1979*c*).

However, the trophic status may alter for other reasons. There is now sufficient palaeolimnological evidence to suggest that extensive changes in the rate of sediment accumulation and in the relative contribution of *autochthonous* particles (organic material produced within the lake) and their biological origins have occurred during the past. Moreover, these changes have occurred in both directions, some of the lakes having become more eutrophic or more oligotrophic at different stages of their development (e.g. Pennington, 1978). These fluctuations can be related directly to climatic changes and their influence on weathering rate of rocks in the catchment, throughflow and leaching of catchment soils and to the nature of the vegetation they support. These processes directly influence the loading of free-nutrients on the recipient lake system. At the same time, changes in the hydraulic retention time (the length of time that elapses before the lake volume can be theoretically displaced by the inflow volume) temper the relative efficiency with which free nutrients can be translated into organic standing crops and, eventually, into permanent sediment, or chanelled through the internal degradation cycles which, potentially, permit nutrients to be used several times over. In this way, the trophic condition of a lake will tend towards an equilibrium with its catchment, in which the nutrient supply is balanced by the morphometric characteristics of the basin itself, but one which can be advanced or retarded through time.

During the present century (and during the last four decades or so particularly) the tropic status of many of the world's lakes has advanced rapidly. Many such cases have been catalogued (for instance, in Vollenweider, 1968; Rohlich, 1969). Almost universally, the relative enrichment has been a direct consequence of social or cultural advances made by growing human populations. Forest clearance and the implementation of agriculture alter terrestrial nutrient cycles in favour of more labile components, proportionately more of which enter drainage water. Ploughing and the use of modern inorganic fertilizers accentuate the trend. Concentration of human populations in large urban settlements, importing nutrients (as food) but discharging them locally (as sewage, whether 'treated' or otherwise) also contributes to net changes in catchment-lake equilibria. In addition, modern detergents based on polyphosphates (which are readily hydrolysed to yield biologically-active phosphates: Clesceri & Lee, 1965) constitute a notorious source of nutrient. Many industrial processes also discharge wastes which may similarly add to the nutrients loaded on to recipient water bodies. Inevitably such changes disturb existing individual catchment-lake equilibria accelerating trophic advance to a greater or lesser extent and bringing about *cultural* or *anthropogenic eutrophication.*

Increased nutrient loading on a water body usually increases its capacity to support greater production and maintain larger standing crops of phytoplankton, by raising the thresholds at which either becomes limited; put at its simplest, the more nutrients the more 'algae'. In reality, the relationship is usually more complex. In the short term, additional free nutrients are added to the 'pool' of transient nutrients that is maintained by natural, internal recycling and so meet the immediate requirements of a larger fraction of the existing biomass (or for all of it, for longer). In

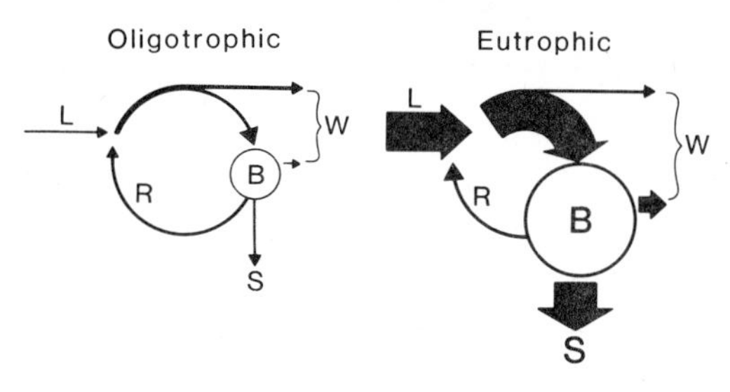

Figure 63. Comparison between the major nutrient pathways in oligotrophic and eutrophic lakes: in the oligotrophic lake, a relatively light nutrient load may sustain a proportionate biomass (B) and a substantial proportion is recycled (R) or washed out of the lake (W); in the eutrophic lake, uptake by planktonic biomass is less dependent upon the recycled nutrient and relatively more of which is in sedimented material (S). (Original.)

the medium term, relatively more of the nutrient resources become partitioned within organisms, whose dependence upon internal recycling is (at least temporarily) alleviated. In the longer term, relatively more nutrient is retained within the lake, ultimately in its sediments (see Figure 63). Thus, the response of the plankton to increased external nutrient loading depends upon the scale of increase relative to the existing internal cycles: the same input of nutrients should have a greater effect on the plankton of an oligotrophic lake than on a eutrophic one (Harris, 1980*b*; see also Reynolds, 1978*c*). The effect of extra nutrients that are definitely not limiting 'algal' growth will be minimal.

Nevertheless, it is the increased average algal standing crop (with attendant decreases in transparency and in hypolimnetic oxygen levels), the sizes of individual maxima, compounded by shifts in species dominance towards more conservative, persistent forms (e.g. many *cyanobacteria*), and the consequent loss of amenity that generate concern about eutrophication of lakes and reservoirs. The eutrophication of potable water supplies has important economic aspects because the offending 'algae' and/or their by-products have to be removed, often at considerable expense. The problems of providing potable water are particularly serious in warm climates.

In turn, this same concern has promoted a great deal of scientific effort towards the formulation of descriptive and predictive models of phytoplankton responses to enrichment with particular nutrients.

The important roles of nitrogen and phosphorus in eutrophication were recognized at an early stage (Vollenweider, 1968). Because the availability of one or other of these two elements places the major constraint on the crop yields of phytoplankton (see §5.1 above), it is not surprising that

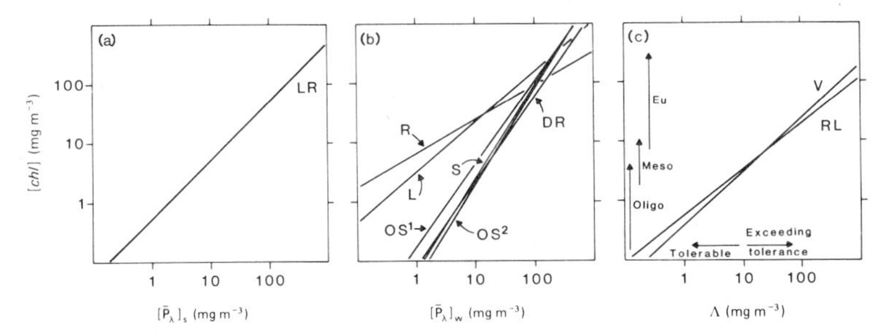

Figure 64. Some relationships between chlorophyll concentration $[Chl]$ in lakes and various measures of the phosphorus supply: (a) is mean summer epilimnetic chlorophyll concentration v. mean summer epilimnetic total P concentration ($[\bar{P}_\lambda]_s$) in Blelham Enclosures (LR: Lund & Reynolds, 1982); (b) *mean* summer chlorophyll concentration in the trophogenic zone (DR: Dillon & Rigler, 1974; OS[1] New York Finger Lakes or, OS[2], deeper stratified examples: Oglesby & Schaffner, 1975; S: Sakamoto, 1966), maximum summer chlorophyll (L: Lund, 1978) or maximum vernal chlorophyll (R: Reynolds, 1978*c*) v. winter-spring concentrations of available phosphorus ($[\bar{P}_\lambda]_w$); (c) mean annual chlorophyll concentration v. annual P-loading factor, corrected for hydraulic loading and residence, $\Lambda = [L(\mathrm{P})\bar{\tau}_w/\bar{z}(1+\sqrt{\tau_w})]$ (RL: Rast & Lee, 1978; V: Vollenweider, 1976). All regression lines significant at 5% level or less; equation as given by the authors or (*) approximated from published figures:

LR: $\log[\overline{chl}]_s = 0.998 \log [P_\lambda]_s - 0.287$
DR: $\log[\overline{chl}]_S = 1.449 \log [P_\lambda]_w - 1.136$
*OS1: $\log[\overline{chl}]_s = 1.333 \log [P_\lambda]_w - 0.809$
*OS2: $\log[\overline{chl}]_s = 1.566 \log [P_\lambda]_w - 1.274$
*S: $\log[\overline{chl}]_s = 1.565 \log [P_\lambda]_w - 1.142$
*L: $\log[\overline{chl}]_{s(max)} = 0.855 \log [P_\lambda]_w + 0.480$
*R: $\log[\overline{chl}]_{v(max)} = 0.585 \log [P_\lambda]_w + 0.801$
RL: $\log[\overline{chl}]_y = 0.76 \log \Lambda - 0.259$
V: $\log[\overline{chl}]_y = 0.91 \log \Lambda - 0.435$

average crops should increase in response to an augmentation in their supply. The essence of quantification of the effect of eutrophication is to determine 'how much phytoplankton' for 'how much nutrient'.

Because phosphorus and chlorophyll *a* each constitute between 0.5 and 2.0% of the ash-free dry weight of healthy phytoplankton cells it can be anticipated that the available phosphorus supports phytoplankton chlorophyll in a ratio of approximate unity. Lund & Reynolds (1982: see also Figure 64a) found a highly significant correlation between the means of weekly estimates of summer chlorophyll concentration and of total phosphorus concentration (which includes fractions believed to be 'unavailable' to 'algae') in the closed limnetic systems of Lund Tubes in Blelham Tarn that were nevertheless subject to very different phosphorus loadings in different years. The regression equation indicated that the chlorophyll:P ratio is consistently about 0.5.

As has been discussed earlier (§5.6), epilimnetic phosphorus concentrations may be low in summer at a time when other conditions (light, temperature) might favour growth. Several authors have sought to relate mean summer standing crops to the phosphorus concentration obtaining at the start of the growing season, i.e. at the winter 'overturn' or in spring, before the vernal increase gets under way. A notable early attempt to establish such a relationship in a series of 12 Japanese lakes was that of Sakamoto (1966). His plots of chlorophyll content (mean of three summer samplings) against early season [P] are redrawn on Figure 64b(S). Dillon & Rigler (1974, 1975) brought together data from several North American lakes as well as one or two European ones and those of Sakamoto's survey; they calculated a similar regression equation (Figure 64b: DR). Oglesby & Schaffner (1975; see also Oglesby, 1977; Oglesby & Schaffner, 1978), working on the 'finger lakes' of New York State, also established a relationship which, at least for the deeper stratified lakes, closely agreed with the previous equations. They infer that the summer chlorophyll content will average about 0.05–0.07 $\mu g\ l^{-1}$ in those waters where the pre-season P content amounts to 1 $\mu g\ l^{-1}$, and between 50 and 100 $\mu g\ l^{-1}$ where it is 100 μg P l^{-1}.

If the amount of phosphorus available to phytoplankton is increased ('eutrophication'), it is important not only to be able to identify the quantities of the nutrient arriving in the lake, but to be able to relate the likely phytoplankton standing crops thereto. Indeed, this is the essence of rational decisions concerning what is or is not an acceptable (tolerable) increase, and this was one of the central aims of the OECD research programmes on eutrophication (e.g. Vollenweider, 1968, 1976; Lee, Rast & Jones, 1978).

Attempts to deal with loading concepts have implicitly recognized that a given input of a given nutrient will have different relative effects on different water bodies. If no biological changes occurred to dissolved nutrient levels, supplied by inflow, then we would expect the water in the recipient system to remain chemically identical. Nutrient uptake by phytoplankton and the subsequent transfer of nutrients through the biological components of the system deplete their concentrations in the recipient water, so that at any given time, the nutrient concentration in the inflow is always likely to exceed that in the water body, and hence tends to enrich it. The loading on the system approximates to the product of the inflow concentration and its volume: its instantaneous impact on a well-mixed system will be 'diluted' out across the entire water area. Thus load is expressed per unit area per unit time. The effect of the loading on concentration is also dissipated through the depth of the water body, requiring that allowance be made for (usually its mean) depth. Moreover, the hydraulic load (the volume of water) must be roughly balanced by outflow, so that the *residence time* (*displacement* or *retention time*, τ) becomes important: it determines the duration that the additional nutrients are available (at short retentions, much may remain unused before being lost in outflow), or from the opposite point of view, the opportunity for nutrients to be absorbed, cycled and sedimented within the system. Thus, the translation of loading rate, $L(\mathrm{P})$, assuming that it can be accurately quantified, into likely chlorophyll biomass requires the introduction of compensatory terms.

Various authors have grappled with this problem (e.g. Oglesby, 1977; Schaffner & Oglesby, 1978), but the expressions developed by Vollenweider (e.g. 1976; see also Vollenweider & Kerekes, 1980) and essentially similar ones by Lee and his co-workers (Rast & Lee, 1978, quoted by Lee *et al.*, 1978; Jones & Lee, 1980) seem to be, potentially, the most useful. These invoke the various components (areal loading, mean depth and hydraulic retention time) mentioned above to deduce the relative retention time of phosphorus within the system and hence predict the mean annual phosphorus concentration $[\mathrm{P}_\lambda]$ in the water body.

$$[P_\lambda] = [P]_q(\tau_\mathrm{p}/\tau_\mathrm{w}) \qquad \text{(37; cf: Vollenweider \& Kerekes, 1980)}$$

where τ_p and τ_w are the mean retention times of phosphorus and the water in the system, and $[P]_q$ is the average phosphorus concentration in the inflow. Vollenweider (1976) showed that, statistically, $\tau_\mathrm{p}/\tau_\mathrm{w}$ approximates to $[1/(1+\sqrt{\tau_\mathrm{w}})]$; surface-area phosphorus loading rate is related to mean inflow concentration as follows:

$$L(\mathrm{P}) = [P]_q \times \text{inflow volume per unit time/lake area}$$
$$= [P]_q \times \text{inflow volume per unit time/(lake volume)/(mean depth)}$$

which is equivalent to:

$$L(\mathrm{P}) = [P]_q \times 1/\tau_\mathrm{w} \times \bar{z}. \qquad (38)$$

Rearranging and substituting (38) in (37)

$$[P_\lambda] = L(\mathrm{P})\tau_\mathrm{w}/\bar{z}(1+\sqrt{\tau_\mathrm{w}}). \qquad (39)$$

Because chlorophyll concentration can be related to $[P_\lambda]$ in the manner of the equations shown in Figure 64(b), the function $[L(\mathrm{P})\tau_\mathrm{w}/\bar{z}(1+\sqrt{\tau_\mathrm{W}})]$ (abbreviated to Λ in Figure 64c) becomes a useful predictor of mean annual biomass ($[chl]_\mathrm{Y}$). Indeed, Vollenweider (1976) and Lee *et al.* (1978)

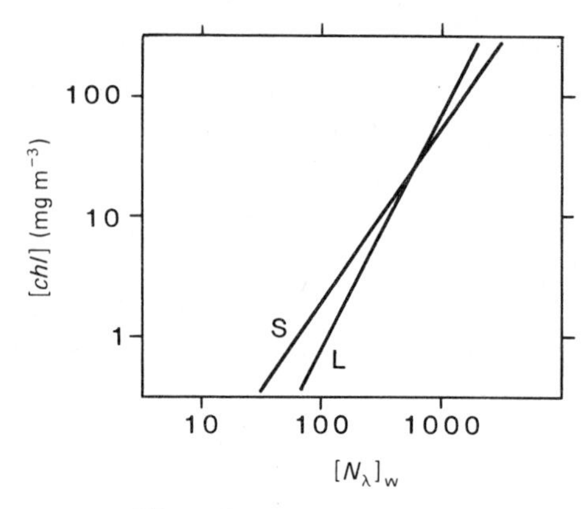

Figure 65. Relationships between mean (S: Sakamoto, 1966) or maximum (L: Lund, 1970) summer chlorophyll concentration and winter nitrate nitrogen concentration $[N_\lambda]_w$. Both regressions are significant at 5% level. Equations, approximated from the author's data:

$$\text{S: } \log \overline{[chl]}_s = 1.44 \log [N_\lambda]_w - 2.62$$
$$\text{L: } \log \overline{[chl]}_{s(\max)} = 1.95 \log [N_\lambda]_w - 4.01$$

confirmed this deduction for various OECD data sets. Using appropriate units [L(P) mg m^{-2} yr^{-1}; τ_w yr and $\bar{z}$ m], Vollenweider calculated that:

$$\log [chl]_Y = 0.91 \log \Lambda - 0.435 \text{ mg m}^{-3}.$$

This deviates little from the Rast/Lee relation (Lee *et al.*, 1978, see Figure 64c);

$$\log[chl]_Y = 0.76 \log \Lambda - 0.259 \text{ mg m}^{-3}.$$

The value of equations such as these is that they enable the effects of projected increases (or decreases) in nutrient loading to be judged in terms of likely new biomass levels and on which terms their acceptability can be based. Scales of oligo-, 'meso-' and eutrophy and the generally recognized critical loading factors for recreational lakes ($\Lambda = 10$ mg P m^{-3}) are inserted on Figure 64 c. These models have been successfully fitted retrospectively to describe the improvement of water quality in Lake Washington following the diversion of all major sewage outfalls into the lake (see e.g. Edmondson, 1972), to certain of the whole lake experiments carried out by Schindler, Fee & Ruszczynski (1978) and to predict the impact of different reductions in the phosphorus loads on the North American Great Lakes (e.g. Lee *et al.*, 1978).

Relatively much less attention has been directed towards nitrogen or other nutrients, mainly because phosphorus has been correctly judged to be the key nutrient in the eutrophication of many lakes. Two examples of regressions relating $[chl]_s$ (mean or maxima) to $[N_\lambda]_w$ are represented in

Figure 65. Their steepness and negative intercepts reflect the relatively greater demands for nitrogen and the greater maximum concentrations in lakes. Equally, it could be argued that the low average chlorophyll concentrations represented when $[N_{\lambda}]_{w} < 300$ mg m^{-3} are actually P-limited, whereas N-limitation is more likely to occur when $[P_{\lambda}]_{w}$ is relatively high.

A point which is stressed by the originators of equations describing average statistical relationships between mean biomass levels and the various measures of nutrient availability (see e.g. Vollenweider & Kerekes, 1980) but which is in danger of being forgotten by those who would apply them is that they are first-order approximations and *do not* predict precisely (within $\pm 50\%$ is a conservative estimate of the error). Moreover, the inaccuracies become progressively larger at higher concentrations. The curves inserted in Figure 64 may not apply throughout their ranges. There is evidence (discussed in Reynolds, 1978*c*) that, at higher phosphorus concentrations, phosphorus is used less efficiently (i.e. $[chl]/[P]$ declines to about one-quarter the level recorded at $[P_{\lambda}]_{w} = 3$ when $[P_{\lambda}]_{w} = 100$). The reason for this is partly that other factors (e.g. nutrients, such as nitrogen, or self-shading in mixed environments) may become limiting. Ultimately, light may well become absolutely limiting (see §4.3.2), when no amount of additional nutrient loading will increase phytoplankton yield.

Another problem which modelling has scarcely tackled is the prediction of the scale and timing of algal maxima. It is rarely the average biomass, as opposed to the short phases of peak abundance, which makes life difficult for water engineers and reservoir managers. Nor can the models discussed here provide much indication as to what species are likely to dominate. Again, the difficulties of contending with a crop yield of 50 mg chl m^{-3} as (say) *Asterionella* differ from those caused by commensurate crops of (say) *Anabaena* or *Cryptomonas*. The seriousness of those difficulties will also vary according to the major function of the water body, whether as a storage reservoir, recreational lake or commercial fishery. The aspects of phytoplankton ecology which influence these phenomena are addressed in the next chapters.

6

Growth and survival

'Be fruitful and multiply and replenish the earth'
Genesis I, v.28

Whereas the previous chapters have attempted to review those aspects of phytoplankton ecology that are common to most planktonic 'algal' species – the need to remain in suspension, to carry out photosynthesis and to obtain nutrients – the concluding ones are essentially directed towards those factors that discriminate *between* species and which contribute to spatial and temporal differences in community composition. The title of the present chapter has been changed more than once during its writing but it is still unsatisfactory. The discussion that I wish to develop here concerns the interspecific differences in metabolic performance in response to seasonally changing environmental conditions, their contribution to periodic changes in community composition and in the means of surviving less-favourable conditions. In short, why do some species grow better than others in certain kinds of water or at certain times of the year?

Let us first make a hypothesis: that any planktonic alga becomes relatively more abundant than its (supposed) competitors (i.e. it becomes the *dominant* species) because (a) it has had the opportunity to do so (i.e. that an '*inoculum*' is already present in the water body concerned) and either (b) it is able to increase its biomass at a faster rate than its competitors under contemporaneous environmental conditions or (c) the initial inoculum is sufficiently great to preserve its advantage over potentially faster-growing competitors, at least until some limiting environmental resource becomes overriding. If we assume the first point (a) to be satisfied, then it is appropriate to begin with a comparison of the rates of increase of different species under optimum conditions and to proceed to a consideration of environmentally imposed constraints thereon.

6.1 Optimal growth rates

6.1.1 *Definitions and units*

In the present context, it is important to define 'increase' and 'growth' and the units of their measurement. By accumulating external supplies of simple nutrients and assembling them into discrete structural components, physiologically-active algal cells will show gain in dry mass, net of respirational losses, which may be equated with 'growth'. As a consequence of its relatively simple organization, the growing cell undergoes a division, preceded by nuclear mitosis and separation (in eukaryotes) or an analogous replication and segregation of chromatin (in prokaryotes), in which the cell material is apportioned between two new daughter cells. In some species, notably among the Chlorococcales and certain colonial members of the Volvocales, further divisions may take place without further gain in mass. Eventually, however, the growing mother cell gives rise to two or more new cells, each potentially able to grow in a similar fashion. Thus, the number of cells present in a population *increases* through a series of doublings. Under ambient steady-state conditions, each doubling is accomplished over similar time intervals, so the increase is *exponential*. While each *generation* of cells attains approximately similar dimensions and mass to the mother, the increases in the total cell volume and dry weight of the population as a whole remain proportional to the increase in the number of cells present.

Under such circumstances, growth and increase may be equivalent. In natural populations, however, observed net increases in cell concentration or dry mass may well lag behind the actual production because cells are constantly lost, by advection, wash-out, sedimentation or other processes (see Chapter 7), so reducing the base for future divisions.

Because both growth and net increase per unit time are non-linear, it is convenient to express them relative to the population existing at a given point in time, t_0. Thus, at time t_1, the population will have increased by a factor, x.

$$\text{i.e. } N_{t_1} = xN_0 \tag{40}$$

The rate of increase is equivalent to:

$$\delta N/\delta t = x = N_{t_1}/N_0 \tag{41}$$

where N is the dry mass of cells or any convenient quantity proportional thereto (e.g. the concentration of cells, their total volume, carbon or chlorophyll content, spectrophotometric absorbance or nephelometric scattering) and x is a fractional rate of increase, which is a constant while the population increases at a constant exponential rate. It may be equally

Table 16. *Maximum exponential growth rates (k', in* ln *units d^{-1}) of selected freshwater phytoplankton in laboratory culture, at the temperatures indicated and, unless otherwise stated, under continuous saturating illumination: based on the review Table of Hoogenhout & Amesz (1965) with some additional entries*

Species	Temp. (°C)	Illumination	k'	Notes
Cyanobacteria				
Microcystis aeruginosa	23	?	1.11	(unicellular culture)
Microcystis aeruginosa	23	sat	0.48	(colonial culture: Reynolds *et al.*, 1981)
Synechococcus sp. (= *Anacystis nidulans*)	25	sat	2.01	
	41	sat	7.97	
Anabaena cylindrica	25	sat	1.56	(Fogg, 1949)
Anabaena flos-aquae	20	sat	0.78	(Foy *et al.*, 1976)
Anabaena flos-aquae	20	sat, 6L:18D	0.35	(Foy *et al.*,
Aphanizomenon flos-aquae	20	sat	0.98	(Foy *et al.*, 1976)
Aphanizomenon flos-aquae	20	sat, 6L:18D	0.27	(Foy *et al.*, 1976)
Oscillatoria agardhii	20	sat	0.86	(van Liere *et al.*, 1975)
Oscillatoria agardhii	20	sat	0.82	(Foy *et al.*, 1976)
Oscillatoria agardhii	20	sat, 6L:18D	0.31	(Foy *et al.*, 1976)
Oscillatoria redekei	20	sat	1.39	(Meffert, 1971)
Oscillatoria redekei	20	sat	1.10	(Foy *et al.*, 1976)
Oscillatoria redekei	20	sat, 6L:18D	0.30	(Foy *et al.*, 1976)
Oscillatoria rubescens	20	sat	0.40	(Meffert, 1971)
Cryptophyta				
Cryptomonas ovata	23	sat, 15L:9D	0.81	(Cloern, 1977)
Cryptomonas erosa	23.5	sat	0.83	(Morgan & Kalff, 1979)
Pyrrhophyta				
Ceratium hirundinella	21	sat, 18L:6D	0.26	(Bruno & McLaughlin, 1977)
Ceratium hirundinella	20	sat, cont	0.21	(G. H. M. Jaworski, pers. comm.)
Chrysophyta–Bacillariophyceae				
Cyclotella meneghiniana	?	sat, 14L:10D	~ 0.85	(Tilman & Kilham, 1976) cf. *Asterionella*
Stephanodiscus hantzschii	20	sat	1.18	
Asterionella formosa	20	sat	1.74	(Lund, 1949)
Asterionella formosa	?	sat, 14L:10D	~ 0.62	(Tilman & Kilham, 1976)
Fragilaria crotonensis	20	sat	1.37	(G. H. M. Jaworski, pers. comm.)
Tabellaria flocculosa var. *asterionelloides*	20	sat	0.76	(G. H. M. Jaworski, pers. comm.)
Chrysophyta–Xanthophyceae				
Monodus subterraneus	25	sat	0.64	
Tribonema aequale	25	sat	0.49	

Table 16. (*cont.*)

Species	Temp. (°C)	Illumination	k'	Notes
Euglenophyta				
Euglena gracilis	25	sat	1.52	
Chlorophyta–Volvocales				
Chlamydomonas spp.	25	sat	2.29–2.91	
Eudorina unicocca	23	sat, 12L:12D	0.62	(M. Rodgers, pers. comm.)
Chlorophyta–Chlorococcales				
Ankistrodesmus braunii	25	sat	1.59	
Chlorella pyrenoidosa	25	sat	2.15	
Chlorella vulgaris	25	?	2.01	
Scenedesmus obliquus	25	sat	1.52	
Scenedesmus quadricauda	25	?	2.84	
Selenastrum capricornutum	24	sat	1.32	(Reynolds *et al.*, 1975)

expressed as a logarithm, either to the base e (k'), to the base 10 (k'_{10}) or to the base 2 (k'_2), which is equivalent to the number of doublings per unit time.

$$x = e^{k'} = 10^{k'_{10}} = 2^{k'_2}$$

k' is the *exponential growth constant*, scaled per unit time (h^{-1}, d^{-1}). Since

$$N_{t_1} = N_0\, e^{k'} \tag{42}$$

$$k' = \ln\,(N_{t_1}/N_0)]/t_1 - t_0) \qquad (\simeq 2.30\,k'_{10};\ 0.693\,k'_2) \tag{43}$$

The *generation time* (t_G) is the time taken for the population to double and is derived from:

$$t_G = \ln 2/k'$$

$$\simeq 0.693/k'; \tag{44}$$

or from $\log_{10} 2/k'_{10}$ ($= 0.301/k'_{10}$) or from $\log_2 2/k'_2$ ($= 1/k'_2$).

Some published results, quoted in one or other of these forms in the literature, are presented (in Table 16) as exponential growth constants (k').

6.1.2 *Specific growth rates in laboratory culture*

Controlled laboratory conditions of continuous saturating illumination, near constant temperature and a plentiful availability of nutrients supplied in artificial media presumably afford the best opportunity for most 'algal' species to realize their highest potential growth rates. Measurements of specific growth rates under such idealized conditions are scattered throughout the literature. However, there have been few systematic attempts to compare the individual performances of a range of species or even to collect published values together in a single review.

Exceptions to both statements are presented by the work of Hoogenhout & Amesz (1965) and to the latter in Fogg (1975). Table 16 is a compilation of selected data for freshwater algae drawn from these and additional cited sources. Although they are not exhaustive, the available data do indicate a very wide variation among maximal growth rates ranging between less than 1 division day^{-1} ($k' < 0.69$) and the 11.5 divisions day^{-1} recorded for *Anacystis nidulans* (= *Synechococcus*) at 41°C by Kratz & Myers (1955), the highest specific rate yet recorded for any 'alga'. This species is evidently thermophilic, whereas many of the others included in Table 16 probably fail to grow much faster, if at all, at temperatures exceeding 30–35° C than they do at 20–25° C (Hoogenhout & Amesz, 1965; see also below). Even so, the optimum rates of growth of *Synechococcus* and of several other small, unicellular (e.g. *Chlamydomonas*, *Chlorella* spp.) or simple coenobial (e.g. *Scenedesmus quadricauda*) algae exceed 2.7 divisions ($k' > 1.9$ d^{-1}) in the latter temperature range. Such rates imply a relatively efficient translation of photosynthetically-fixed carbon into new cell material. Light-saturated specific photosynthetic rates that may be attained at 20–25° C, in the order of 20 mg O_2 (mg chl*a*)$^{-1}$ h^{-1} (or about 0.15 mg C fixed per mg cell carbon per hour, assuming 1 molecule of carbon to be equivalent to 1 molecule of oxygen and a carbon: chlorophyll ratio of 50; see §§4.2.2, 4.2.6) should theoretically permit the cell to double its mass in under seven hours and to undergo up to 3.6 divisions per day ($k' \sim 2.5$ d^{-1}). In contrast, many larger, filamentous and (especially) colonial 'algae' grow relatively more slowly.

Indeed, although growth rates are undoubtedly influenced by a variety of physiological and metabolic factors, they are widely recognized to be generally coupled to size and structural organization (e.g. Belcher & Miller, 1960; Findenegg, 1966*a*; Fenchel, 1974; Fogg, 1975; Laws 1975; Banse, 1976). The plot of growth rates (from Table 16) against mean unit volume (from Table 3) for those species for which data are available shows a negative correlation (Figure 66). The implication that habit influences growth rate is strengthened by the available data for the growth of *Microcystis aeruginosa*. In most cultures, this species loses the colonial organization that characterizes natural populations (for a discussion, see Reynolds *et al.*, 1981). Reynolds & Jaworski (1978) succeeded in maintaining colonial strains in a modified ASM-1 medium, none of which attained the rates of growth reported elsewhere (e.g. McLachlan & Gorham, 1961; Krüger & Eloff, 1977) for cultured (unicellular) *Microcystis* strains.

Size and shape may have two separate influences upon metabolic activity essential to growth. One is that the distance between sites of uptake and nutrients and of their assembly into cellular components will tend to be

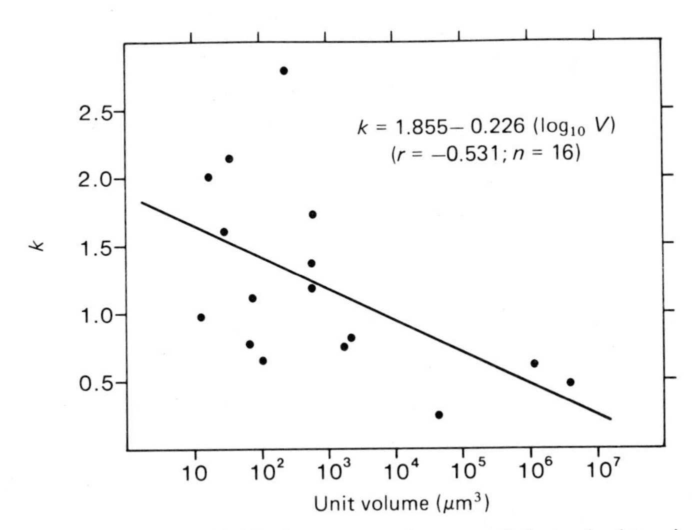

Figure 66. Maximum growth rates of phytoplankton in culture (θ: 20–25°C) plotted against cell or (for colonial species) unit volume. Data from Tables 3 and 16 for species entered in both. (Original.)

greater in large cells, or in cells embedded in mucilaginous colonies. The other is that the smaller is the cell, then the relatively greater is its surface/volume ratio (see Chapter 1); potentially, so is its relative rate of nutrient uptake. Irregularities in cell shape, which contribute to the 'conservation' of SA/V ratio (also discussed in Chapter 1), help to overcome the limitation on size *per se*. Nevertheless, there is a general, positive relationship between k' and (SA/V) among planktonic species (see Figure 67), that accounts for about 18% of the variation in growth rates. This was the conclusion of Foy's (1980) investigation of growth rates in 22 strains of planktonic cyanobacteria (the fastest of which are included in Figure 67); Foy also commented that cultures of nominally the same species often showed wide variations in cell morphology and growth rates. In diatoms, and other cells with relatively large intracellular vacuoles, the surface available for exchange of metabolities may be proportionately increased (Paasche, 1960).

Specific maximal growth rates, like most physiological processes, respond directly to temperature changes provided that other factors continue to be saturating. Examples of temperature effects on the growth rates of 'algae' in cultures are represented in Figure 68: supposed hyperbolic curves are normalized by plotting k' semilogarithmically against temperature on an Arrhenius scale ($1/\theta$K). Although their growth rates generally increase with increased temperature, the specific responses vary significantly. According to data in Lund (1949) and Foy, Gibson & Smith (1976) growth

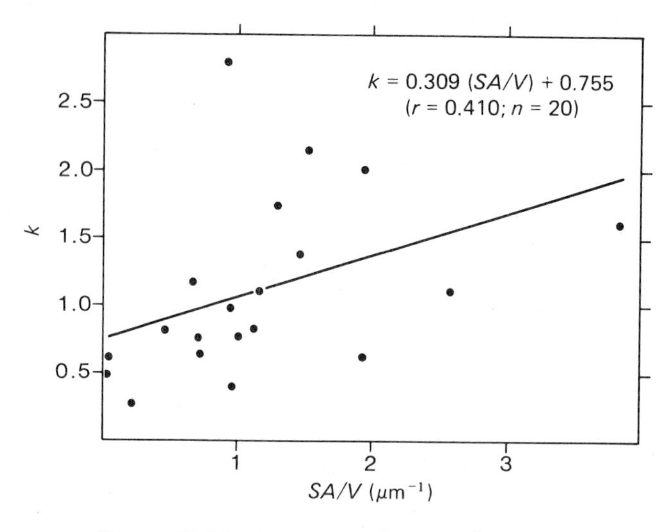

Figure 67. Maximum growth rates of phytoplankton plotted against *SA/V* ratio of individual cells or (for colonial species) units. Data from Tables 3 and 16 for species entered in both, and four points from Foy (1980). (Original.)

rates of *Asterionella*, *Anabaena*, *Aphanizomenon* and *Oscillatoria* increase between 1.8- and 2.9-fold over the range 10°–20°C, while the average Q_{10} for Kratz & Myers (1955) *Synechococcus* between 25° and 41°C is ~ 2.4. Cloern's (1977) data from *Cryptomonas ovata*, however, indicated that its growth rate increased about nine-fold between 8° and 20°C (Q_{10} ~ 6.1). Above 20°C growth rate increased more slowly to an optimum at about 23.5°C, above which it decreased significantly. On average, however, the Q_{10} values of temperature-limited growth seem generally higher than those of (say) photosynthesis. Most 'algae' probably reach their optimal growth rate in the range 20–25°C, but there are many thermophilic species (e.g. the *Synechococcus* of Kratz & Myers, 1955) whose growth rate may continue to increase with temperature up to 35–45°C.

There are remarkably few data available to describe 'algal' growth at temperatures much below 5°C, even though the temperatures of many lakes in high latitudes remain well below this level for a large proportion of the year. Nor may the curves drawn in Figure 68 be extrapolated, as the growth of many species in natural waters seems to commence abruptly (see §6.2). This may be due to other factors which alter coincidentally with increasing temperature, but nothing can be established without some experimental evidence. This might also help to explain the tendency for diatoms to dominate vernal maxima in temperate lakes ($\theta < 8$–10°C), with

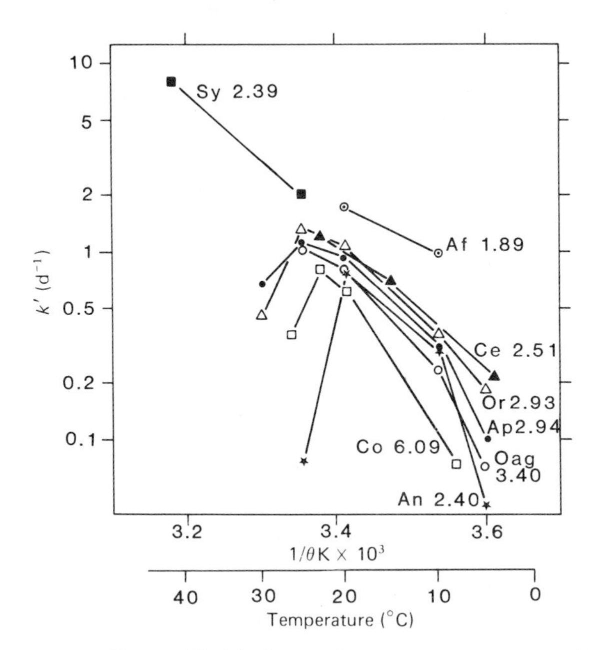

Figure 68. Algal growth rates v. temperature (shown on Arrhenius plot, with scale of customary temperature below). Individual species are identified by symbols. Approximate Q_{10} values for the 10°–20°C are given alongside. (Af, *Asterionella formosa*; An, *Anabaena flos-aquae*; Ap, *Aphanizomenon flos-aquae*; Ce, *Cryptomonas erosa*; Co, *Cryptomonas ovata*; O ag, *Oscillatoria agardhii*; Or, *O. redekei*; Sy, *Synechococcus*. (Original presentation of data from Lund (1949), Kratz & Myers (1955), Hoogenhout & Amesz (1965), Foy *et al.* 1976), Cloern (1977), and Morgan & Kalff (1979).)

little competition from the cyanobacteria and chlorophytes which may succeed them later in the year.

Excepting these upper and lower temperate extremes, most planktonic algae are able to grow within the range 10–25°C; few *stenotherms* (species with narrow temperate limits) are known to exist, although it has been suggested that *Oscillatoria rubescens* is a cold-water stenotherm, based on its ecology in central European lakes (Findenegg, 1947), and Reynolds (1973*a*) commented that 'resting' cells of *Melosira granulata* fail to 'germinate' when water temperatures are low. Again, the involvement of other factors may well account for these effects.

The importance of light, through its influence on photosynthesis, is not to be doubted. Light intensities which are insufficient to saturate immediate photosynthetic requirements are scarcely likely to support maximal growth rates. The same is true of inhibiting irradiance levels, though in either case, adaptive responses may eventually improve growth performance. It is

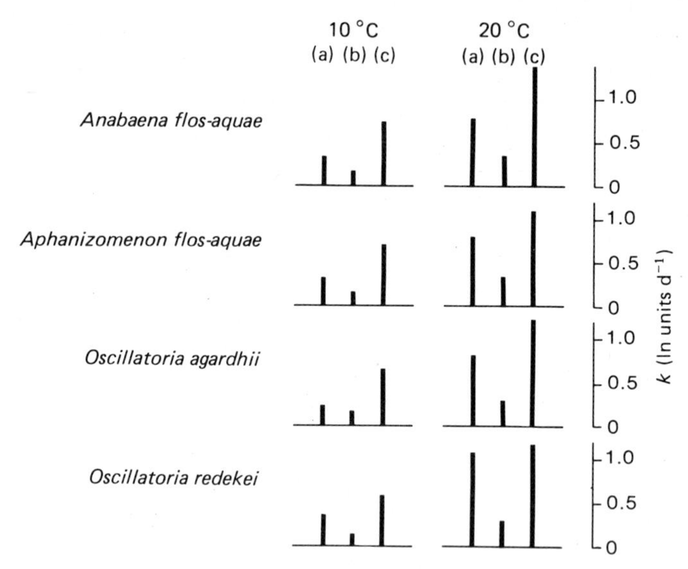

Figure 69. Growth-rate responses of cyanobacteria in culture to changes in temperature (10°, 20°C) and photoperiod: (a) in continuous light (~ 27 μE m^{-2}); (b) under 6-h L:18-h D cycle; (c) as (b) but expressed per 24-h light period. (Drawn (with conversion of units) from data of Foy *et al.*, 1976.)

significant that when enclosed cultures of diatoms have been exposed to the natural light-field of lakes, there has been little evidence of near-surface photo-inhibition (cf. §4.2.3) the fastest growth rates and greatest yields being obtained at depths of 0.5 m (Lund, 1949; Reynolds, 1973*a*).

'Algal' growth rates are equally likely to respond to the *photoperiod* as to the intensity of irradiance. Foy *et al.* (1976) attempted to simulate the natural growth conditions of cyanobacteria in Lough Neagh by subjecting some of their experimental cultures to saturating irradiances for only 6 h out of every 24 h; their results are represented in Figure 69. Although growth rates were universally lower than those of continuously-illuminated cultures at the same temperature, the number of divisions per light hour (extrapolated for a 24-h day) was increased by between 9% (in *Oscillatoria redekei*, at 20°C) and 130% (in *Anabaena flos-aquae* at 10°C). They also noted that all the species they considered were relatively greener when grown on 6-h L/18-h D cycles than on 24-h L cycles and that, with the exception of *Oscillatoria redekei*, showed a marked increase in their light-saturation requirement. The results suggest adaptation to fluctuating light regimes, enabling the organisms to attain higher photosynthetic efficiencies during the light period (see §4.2.2).

It is equally clear that the specific responses of growth rate to temperature and light fluctuations will interact with each other. For instance as metabolic activity increases with temperature, so the irradiance levels required to saturate growth rates might be expected to increase. Morgan & Kalff (1979) showed that such a relationship holds in laboratory cultures of *Cryptomonas erosa*: light saturation of growth increased from the equivalent of 8 $\mu E\ m^{-2}\ s^{-1}$ at 4°C to 30 at 15°C and 138 at 23.5°C. Similarly, the data of Foy *et al.* (1976) show that the growth rate of all four species of cyanobacteria they investigated was saturated at a light intensity equivalent to 17.5 $\mu E\ m^{-2}\ s^{-1}$ at 10°C when grown in continuous light; both *Aphanizomenon flos-aquae* and *Anabaena flos-aquae* showed higher light-saturation requirements at 20°C ($\sim$ 28, $\sim$ 40 $\mu E\ m^{-2}\ s^{-1}$ respectively) but no such increase was observed in *Oscillatoria agardhii* or *O. redekei* cultures. This difference in the behaviour of *Oscillatoria* spp. recalls Harris' (1973, 1978) observations concerning the different temperature-dependent effects on the onset of light saturation (I_k) of photosynthesis of populations dominated by *Oocystis* and by diatoms (see Figure 47, p. 131).

These findings support the postulated existence of two separate but overlapping reactions to temperature change (Tamiya *et al.*, 1953). One (L_d) is also light-dependent, regulated primarily through photosynthetic rate, with a Q_{10} of about 2. The other (L_i) is the light-independent assimilation of photosynthate into proteins and new cell material, which is rather more sensitive to temperature ($Q_{10} > 2.4$) and, arguably, to cell size and SA/V ratio. At low temperatures, growth is more likely to be L_i-limited, excess photosynthesis being liable to excretion or photo-respiration. At higher temperatures L_i 'catches up' with L_d until the latter becomes limiting. Some examples of cyanobacteria whose growth responses to temperate variation differ from those of photosynthetic rate are given in Konopka & Brock (1978).

Under a fluctuating light regime, the L_i reaction can continue through the dark periods and, the longer they are, the more growth limitation moves towards the L_d process. Here, size may be advantageous in providing a relatively greater capacity for the retention of L_d products or a lower rate of respiratory or photorespiratory loss from the cells (cf. Laws, 1975). It is significant that Foy *et al.* (1976) found that, over the 10–20°C range, the Q_{10} values for the specific maximal growth rates of the four cyanobacteria grown under 6-h L/18-h D cycles were in the range 1.5–2.1 (compared with 2.4–3.4 under continuous light) and that those rates were absolutely higher in the larger (low SA/V) species than in the smaller ones.

Another possible effect of the alternation between light and dark exposure is on the timing of cell division. It is well known that division in algal cultures frequently becomes *synchronized* (i.e. division occurs at regular intervals and approximately simultaneously) to strict 24-h L/D cycles, but the effects of altering the photoperiod on the phasing are generally unclear, varying widely among different taxonomic groups and apparently interacting with temperature and nutrient limitation. Most of the available data (reviewed by Rhee, Gotham & Chisholm, 1981) apply to marine species; it is of interest that division among marine diatoms, especially, seems to be more closely coupled to the supply of light energy *per se* rather than to any endogenous day-night rhythm. This would be advantageous under a fluctuating light regime, characterized by short photoperiods.

Although these aspects of phytoplankton ecology are still under-researched, the above cases exemplify possible mechanisms whereby the competitive advantage conferred by a faster rate of growth moves between species in response to both the short-term imposition of high-frequency (e.g. light in mixed columns) and to more gradual, low-frequency (e.g. seasonal temperature changes) scales of environmental variability. Collectively, 'algae' do possess what Harris (1980*a*) described as a 'suite of physiological responses which serve to buffer the growth rate of the cell against fluctuations in the environment'. Harris (*loc. cit.*) also argued that the scale of environmental perturbation determined the scale of the response: whereas small or brief stimuli may be met by purely physiological responses, larger or more persistent ones invoke higher-order responses, manifest as changed growth rates and, ultimately, altered composition of the community. These possibilities are examined in subsequent sections.

6.2 Increase in natural waters

Rates of growth in natural habitats are rarely directly measurable owing to the cumulative effects of the various loss processes that deplete the standing population. In many instances, it is possible to assess the *exponential rate of increase* (k_n), net of losses, provided that the standing populations can be adequately described and the sources of error can be reliably estimated. Satisfaction of this condition demands the adoption of sampling designs appropriate to the questions that are posed (see e.g. Sournia, 1978): considerable sampling effort, both in space and in time, is necessary to determine rates of change in phytoplankton populations. If it is also possible to estimate the simultaneous rates of loss (k_L) attributable to washout, death, sedimentation and consumption by animals, then specific growth rates may be approximated by summation ($k' = k_n + k_L$). This has been attempted for a number of phytoplankton populations husbanded within experimental limnetic enclosures (the

'Lund Tubes' in Blelham Tarn, Cumbria: see Lund & Reynolds, 1982). In Table 17, maximum observed rates of increase (k_n), together with deduced rates of growth (k'), where available, of some of the species that have developed large populations in these enclosures, are compared with k_n values derived by Sommer (1981) for populations in the Uberlinger See, an arm of Lake Constance (Bodensee). In comparison with the growth rates under idealized conditions (Table 16), *in situ* increase is usually much slower, even when temperatures are approximately equivalent. However, the relative reduction varies with species, it being less among the large-celled and low SA/V colonial forms (e.g. *Cryptomonas ovata, Ceratium, Microcystis, Eudorina*; 'reduction' being by 25–50%) than among smaller, high SA/V forms (e.g. *Asterionella, Fragilaria;* 'reduction $> 75\%$). This is what would be expected in a fluctuating light regime. Of course, other factors contribute to the observed effect, not least that the smaller cells are more vulnerable to *in situ* losses (see Chapter 7). Moreover, the depth of mixing and, hence, the frequency of fluctuations at the time of the observations would not have been identical in every case.

In natural waters, there are many environmental variables to which algal growth rates might respond, e.g. temperature, average light intensity and period, nutrient availability. The potential individual responses of the various species depend in large measure upon which factor or group of factors is critical at the time. Equally they are likely to alter as the critical limiting condition switches between factors. For instance, temperature and/or light may be limiting early in the year, whilst nutrients are more likely to be critical during the summer.

These seasonal shifts in emphasis are generally accepted to influence the sequence in which the various species will each be able to grow and, perhaps, to dominate the plankton assemblage. Although periodic change in dominance ('the seasonal succession') has been investigated and described for a large number of (mainly temperate) water bodies and the sequences have been shown often to conform to one or other of a relatively small number of generalized patterns (e.g. Figure 43), few explanative theories to account for them have yet emerged and, certainly, none has won general acceptance. One approach that promises some progress in this direction is to define the ranges of supposedly critical environmental variables under which different species are known to grow in natural waters. Some pointers can be given here.

Periods of finite net increase in the standing populations of selected 'algal' species in the plankton of the Blelham Enclosures during each of three calendar years, in relation to seasonal variations in surface water temperature, mixed depth, Secchi-disk extinction depth and nutrient

Table 17. *Maximum observed rates of net population increase* (k_n) *and derived (see text) growth rates* (k') *for selected phytoplankton species in Lund Tubes (from Reynolds et al., 1982b, and unpublished) and in Bodensee (from Sommer, 1981); the times of year of the maxima cited are indicated (V = vernal prestratified period; ES = early, MS = mid- and LS = late-stratification)*

	Period	Lund Tube k_n	Lund Tube k'	Bodensee k_n
Microcystis aeruginosa	MS	0.24	0.24	—
Anabaena flos-aquae	MS	0.26	—	0.41
Aphanizomenon flos-aquae	MS	0.23	—	0.43
Oscillatoria agardhii v. *isothrix*	V	0.06	—	—
	LS	0.33	—	—
Cryptomonas ovata	V	0.15	> 0.15	0.46
	ES, MS	0.49	> 0.61	0.89
Rhodomonas minuta	V	0.17	—	0.58
	ES	0.71	—	0.56
Ceratium hirundinella	ES, MS	0.13	—	0.17
Peridinium cinctum	MS	0.16	—	0.18
Dinobryon spp.	ES, MS	0.27	—	0.45
Uroglena cf. *lindii*	MS	0.23	—	—
Chromulina sp.	MS	0.59	> 0.88	—
Melosira granulata	MS	—	—	0.43
Asterionella formosa	V	0.15	> 0.16	—
	MS	0.34	> 0.40	0.36
Fragilaria crotonensis	V	0.10	0.10	—
	MS	0.24	0.24	0.27
	LS	0.10	—	—
Tabellaria flocculosa v. *asterionelloides*	V	0.13	—	—
	MS	0.33	—	—
*Eudorina unicocca**	ES	0.43	> 0.43	—
Pandorina morum	MS	—	—	0.52
Ankyra judayi	MS	0.50	> 0.91	—
Sphaerocystis schroeteri	ES, MS	0.43	—	0.64
Closterium aciculare	LS	—	—	0.18
Staurastrum pingue	LS	0.28	—	0.13

* Referred to as *Eudorina* sp. in Reynolds *et al.* (1982*b*; see also Lund & Reynolds, 1982).

concentration, are represented in Figures 70–72. Quite different fertilities were maintained in each of the three years: Enclosure A was left unfertilized throughout 1977 (Figure 70) and quickly became deficient in available phosphorus (and later, also in silicon and, apparently, in iron: Reynolds & Butterwick, 1979); in 1978, the same enclosure was fertilized each week in order to restore the dissolved nutrient concentrations to predetermined arbitrary levels, such that none should be severely limiting

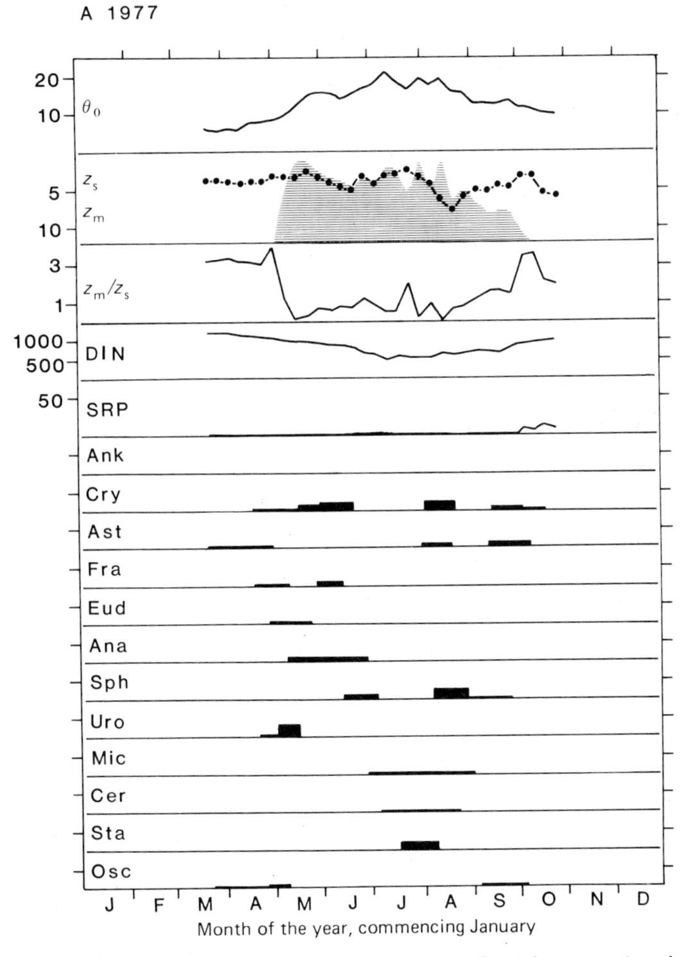

Figure 70. The periodicities and rates of net increase (vertical scale: datum line, O; tag ≡ $k' = 0.5\ \mathrm{d}^{-1}$) of selected phytoplankton species in a Blelham Enclosure, left unfertilized after isolation in March, in relation to temperature, Secchi-extinction (z_s) and mixed depths (z_m) the ratio between them (z_m/z_s) and the concentrations of SRP and DIN. Ank, *Ankyra judayi*; Cry, *Cryptomonas* spp.; Ast, *Asterionella formosa*; Fra, *Fragilaria crotonensis*; Eud, *Eudorina unicocca*; Ana, *Anabaena flos-aquae* (and occasional *A. circinalis*); Sph, *Sphaerocystis schroeteri*; Uro, *Uroglena* sp(p).; Mic, *Microcystis aeruginosa*; Cer, *Ceratium hirundinella*; Sta, *Staurastrum pingue* (and other spp.); Osc, *Oscillatoria agardhii* var. *isothrix*. (Original, from data in Reynolds & Butterwick, 1979.)

(Figure 71: Reynolds & Butterwick, *loc. cit.*); Enclosure B was heavily fertilized during the vernal pre-stratified period, but left unfertilized thereafter (Figure 72). The mean levels of 'algal' biomass were commensurate with nutrient loads applied in each of the three years (Lund &

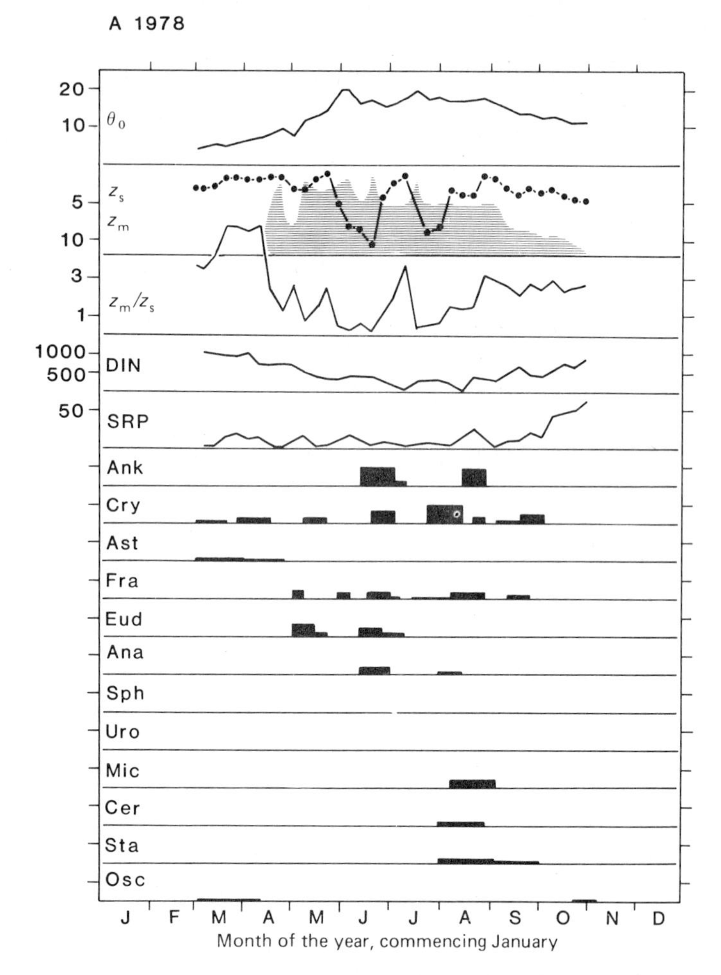

Figure 71. Periodicities and net rates of increases of selected phytoplankton in a Blelham Enclosure, fertilized each week throughout March–September in relation to physical and chemical factors. Construction as in Figure 70. (Original, from data in Reynolds & Wiseman, 1982, and Reynolds *et al.*, 1982*b*.)

Reynolds, 1982). The annual sequences of the dominant species were respectively analogous to those of nearby mesotrophic lakes, of a highly eutrophic system, and of Blelham Tarn itself (for details, see Reynolds 1980*a*; Lund & Reynolds, 1982). In Figure 73, specific rates of increase are collected together and plotted, in turn, against contemporaneous evaluations of mixed depth, of mixed depth/Secchi-disk depth ratio (an index of the frequency of fluctuations in saturating light intensity), of temperature, of soluble reactive phosphorus (SRP) and dissolved inorganic

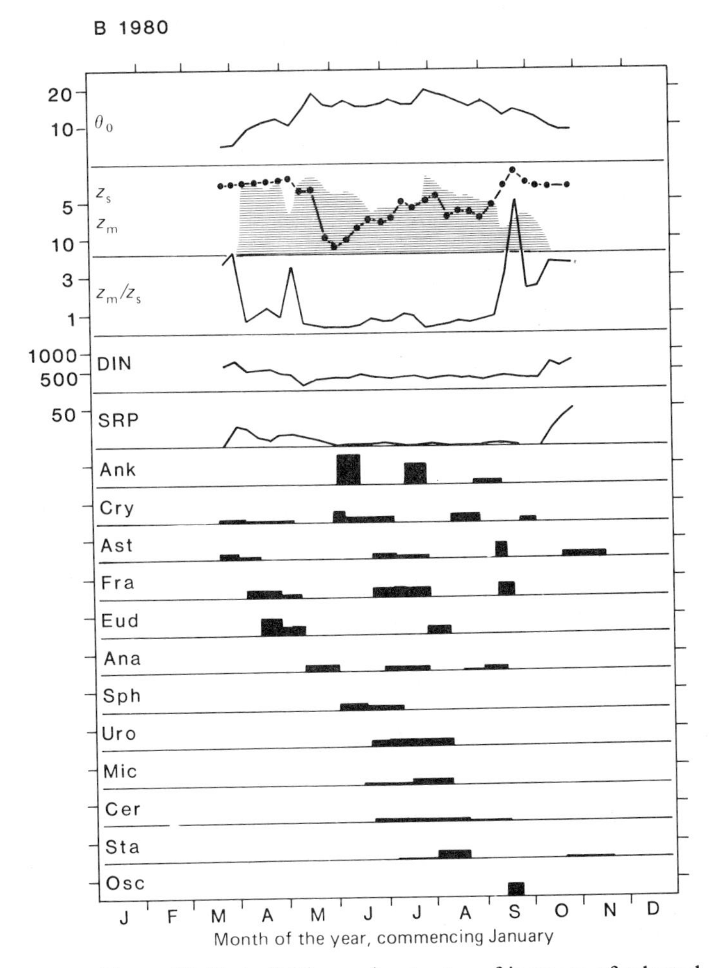

Figure 72. Periodicities and net rates of increase of selected phytoplankton in a Blelham Enclosure, heavily fertilized until the onset of stratification in April but then left unfertilized through the summer, in relation to physical and chemical factors. Construction as in Figures 70 and 71. (Original, based on unpublished data.)

nitrogen (DIN = $NO_3' + NO_2' + NH_4'$. N). In few cases does there seem to be any evidence that net increase rate is directly correlated with a given environmental variable, but individual points are frequently grouped within definable areas of the two-dimensional representations. Moreover, the distributions against given variables differ among the species. Net increase in *Asterionella*, *Uroglena* and *Oscillatoria* tended to coincide with the higher values of z_m and z_m/z_s; *Sphaerocystis*, *Ceratium* and *Staurastrum* with lower ones. *Asterionella*, *Oscillatoria* and *Cryptomonas* grew over a wider range of temperatures than did *Eudorina*, *Anabaena* or *Microcystis*.

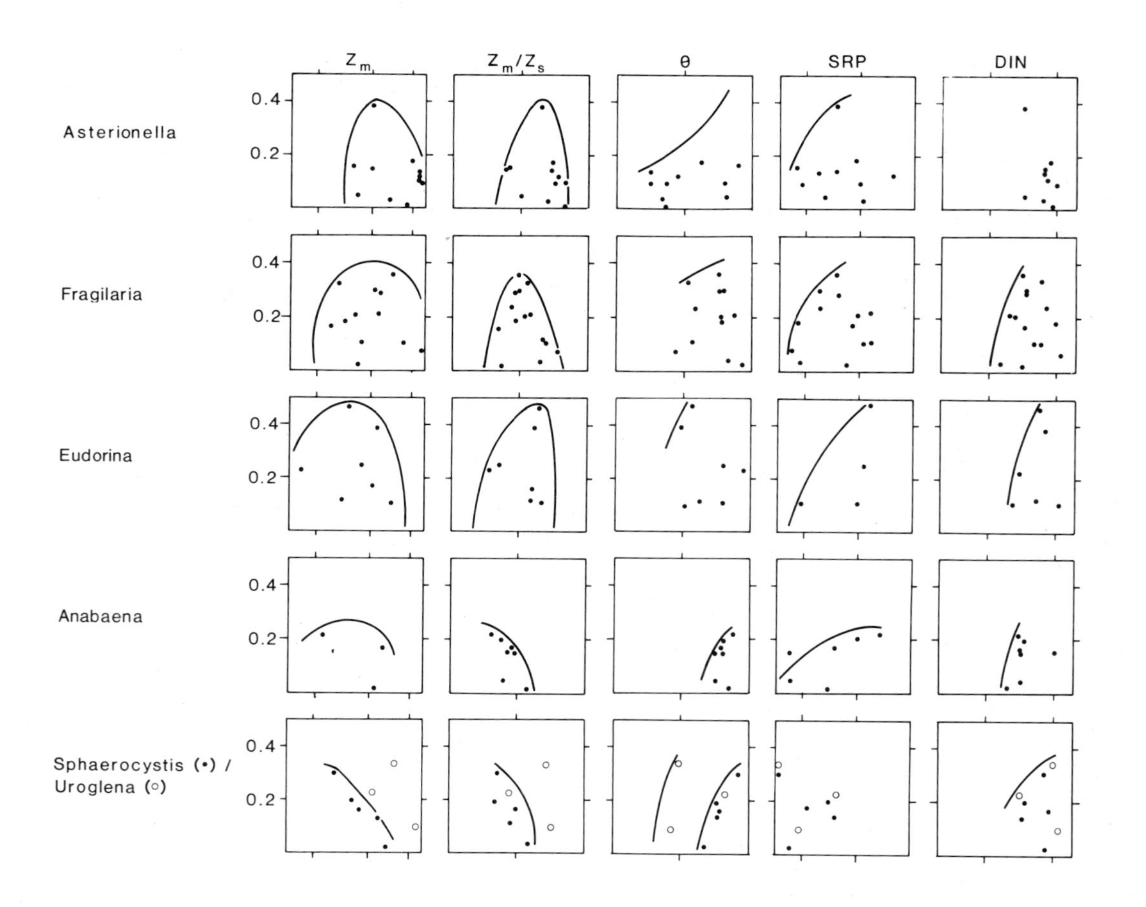

Z_m
Z_m/Z_s
θ
SRP
DIN
Asterionella
Fragilaria
Eudorina
Anabaena
Sphaerocystis (•) /
Uroglena (○)
0.4
0.2

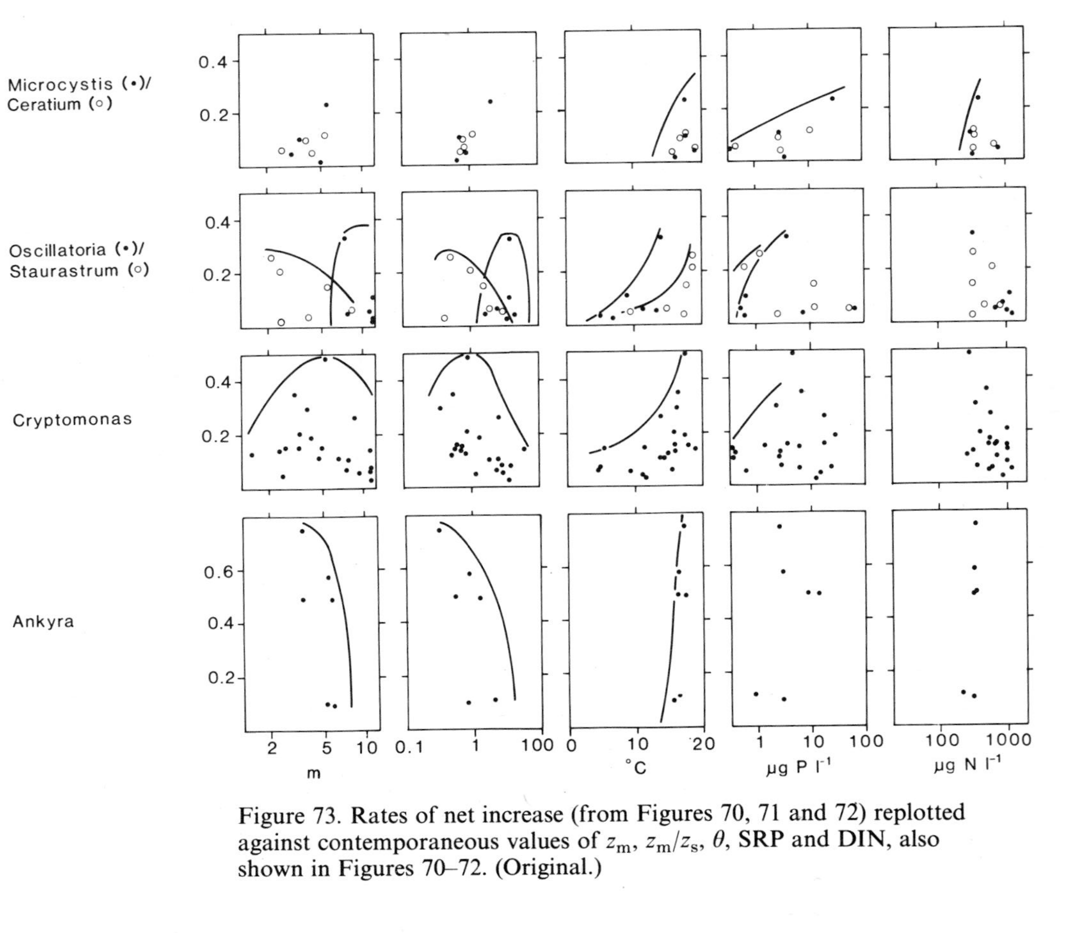

Figure 73. Rates of net increase (from Figures 70, 71 and 72) replotted against contemporaneous values of z_m, z_m/z_s, θ, SRP and DIN, also shown in Figures 70–72. (Original.)

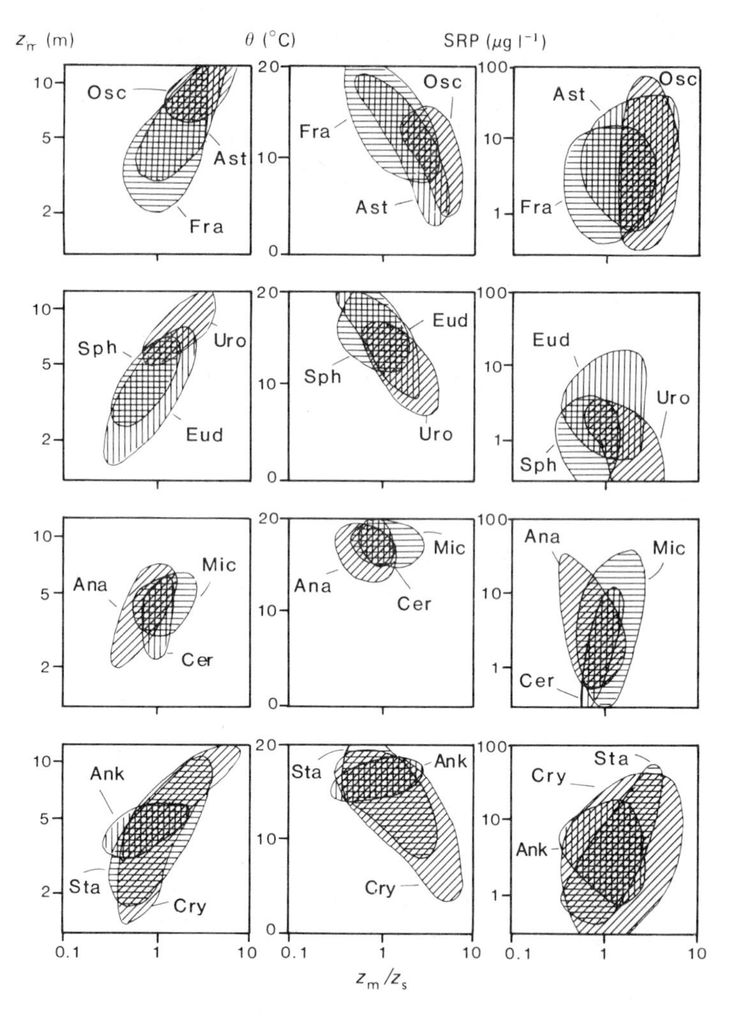

Figure 74. The scatter of points representing specific increase (as shown in Figure 73) but plotted relative to given pairs of variable factors (z_m, θ, SRP, v. z_m/z_s). Shaded areas enclose all points for the same species (identified in Figure 70). (Original.)

Sphaerocystis and *Uroglena* apparently tolerated lower external concentrations of SRP than *Eudorina*. That nitrogen was probably not a significant limiting factor in any of the three years may explain the general similarity of distribution of points against the DIN-axis.

These apparent differences in performance are accentuated in the representations in Figure 74. Here, the *rate* of increase is ignored, but each point signifying a positive increase in the population of each species has been replotted in turn on axes representing mean mixed depth, mean surface temperature and mean SRP concentration against the optical depth/ stability correlative, z_m/z_s. The individual points have been omitted, for clarity, but the 'thumbprints' enclose all plotted points for the species named. Moving down the first column (z_m v. z_m/z_s) it may be seen that, although there is considerable overlap, there is a progression from species which grew in well-mixed, optically deep columns to those which grew under relatively less-mixed columns, with decreasingly frequent oscillations in light exposure. Only *Cryptomonas* was observed to increase across the whole spectrum. The pattern broadly corresponds with morphological and physiological characteristics established in earlier chapters: those species which are reliant on water movements for suspension are apparently confined to deep-mixed columns; those which are small, motile or buoyant tolerate weaker circulation patterns; equally, species which tend to photo-inhibit more readily, such as diatoms, tend to be grouped against intermediate z_m/z_s ratios.

In the second column, a sharp differentiation is evident between the species associated with stably stratified conditions in summer (when water temperatures were higher: e.g. *Anabaena*, *Ankyra*) and those which tolerated lower water temperatures or grew in well-mixed columns at the higher ones (e.g. *Oscillatoria*, *Asterionella*). Again, these distributions match the physiological findings, the high SA/V forms, whose growth at low temperatures tends to be limited by photosynthetic rate, performing relatively better than the low SA/V forms when the water is cold, or at longer photoperiods when the water is warm.

In the third column, the patterns differentiate perceptibly between species whose growth was maintained despite very low external concentrations of phosphorus and those whose capacity of increase became inhibited. Here, the published literature cannot be directly invoked to support the observations beyond that species like *Eudorina* and *Anabaena* are most frequently associated with nutrient-rich eutrophic lakes while *Sphaerocystis* and *Uroglena* are commonly encountered in nutrient-deficient lakes. These species might make good subjects for future investigations of the principle of resource-based competition (see §5.6.1).

Moss (1973*b*) has suggested that, weight-for-weight, intrinsically faster-growing species make a heavier demand on available nutrient resources per unit time than do slower-growing ones, and so the latter are better equipped to maintain growth at such times as when the nutrient supply is maintained principally through internal recycling. If nutrients are more freely available, they will be out-competed by the faster-growing species. The second part of this statement is always likely to hold, though the data in Table 17 suggest that *Sphaerocystis*, at least, potentially increases at a rate comparable to the fastest rates observed for *Eudorina*. The first part apparently ignores the fact that nutrient-limited growth of most species reduces cell quotas of nutrients, although it is possible that the half-saturation coefficient (K_s) of nutrient-limited growth might be significantly lower in *Sphaerocystis* than in *Eudorina* (see Chapter 5). With this condition in mind, Moss's very interesting suggestion, or something like it, might well explain the common observation that some species are characteristically associated with nutrient-rich systems and others with nutrient-poor ones. It is of relevance in this context that both *Eudorina* and *Sphaerocystis* featured in the periodic sequence of phytoplankton growth in Tube B, 1980 (Figure 72), against a background of declining nutrient (phosphorus) availability.

It must be emphasized that the diagrams in Figures 73 and 74 merely *describe* the conditions under which species were observed to grow and do not, by themselves, *explain* them. It is possible, for instance, that the temporal sequence of species coincides with the series of suggested critical events associated with seasonal increases in temperature, radiation income and structural stability. Nevertheless, the sequences in natural lakes, essentially imitated in the Lund Tubes, do seem to share similar correlations with the cycles of temperature change, stratification and epilimnetic nutrient depletion (see Figure 43 for references). Moreover, out-of-season perturubations of structural stability, in natural lakes (Reynolds, 1976*a*, 1980*a*; Lewis 1978*b*; Harris & Piccinin, 1980) and both natural (Figure 72) and artifically-induced ones in the Lund Tubes (Lund & Reynolds, 1982; Reynolds *et al.*, 1983*b*), are invariably followed by 'shifts' in community composition brought about by the renewed or enhanced growth of species favoured by well-mixed conditions normally obtaining earlier (or later) in the annual cycle. Subsequent recovery periods are marked by further changes in relative growth rates and community composition which, to a greater or lesser extent, recapitulate the natural seasonal sequence (Reynolds, 1980*a*).

Equally, water-bodies whose limnological characteristics tend to generate or prolong conditions representing extremes in one or other of the criteria of the 'normal' cycle, described in Figure 74 (i.e. they are usually well-mixed and optically deep or are ultra-stable, are always cold or warm, are extremely oligotrophic or hypereutrophic) tend habitually to support communities resembling those which occur episodically, under the appropriate conditions, in 'normal' seasonal sequences. A few examples may be cited to support this statement. The planktons of the well-mixed, frequently turbid and nutrient-rich Lough Neagh (Gibson *et al.*, 1971) and of some rich relatively shallow, exposed polder-lakes in the Netherlands (e.g.

Veluwemeer, Wolderwijd: Berger, 1975) tend to be dominated through much of the year by diatoms (*Melosira italica*, *Stephanodiscus* spp., *Diatoma*) or *Oscillatoria* spp. In relatively exposed but nutrient-poor large Scottish lochs, whose summer epilimnia may be 8–17 m deep for long periods, the vernal diatom (*Melosira italica* – *M. islandica*) – *Oscillatoria* phase extends (and peaks) well into summer, when populations of *Coelosphaerium*, *Staurastrum* and *Oocystis* spp. develop (Bailey-Watts & Duncan, 1981). Lakes at high latitudes may be ice-covered for eight or more months each year and their ice-free waters remain cold by even temperate standards. Char Lake and Meretta Lake in Northern Canada (latitude *ca.* 74°N) support some vernal flagellate (*Peridinium*, *Gymnodinium*) production under snow-free ice but diatoms (*Cyclotella glomerata*), *Ochromonas* and *Dinobryon* spp. develop in the ice-free summer period, peaking in July and August when the water temperature reaches 6 or 7°C (Schindler *et al.*, 1974). In Mývatn, Iceland (latitude *ca.* 65°N), where the ice-free period extends from May to September, and maximum water temperatures are about 14–15°C, a truncated diatom → *Dinobryon*/*Uroglena* (and *Anabaena* in the richer north basin) sequence is evident (Jónasson & Adalsteinsson, 1979). At the other extreme, the distinctive sequence of specific growth optima in the tropical Lake Lanao, Philippines (*Melosira granulata* → *Cryptomonas* → various green algae → *Anabaena* → dinoflagellates) shares obvious affinities with the sequences in temperate lakes but may be completed or partially completed on cycles of rather less than one calendar year's duration, in response to seasonal fluctuations in structural stability (Lewis, 1978*b*). The tendency of certain colonial species of Chlorococcales (*Pediastrum*, *Scenedesmus*, *Oocystis* spp.) to succeed diatoms in certain hyper-eutrophic, physically variable lakes (e.g. Onondaga Lake, U.S.A.: Sze, 1980; Arresø, Denmark: Olrik, 1981), small ponds and meres (Reynolds, 1978*b*) and, perhaps, certain lowland rivers may represent an extension of their periodic growth in some eutrophic lakes (Reynolds, 1980*a*). Haffner *et al.* (1980) showed that the growth of diatoms (*Stephanodiscus hantzschii*, *Fragilaria construens*) alternated with the growth of *Oocystis borgei* and *Ankistrodesmus* spp., in tune with mixing and more stable phases, respectively, in Hamilton Harbour, Canada. As already discussed in Chapter 3, the diel alternations of destratification and restratification in equatorial Lake George, Uganda (Ganf, 1974*a*) are too rapid to invoke a community response. The physiological and behavioural flexibility of *Microcystis* evidently accommodates to these extremes better than most, potentially faster-growing species.

It would be presumptive to assume that the factor-interactions so far considered might contribute to the explanation of all distinctive sequences in phytoplankton communities. Indeed, extremes in other dimensions may be of far greater selective importance in the habitats where they occur. Thus, the floras of water which are influenced by geothermal flows, high salinity, high alkalinity or high acidity are usually very different in character from those so far considered. A remarkable case of a change in the plankton flora in response to pH fluctuation has been documented by Swale (1968) and Reynolds & Allen (1968). The plankton of Oak Mere, a small, acidic lake (pH 4.5), was dominated by various chlorophytes (*Ankistrodesmus*, *Lagerheimia*, *Closterium* and *Chlorella* spp.) until after the introduction of base-rich borehole water had raised the pH to 6.5, when *Asterionella*, *Pediastrum*, *Anabaena* and *Microcystis* became the dominant forms.

Excepting these specialized cases, there does seem to be sufficient circumstantial evidence to propose that the environmental control over phytoplankton growth and periodicity is exercised predominantly along three gradients: one of increasing temperature, one of increasing stability and decreasing frequency of light fluctuations, and one of decreasing nutrient availability. The 12 species whose growth in the Lund Tubes is evaluated against each of these variables in Figures 73 and 74 are arranged along each of the three axes shown in Figure 75 in a 'sub-sequence' corresponding to their apparent specific tolerances. For example, *Asterionella* is placed nearer to the origin of the axis of increasing temperature than *Anabaena*, because the growth of *Anabaena* is more severely limited by lower temperatures. *Asterionella* is placed nearer to the origin of the axis of decreasing nutrients than *Uroglena* because its growth is apparently more sensitive to falling concentration of nutrients (in this case, phosphorus). We may envisage periodic sequences starting at the origin and proceeding along whichever gradient is exacting the most powerful constraint on 'algal' growth at a given point in time. Thus, if temperature is low, but nutrients are high, the progression might follow the temperature axis. As the lake stratifies and the water temperature stabilizes the progression will switch to an appropriate point on the stability axis and move outwards or inwards in accord with changes in physical variability. As nutrients fall to low concentrations, the progression is more likely to follow the nutrient axis. The arrow corresponds to hypothesized switches between the sub-sequences in response to advancing season. Many more species could be inserted on the axes, and other axes could be added. An obvious candidate might be one describing daily light-energy income, since it is probable that it is low light, rather than low temperature, that limits winter growth in temperate lakes. This dimension is not wholly accommodated by the stability axis, by definition, until the energy flux to

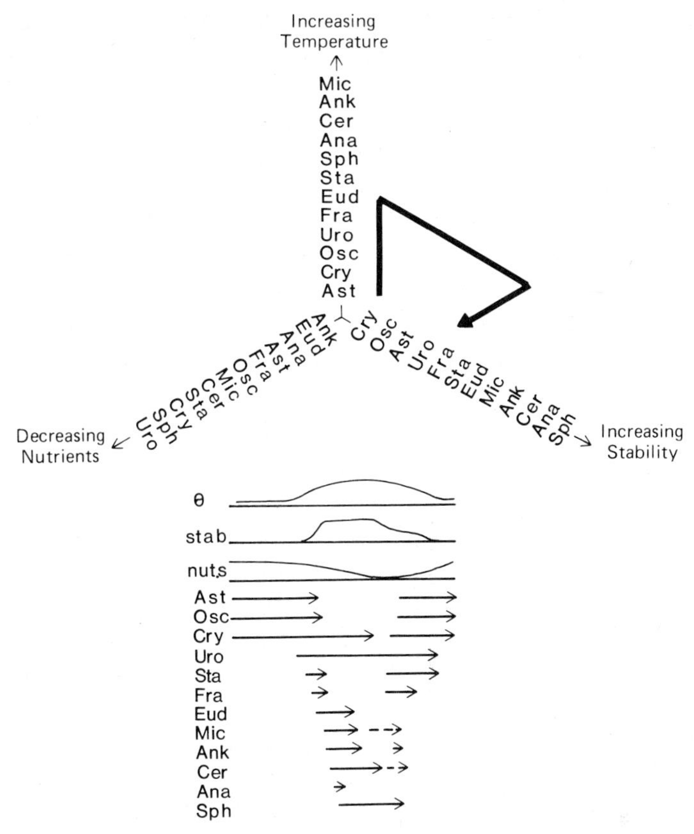

Figure 75. The growth of the twelve species of phytoplankton (Figures 70–74) ranked along axes representing three supposed limiting-resource gradients (in ascending order of their temperature thresholds and descending order of sensitivity to thermal stability and dissolved nutrients). The broad arrow traces the likely direction of selection in a lake, through its vernal heating, onset of stratification and its eplimnetic nutrient depletion (show below). Arrows against species identifications cover those parts of the year in which the combination of the three gradients should be conducive to good growth performance. (Original.)

the lake surface passes certain, possibly species-specific, critical thresholds of intensity and duration.

For the sake of simplicity, let us assume that the three axes shown in Figure 75 exemplify several such axes representing environmental variables. A given species will be able to grow under a given combination of environmental conditions provided the empirical valuation of each variable encompasses the critical limit of that species. For example, *Fragilaria* could be expected to grow once its minimal requirements of temperature, physical mixing and nutrients were simultaneously satisfied. Then the *same* set of conditions might be adequate to support the growth of *Cryptomonas* or *Uroglena*, for example, but would possibly be inadequate for the growth of *Anabaena*. Equally, were the nutrient concentrations alone to fall below the critical limit for *Fragilaria* the continued growth of (say) *Uroglena* need not be impaired to the same extent.

These principles are embodied in the lower part of Figure 75. Hypothetical, year-long gradients in the three major environmental variables are shown above arrows to represent the periods of the year in which all three satisfy the conditions for net increase of each species. It is apparent that none is sufficiently flexible in its requirements to enable it to maintain growth throughout the year, though some come relatively closer to that ideal than others. Increasing stability apparently selects against the diatoms and *Oscillatoria*. Increasing temperature selects for *Sphaerocystis* and *Anabaena* while decreasing nutrients selects against *Eudorina*. Only *Cryptomonas* seems potentially able to maintain its growth through most of the year, while *Microcystis* and *Ceratium* may be able to tolerate low nutrient concentrations in summer by virtue of their extreme vertical mobility which could enable them to exploit hypolimnetic nutrient reserves unavailable to their competitors.

At other times, the environmental resources often meet the requirements of several species simultaneously (that is, their ranges overlap). Indeed, they may be found to be growing simultaneously. We may propose that, other factors being equal, the species that is most likely to become dominant will be the one, *of those available*, that is able to sustain the fastest rate of net increase (cf. Table 17). However, as its growth proceeded, light penetration would fall and nutrients would be consumed, so the 'spectrum' of resources, defined by the axes in the upper part of Figure 75, is perceptibly shifted. As the combinations of environmental factors evolve seasonally or are suddenly disturbed (for instance, by increased wind mixing), so the selective advantage should move among the various species. Indeed, such a mechanism could be the basis for phytoplankton periodicity.

That these representations provide a reasonable description of the sequences they are intended to model can be judged by comparing them with Figures 70–72. Although good general agreement is evident, several points of difference from the model require emphasis. One is that the actual variations in z_m/z_s and nutrient concentration are less regular than the model implies, and these may be held directly responsible for irregularities in the temporal phasing of (for instance) diatom growth. Another point is that fewer of the selected species were actually increasing simultaneously than might be 'predicted' from Figure 75, 'Availability' (i.e. being present to take advantage of the supposedly favourable conditions) may be a contributory factor. Equally, it is likely that the requirements of some potential competitors were not met in the way the model suggests. For instance, *Eudorina* evidently has a high demand for nutrients which, presumably, was met in 1978 (Figure 72) but scarcely so in 1977 (Figure 71). In the latter case, *Uroglena* dominated. Finally, although *Cryptomonas* showed frequent bursts of net increase in all three years, it was never sustained in the manner suggested by the model. Nor did its rate of increase approach its potential growth rate, which presumably would have enabled it to dominate the plankton for long periods in all three years. The role of processes which detract from rapid increase among actively growing species and the factors which contribute to 'availability' of inocula are discussed in Chapter 7.

6.3 Perennation

The ability of a species both to survive periods when environmental factors are unsuitable for growth and to maintain a potential 'inoculum', poised to exploit the return of favourable conditions, is of fundamental importance to the planktonic way of life. It is manifest in quite different ways among the various groups.

Perhaps the most striking and best-documented example of alternation between extant vegetative populations and distinctive resting phases is provided by the dinoflagellate, *Ceratium hirundinella. C. hirundinella* occurs in a wide trophic range of (usually) deeper, temperate lakes (Höll, 1928), where it frequently builds up its populations through the summer period. Later in the summer or autumn, an abrupt decline in numbers of vegetative cells takes place, with a concomitant production of distinctive cysts. These have thick resistant walls of cellulose and electron-dense granules and their protoplasts usually contain conspicuous reserves of starch and lipid, accumulated during a phase of stationary growth (Chapman, Dodge & Heaney, 1980). The cyst wall is laid down within the theca of the vegetative cell; the mature cyst is released by dehiscence of the theca. On release, the dense, non-motile cysts rapidly settle out to the sediments, where they are presumably equipped to survive the low-temperature, low-light conditions characteristic of high-latitude lakes, and whence they are presumably able to reinfect the water column the following year. Because the cysts are lost rapidly from suspension, it is not easy to assess the proportion of cells which encyst. However, recoveries of cysts in sediment traps placed near the bottom of the water-columns of Lund Tubes suggest that they account for 35–42% (Reynolds *et al.*, 1983*b*) or more (Reynolds' unpublished data) of the maximum standing crop of vegetative cells.

'Germination' (excystment) has been described by Huber & Nipkow (1922) and more recently by Chapman *et al.* (1981). A naked flagellate cell, or gymnoceratium, emerges through an exit slit and soon develops the thecal plates of the vegetative form. Heaney, Chapman & Morison (1981) have shown that a sharp recruitment of vegetative *Ceratium* cells to the plankton of Esthwaite Water occurs in late winter, when the water temperature rises to 5°C, and coincides with an abrupt increase in the proportion of empty cysts recoverable from the sediments of the lake. There follows a period in which the population remains nearly static or declines slightly, until in April or May, cell division and exponential increase in the population become established.

Nevertheless, Heaney *et al.* (1981) remarked that relatively few of the cysts produced the preceding autumn are required to 'germinate' in order to establish the basis of a dominant summer population. There may be high mortalities among the overwintering cysts (brought about by parasitic attack or consumption by browsing benthic protozoa and crustacea) or among the gymnoceratia. Nevertheless, an enhanced spring recruitment of *Ceratium* to the plankton of Crose Mere generally follows prolific cyst production the previous autumn (Reynolds, 1978*b*). The precise location of cysts on the sediment may be another important factor: Lund & Reynolds (1982) related the apparently poor development of *Ceratium* populations in the Lund Tubes, compared with those of the outside tarn, not to an inferior growth rate but to very low initial inoculations of cysts. Being hydrologically isolated at the time when cysts are produced externally means that the tube populations must be dependent upon the absolutely low production of cysts within the tubes and upon the delivery of resuspended viable cysts deposited within the enclosures during winter openings. Another, as yet unquantified, factor is that *Ceratium* spp. produce at least two types of cyst, one of which is apparently less resistant than the other (Chapman *et al.*, 1980).

Asexual spores, or *akinetes*, are produced by many species of Nostocales (*Anabaena*, *Aphanizomenon*, *Gloeotrichia* spp.). These develop directly from vegetative cells usually singly, occasionally in groups of two or three. In *Anabaena solitaria* and *Gloeotrichia echinulata*, akinetes occur adjacent to, or in close proximity with, heterocysts (when present). Akinetes are usually wider and (especially in *Aphanizomenon*) longer than the vegetative cells. Their thick-walled appearance suggests a perennating function, though the temporal phasing of their formation is not always so clearly associated with the demise of vegetative populations as is that of encystement in *Ceratium*.

In their study of sporulation in White Mere and Kettle Mere (Shropshire), Rother & Fay (1977) related the induction of akinete formation to exposure to extreme conditions at the water surface during bloom formation rather than to any nutrient deficiency. Neither were they able to show conclusively that akinetes provided a means of overwintering, nor that their germination in spring necessarily provided the inoculum for vegetative growth the following year (cf. Roelofs & Oglesby, 1970; Wildman, Loescher & Winger, 1975; Reynolds, 1975*a*). On the other hand, they found that akinete germination took place sooner rather than later and that this contributed to an overwintering stock of vegetative trichomes. They suggested that akinetes served more as a means of surviving summer blooms than as overwintering propagules.

Rother & Fay (1979*a*) later investigated physiological and biochemical characteristics of sporulating *Aphanizomenon* populations. Intracellular phosphorus reserves remained high during bloom formation but declining levels of net photosynthetic production and nitrogenase activity, with a concomitant fall in cellular C/N ratio, were detected in sporulating populations. They attributed the induction of sporulation to an imbalance in the carbon and nitrogen metabolism with respect to normally-growing vegetative cells. They were able to support these findings experimentally under controlled laboratory conditions (Rother & Fay, 1979*b*).

However, akinetes have been shown not to be exclusively short-lived propagules; Livingstone & Jaworski (1980) recovered akinetes from a 1-m sediment core from Rostherne Mere at depths of up to 270 mm below the mud surface and deposited up to 64 years previously. Laboratory incubation under continuous illumination (40 μE *PhAR* $m^{-2} s^{-1}$) at 18°C induced germination within 20–30 days and sustained vegetative growth (one of these is still in culture: G. H. M. Jaworski, personal communication). The significance of this result is not that the akinetes were ever likely to have germinated naturally so much as an illustration that they had not been previously exposed to suitable conditions for germination and were still viable. Whilst the work of Rother & Fay (1977, 1979*a*) leaves little doubt that sporulation is triggered by bloom conditions, akinetes may nevertheless persist and germinate when conditions are suitable. Reynolds (1972) observed that *Anabaena* akinetes were episodically resuspended into the open water of Crose Mere and germination eventually occurred in the water column, presumably after certain minimum temperature and/or insolation requirements had been satisfied.

Other morphologically-distinctive resting propagules produced by phytoplankton include the sexually-produced zygotes of *Eudorina* and *Volvox* (which apparently are produced at the end of vegetative growth, and which, unlike vegetative cells, survive for several months, at least, after being deposited on the bottom, still within the colonial mucilage: Reynolds & Wiseman, 1982) and the siliceous cysts of Chrysophyceae, such as *Dinobryon*, *Mallomonas*, *Synura* and *Uroglena* spp. (many of which can only be reliably identified with light microscopy from their spores: Bourrelly, 1968).

Microcystis overwinters as vegetative colonies that reside on the lake bottom where they can survive for several years, apparently without light or oxygen (Reynolds *et al.*, 1981). Some centric diatoms, particularly of *Melosira* spp. (Lund, 1954, 1966*a*), *Stephanodiscus* and, perhaps, other genera can survive for long periods (months to years) on the sediments of lakes. There, the vegetative protoplasts contract into physiologically-resting globules still within the frustule. Those of *Melosira* spp. (*M. italica*, *M. granulata* a.o.) readily 'germinate' when resuspended into the water column and suitable growth conditions obtain (e.g. Lund, 1966*a*; Reynolds, 1973*a*). No such resting stages have been discerned among the pennate diatoms, though *Asterionella* and *Fragilaria* often settle out live and may remain capable of revived vegetative growth if resuspended within 1–3 months. In the main, however, these diatoms, together with many small planktonic species of cryptomonads and chlorellids have no known means of perennation save that of being capable of attaining relatively fast rates of growth. They can quickly increase their numbers to detectable proportions when exposed to suitable growth conditions (e.g. Lund & Reynolds, 1982). Incidentally, this impinges upon one of the most remarkable features of the freshwater phytoplankton, that is, how it is transferred from one water body to another, even when there is not always direct hydrological communication between them. Transport on the bodies of migrating waterfowl has been suggested to be important, but it would be expected that the 'algae' must also be capable of withstanding desiccation. The work of Atkinson (1980) suggests that faeces may be more important than feathers as a means of dispersal; although viable planktonic algae have been recovered from the faeces of ducks, the long-distance dispersal of plankton 'algae' by wildfowl has not yet been proved.

In relation to the algal periodicity of particular lakes, the production and maintenance of viable propagules represents a forceful contribution to the constancy of annual periodic cycles, since it preserves the 'availability' of a species to provide an inoculum when conditions are suitable. It also confers an advantage over potential competitors compelled to start from a smaller base or to reinfect the water body from outside. As has been

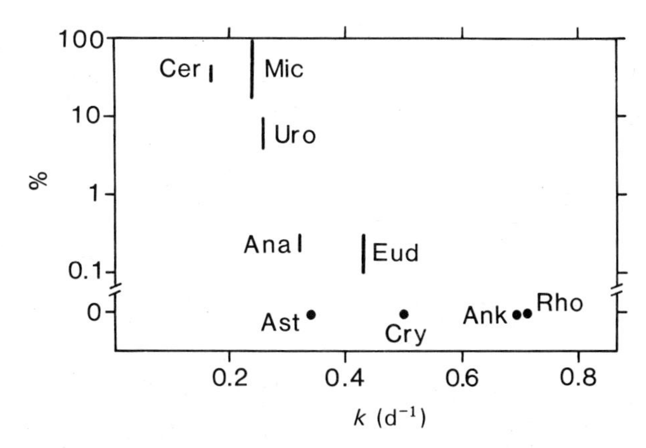

Figure 76. Total observed propagule formation (ranges) during the decline stages of specific populations, expressed as a percentage of the preceding maximum population of vegetative cells, plotted against maximum *in situ* increase rates (from Table 17). Abbreviations to 'algae' as in Figure 70. Original; data for *Microcystis* and *Eudorina* from Reynolds & Wiseman (1982); for *Ceratium* and *Anabaena* from Reynolds *et al.* (1983*b*); for *Uroglena* (author, unpublished).

pointed out, non-perennating species must rely on being able to maintain a relatively larger stock of vegetative cells (for instance, by maintaining growth over a wider range of environmental conditions) and/or capitalizing on favourable conditions by a rapid rate of replication. The latter options are more likely the preserve of smaller-celled algae. It is interesting to note that an apparent negative correlation can be deduced between the provision of morphological or physiological propagules and growth rate amongst phytoplankton (Figure 76); this, in itself, testifies to the existence of more than one strategy for the growth and survival of planktonic 'algae' in lakes.

6.4 Growth and survival strategies

The interrelationships among specific growth rates, morphologies, and optimal ranges with respect to environmental resource 'spectra' and means of perennation are indicative of distinct trends in the evolutionary adaptation of 'algae' to a planktonic existence. It is possible, for instance, to discern a broad polarization between, on the one hand, species that have a high growth rate and can respond quickly to the availability of environmental resources but do not or cannot maintain their maximum density for long and, on the other, species with lower growth rates, slower responses and an ability to tolerate or accommodate to periods of resource-stress. The former may be recognized as colonizing, opportunistic, 'weedy', or '*r*-selected species', the latter with constant, autochthonous, equilibrium or '*K*-selected species'.

The terms *r*-selection and *K*-selection are particularly useful ones, but a clear understanding of what they convey is essential. The concept owes to MacArthur & Wilson (1967) and is based upon the logistic equation of growth, *r* being the intrinsic growth rate and *K*, the upper asymptote of the hyperbolic growth curve (i.e. the point at which a resource or resources became absolutely limiting). They recognized that, in natural environments, there is competition both for the space available, where rapid colonization and rapid growth are (*r*-) selectively advantageous, and for the available resources, where the ability to operate close to the environmental carrying capacity will be (*K*-) selected. These principles apply in all ecosystems, where they contribute to the maturation of community organization (or 'succession': see also Chapter 8). An unoccupied habitat may be first colonized by pioneer (*r*-selected) species which are later displaced by (*K*-selected) species that make better use of the resources available. Subsequent development of the original concepts of *r*- and *K*- selection (Hairston, Tinkle & Wilbur, 1970; Pianka, 1970, 1972; Margalef, 1977) has firmly established them in theoretical ecology, where it has become usual to refer to organisms as being either *r*- or *K*-strategists: the *K*-strategists grow more slowly but get further than the *r*-strategists.

The application of *r*- and *K*-selection to phytoplankton ecology has been attempted by Kilham & Kilham (1980) and Sommer (1981). Both studies introduced an element of adaptation to the original theory of *K*-selection to accommodate it to the relative unstable and short periodic cycles of plankton communities, which terrestrial biologists would consider to be universally *r*-selected (Kilham & Kilham, 1980). Both they and Sommer demonstrated (as I hope I have also succeeded in doing in this book) that a full range of life-history strategies exists among phytoplankton, in which the allocation of time, energy and matter differs between species in ways that can be analogized to the principles of *r*- and *K*-selection. The theoretical approach developed by Kilham & Kilham distinguished *r*-selection as favouring a relatively greater investment in reproductive effort, whereas under *K*-selection, proportionately more resources are diverted to non-reproductive activities (for instance, vertical migration), in order to maximize the exploitation of available resources. Uptake and retention of nutrients by *K*-strategists make them less available to other species. Though different, both strategies can be seen to increase the survival prospects of the gene pool. They also recognized that *r*-selection would be favoured in habitats where existing population densities are well below the carrying capacity, where supply of resources (*S*) is in excess of their demand

(D). Such situations might be expected to exist early in the year, or, to some extent, after density-independent external ('allogenic') perturbation of the existing physical structure (such as increased wind-mixing: Reynolds, 1980*a*). Under such circumstances, the species whose biomass is likely to be better represented will be the faster-growing species. As resources are depleted ($D/S \rightarrow 1$), however, the emphasis moves towards increased competitive ability, i.e. towards K-selection. The controlling resource may be a nutrient, light, temperature, turbulence or, perhaps, a biotic variable (such as predator or parasite populations). Kilham & Kilham (1980) discussed a number of examples, drawn from both marine and freshwater habitats, which seem generally to show parallelism between declining growth rate, increasing cell size and decreasing niche-overlap (i.e. favoured species become more specialist in their requirements) and an increasing D/S ratio. These trends are consistent with a progressive shift from r- to K-selection.

Sommer (1981) related seasonal changes in the phytoplankton of Bodensee (Lake Constance) to the morphology and performance of individual species. He classified the abundant species according to their rate of increase ($k > 0.4\ d^{-1}$; 0.2–$0.4\ d^{-1}$; $< 0.2\ d^{-1}$) and then according to their size (cell of colony units $< 30\ \mu m$ in length and width; units $> 30\ \mu m$ in length but $< 30\ \mu m$ in width and easily fragmented filaments; units $> 30\ \mu m$ in length and width). He found that species of the same growth- and size-classes tended to be associated with each other during the annual cycle, the small, fast-growing species occurring in spring and the large, slow-growing forms in mid-summer. Sommer equated these periods as being respectively indicative of predominant r- and K-selection and their ranking in time as evidence of a trend towards increasing K-selection. By mid-summer, phosphorus and silicon had become severely limiting. Towards the end of summer, when thermal stratification began to break down, nutrients became more plentiful but average light levels began to fall, there was a shift back towards the more r-selected species.

It is interesting to note that, both in the sea (Kilham & Kilham, 1980) and in lakes (Sommer, 1981), large dinoflagellates (*Ceratium* spp.) apparently represent the ('most') extreme K-strategists. This conforms with their acknowledgedly complex characteristics of high mobility, of slow growth and of relative immunity to grazing animals (Heaney & Talling, 1980*b*; *q.v.* for references). Their versatile and adaptable behavioural responses undoubtedly equip them to live to resource limits which few other 'algae' tolerate. Possibly the facultative ingestion and assimilation of bacteria and other small detrital particles ('phagotrophy': Dodge & Crawford, 1970) may aid survival when nutrients are scarce. In tropical oceans, characterized by a high level of environmental constancy and generally low nutrient levels, *Ceratium* spp. are usually conspicuous dominants. In temperate lakes, seasonal variations restrict their abundance to two or three months per year. If the production of cysts (and indeed of resting propagules by other species) is regarded as a natural alteration of status in the natural life-cycle (Sommer, 1981), then the overwintering stage could be seen as yet another refinement in the life of a planktonic *K*-strategist. In (warmer) lakes at lower latitudes, the place of *Ceratium* 'is taken' sometimes by another dinoflagellate (*Peridinium* sp.), sometimes by *Microcystis*. The overwhelming dominance of *Microcystis* in the dielly-mixed Lake George, Uganda, throughout most of the year and the evident inability of any other species to compete seriously with it is, again, what might be expected of a *K*-strategist in an ecologically stable environment. Its frequent seasonal dominance in eutrophic temperate lakes, sometimes to the apparent exclusion of *Ceratium* (Reynolds, 1978*d*), suggests that *Microcystis* is another example of an extreme *K*-selected phytoplankter.

7

Loss processes

'Lose, and start again at your beginnings'
KIPLING, *If*

7.1 What is a loss process?

It is almost axiomatic that much of the specific biomass of phytoplankton elaborated in growth is eventually destined to 'disappear', or be lost, from the open waters. Apart from the provision of perennating propagules, the loss from suspension at a given time and place may often be attributed to mortality, whether it is caused by an inadequate source of light energy or nutrient or some other physiological limitation, or whether it is brought about by disease, attack by parasitic organisms or consumption by animals. Sometimes the removal may be due to entirely physical processes – sedimentation to the bottom or physical displacement by hydrological flow – though it may eventually be followed by biological death.

Populations may thus be said to be subject to 'losses', the magnitude of which can exceed that of anabolic production. There is no reason to suppose that the final total 'loss' is sustained at the end of a population. Rather, losses are sustained during the entire course of a population, although they will usually increase, relatively as well as absolutely, as it progresses. In the end, the population maximum should be regarded as the point when the sum of the rates of losses balances the rate of growth for the first time (either because the former increases or the latter decreases or both). Prior to that point, the rates of loss (Σk_L) detract from the true growth rate (k') but nevertheless fail to prevent a positive rate of increase ($+k_n$). After the maximum, $\Sigma k_L > k'$ and k_n is negative. It should be noted, however, that population decline is not synonymous with zero growth.

That a full understanding of phytoplankton population dynamics will come only when the factors influencing both the waxing and waning (i.e. both growth and mortality) can be ascertained has been recognized by several reviewers (e.g. Kalff & Knoechel, 1978; Le Cren, 1981). Indeed various attempts to model species succession have strongly suggested that the rates of loss are essential driving variables (O'Brien, 1974; Lehman *et al.*, 1975*b*; De Pinto, Bierman & Verhoff, 1976). However simple is the concept of several conflicting processes acting simultaneously to effect observable net rates of change, its practical demonstration is fraught with considerable difficulties. Firstly, the measurement of true growth *in situ* is far from being a straightforward matter. The direct measurement of photosynthetic carbon fixation is a well-tested, relatively simple and therefore attractive method for assessing production. However, respirational and excretory losses of carbon from physiologically active cells, especially when nutrient-limited, detract from the direct translation of carbon fixed per cell into an equivalent production of new cells. Few other variables offer an independent assessment of cell production, though silicic acid uptake in the special case of diatoms is, potentially, an obvious exception (see §7.3 below).

The second difficulty is the paucity of good data relating to the rates or magnitudes of the losses themselves. Many of the dynamic models of plankton biomass changes (reviewed in Patten, 1968; Jassby & Goldman 1974*a*) have inserted approximate terms to account for the processes assumed to be important. Many have incorporated a factor for the rate of removal by herbivores; allowance for the rate of removal by sinking and vertical transport has been made less frequently (but see Riley, Stommel & Bumpus, 1949); horizontal transport has been virtually ignored (Jassby & Goldman, 1974*a*) although much theoretical work now exists (Chapter 3). Other sources of loss (cell mortality, due to parasitic attack by viruses, bacteria and fungi, and due to exposure to physiological extremes of light, temperature, nutrient concentration and toxic substances) have rarely been incorporated, although some important advances have been made in these fields.

An alternative approach to the problem is to attempt to quantify the losses and loss rates in experimental systems. A notable conceptual advance in this direction was contributed by Uhlmann's (1971) work with batch cultures of planktonic algae. Much the same approach was used in the work on the dynamics of change in phytoplankton populations husbanded in Lund Tubes, with which I have been involved (see, for instance, Reynolds, Thompson, Ferguson & Wiseman, 1982*b*; Reynolds; 1982; Lund & Reynolds, 1982), and the findings of which contribute much to the following chapter. Before proceeding, however, it is necessary to clarify exactly what is considered to constitute a 'loss process' and to define the units of its measurement.

The term 'loss processes' has crept into the literature without, so far as I am aware, any formal definition. In terms of population dynamics, it is any process which actively removes biomass from the part of a water body under consideration and therefore depletes the potential stock of growing organisms. To some extent, they are all instantaneously density dependent and can be expressed in the same rate terms as growth (i.e. in fractional or natural logarithmic units).

Thus, $\delta N/\delta t = \mathrm{e}^{k_n} = \mathrm{e}^{(k' - \Sigma k_L)}$ (45)

where N is the biomass and k_n, k' and Σk_L are respectively the exponential rate constants of net change, growth and the summated loss rates. The units of biomass may be given in terms of carbon, dry weight, cell number or volume. Jassby & Goldman's (1974*a*) study of losses in Castle Lake, California, took photosynthetic carbon fixation as the basic unit of biomass and compared it with the actual (net) change in biomass per unit area of the lake. Loss rates were calculated as the difference, and specific loss rates (loss rate per unit biomass) as loss rate/phytoplankton carbon. Seasonal variations in specific loss rate ranged between 0.2 and 0.8 d^{-1}, which they were rarely able to attribute to the combined effects of water transport, sinking and grazing. They assumed that cell mortality and decomposition made up the unaccounted fraction of the inferred loss rate, and that this was more likely to be due to environmental tolerances of the planktonic algae being exceeded rather than to pathogenic attack. They did not, however, furnish any additional evidence that the shortfall of carbon had ever constituted, even briefly, new cells. In the work of both Uhlmann (1971) and Reynolds *et al.* (1982*b*), cell concentration was used as the basic unit, changes in which were related to known or *in situ* measurements of loss rates. The calculation of true growth (= net increase + net losses), however, makes the same assumption that the major losses were all accounted. Neither regarded death rate as being often significant. Reynolds *et al.* (1982*b*) substantiated this view, in part, through a consideration of the dynamics of the diatom species present, for which growth rate was monitored independently; they also cited results from a parallel study of decomposition in the same experimental system (reported by Jones & Simon, 1980) which showed that almost all the measurable decomposition occurred at the sediment or near the bottom (> 11 m) of the experimental enclosure. By excluding hydraulic washout (by hydrological isolation of the experimental system) and by reducing any likelihood of nutrient limitation of plankton growth (by frequent replenishment of known amounts of nutrient) and in assuming that horizontal distributional variations were unlikely to have been consistent in any one direction, they were able to show that sedimentation (and *subsequent* decomposition) and grazing were the major loss processes operating. The application of the findings to the wider context of phytoplankton ecology was also discussed (Reynolds *et al.*, 1982*b*; Reynolds, 1982) and, together with a washout and pathogenic or physiological death *of existing cells*, sedimentation and grazing were considered to be the most important loss processes affecting phytoplankton.

i.e. $\Sigma k_L \simeq k_w + k_s + k_d + k_g$

where k_w, k_s, k_d and k_g are respectively the exponential rate constants for hydraulic washout, sedimentation, death and grazing losses. Each is considered separately in the following sections (in order of ascending complexity); comparative and overall effects are considered in a final section (§7.6).

7.2 Hydraulic washout

The slow displacement of water supporting phytoplankton through an outflow by inflowing waters devoid of planktonic organisms approximates to a dilution function, provided the water body itself delays the flow rate for long enough for the inflow to become fully integrated and the distribution of planktonic organisms remains approximately uniform. The less this condition applies, the greater is the departure from straightforward dilution. The departure will be greatest in rivers, subject to strong unidirectional flow, where the phytoplankton-bearing water is almost exclusively displaced downstream. The removal of phytoplankton from a section of river, represented in Figure 77a, will be in constant and direct proportion to the discharge (q) relative to the volume, V. The time taken to flush the population from the section (t') is given by V/q; in the example illustrated, $t' = 3$ time units of length t. The fraction of the original population (N_0) removed at any given intermediate value of t will be in the proportion t/t'; the proportion remaining (N_t) is given from:

$$\begin{aligned} N_t &= N_0 - N_0(t/t') \\ &= N_0(1-t/t') \\ &= N_0(1-qt/V) \end{aligned} \qquad (46)$$

In contrast, Figure 77c, shows the corresponding fractions of a similar initial population remaining at similar time intervals after they have been fully randomized at each interval of t. At t_1, the population corresponds to:

$$\begin{aligned} N_{t_1} &= N_0 - N_0 q t_1/V \\ &= N_0(1-qt/V) \\ \text{At } t_2,\ N_{t_2} &= N_{t_1} - N_{t_1}(1-qt/V) \\ &= N_0(1-qt/V)(1-qt/V) \\ \text{At } t',\ N_{t'} &= N_{t_3} = N_0(1-qt/V)^{t'/t} \end{aligned} \qquad (47)$$

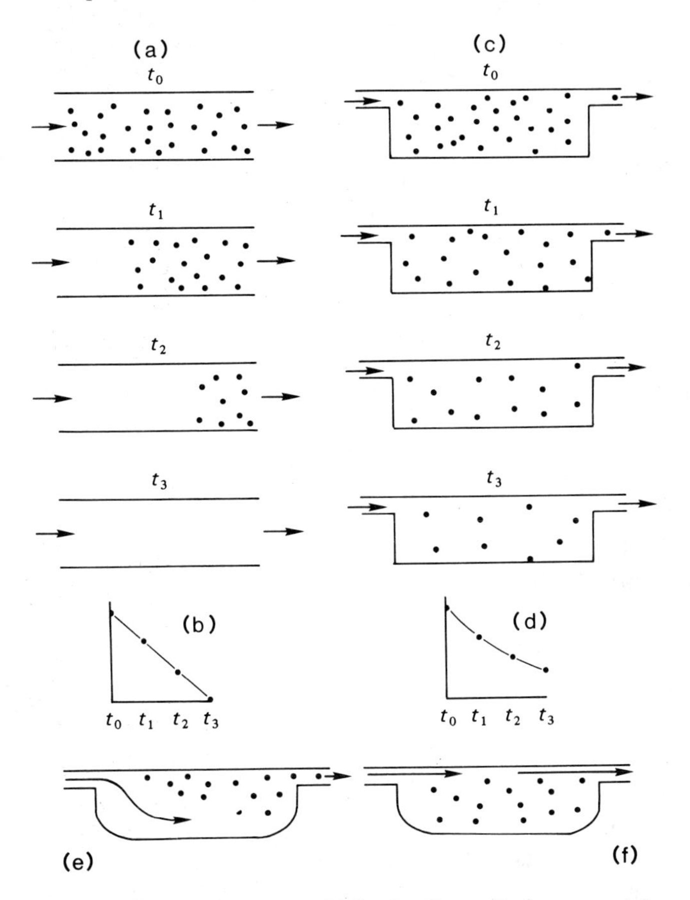

Figure 77. Successive stages (a) in the direct displacement of particle-rich water by clear water and (c) assuming full spatial randomization of particles remaining at times t_1, t_2, t_3. The falls in the total numbers of particles remaining in the section under either treatment are respectively shown in (b), (d). The lowermost figures show two possible variations where (e) the inflow forces under a less-dense lakewater and (f) where inflow 'slides' over a more-dense lakewater. (Original.)

Empirically this corresponds to

$$N_{t'} = N_0\,(1-1/3)^3$$
$$= 8/27\,N_0$$

The series is hyperbolic and is asymptotic to zero (Figure 77d). In a continuously mixed system, in which more, smaller time units, t/t' are accommodated within t', the series rapidly tends to:

$$N_{t'} = N_0\,(1/\mathrm{e}).$$

This solution is analogous to that for dilution by sinking, derived in §2.3 (equations 5 to 9). The population remaining at any given value of t is thus:

$$N_t = N_0\,(1/\mathrm{e}^{t/t'})$$
$$= N_0\,\mathrm{e}^{-(qt/V)} \qquad (48)$$

The rate of net change (k_n) in the population is easily solved, since:

$$N_t = N_0\,\mathrm{e}^{k_n t}$$
$$k_n = -q/V \qquad (49)$$

It may be seen that for the population to remain constant, it must sustain a growth rate (k') such that $k_n = k' - k_w = 0$, i.e. $k' = q/V$ (in the example, $q/V = 0.33$). No amount of growth, however, can offset the effect of displacement.

A further difference between the hypothesized systems represented in Figure 77 concerns the perceived population density at a given point. In the integrated, diluting system, the instantaneous concentration of planktonic organisms is, at least, theoretically uniformly distributed so that, other factors being equal, the concentration will be instantaneously identical, at any two horizontally separated stations. Equally, the concentration at any given point will be observed to decrease with time, in accordance with the general dilution model. At a given point in the system which is being linearly displaced by clear inflow water, the concentration will apparently remain constant for a time, before decreasing abruptly to zero. This effect would be overridden if the inflowing water itself contains the same species of planktonic organisms. Thus, if river phytoplankton is sampled frequently at a fixed point, the effect of outwash is countered by the import of plankton from upstream. One would therefore be effectively monitoring a sequence in the downstream transport of plankton and not necessarily periodic changes in the population dynamics of organisms in the river as a whole. It is generally true, however, that a true river phytoplankton is a characteristic of larger, longer rivers in which 'algae' have an opportunity to grow and develop during downstream passage. Conversely, small headwater streams, fed directly or indirectly by precipitation rarely carry a planktonic element. Intermediately, the population ecology of river plankton is strongly dependent upon changes in discharge (precipitation input), developing best during low flows and declining during flood periods. Of course, these effects are conditioned by the presence of lakes, lagoons or artificial impoundments along the water course. Events in those water bodies are always likely to influence the 'algal' flora in the river and, to some extent, in similar water bodies downstream of them.

The dynamics of growth and loss in river plankton are clearly something of a special case, the study of which requires appropriate sampling strategies and approaches to interpretation. Nevertheless, an appreciation of the factors involved in such extreme cases aids the interpretation of the effects of throughflow and outwash in lakes. Some specific cases can be considered briefly.

Crose Mere is a small shallow lake (area: 15.2 ha; volume: 0.73×10^6 m^3) that is fed and maintained largely by ground-water influx, which we will assume imports no live phytoplankton cells, and is drained by a small outflow ditch discharging between 100 and 1 700 m^3 d^{-1} (Reynolds, 1979*b*), that is between 0.01 and 0.23% of the lake volume per day; $-k_w \leqslant 0.0023$. It may be seen immediately that the rate of loss is minuscule, unlikely either to prevent the plankton from becoming fully integrated throughout the lake as a whole or ever to be, by itself, a critical factor limiting the development of a phytoplankton or, for that matter, in its post-maximal removal.

Grasmere (area: 64.4 ha; volume: 4.99×10^6 m^3; Ramsbottom, 1976) is located near the headwaters of the River Rothay (one of the major water courses feeding Windermere) which, together with several smaller streams, flows directly into the lake. The catchment (area: 27.9 km^2) is largely open fell grassland, based on thin soils overlying unyielding rocks and includes several small tarns; annual precipitation (in the range 2.3 to 3.3 m) net of evapotranspiration (about 0.5 m annually) generates a mean annual flow of between 50 and 78×10^6 m^3, that is enough to displace the volume of Grasmere between 10 and 15 times per year. Thus, the mean retention time of Grasmere is between 24 and 37 days but, owing to the nature of the catchment and the variability in intensity and seasonal distribution of its rainfall, the variation in instantaneous retention time is a good deal wider (estimated to be 9 to 65 days, i.e. between 1.5 and 11.1% of the lake volume may be removed each day; $-k_w \simeq 0.015$ to 0.111 d^{-1}). During droughts, the effect of outwash on the dynamics of the phytoplankton may be relatively small but a slow dilution may be expected whenever growth rate is limited by other factors. At high discharges, however, especially in the winter, loss rates can exceed the growth rates of many phytoplankton species and, if sustained, lead to severe depletion of the suspended stock. Indeed, winter discharges may be such as to delay the onset of the spring growth and its eventual maximum, even though nutrients and light availability may be adequate to support growth earlier in the year, simply because 'recovery' of a phytoplankton to that capacity may commence from very small inocula. The effect can be compounded by increased turbidity associated with floods, which impedes the penetration of light,

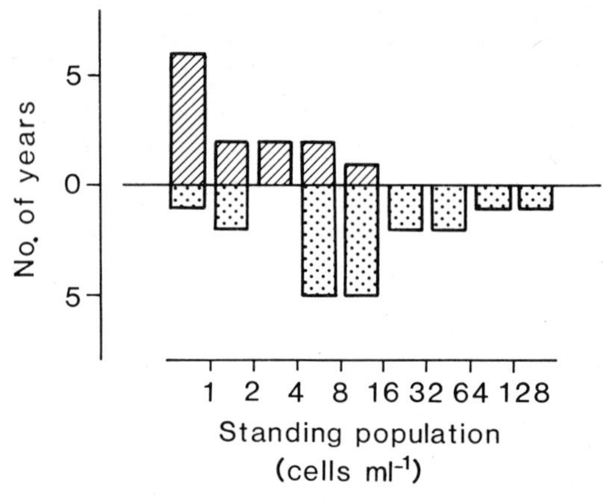

Figure 78. The numbers of years, between 1945 and 1976 (inclusive) in which the standing population of *Asterionella* cells in Blelham Tarn, on the nearest weekly sampling date to 26 January fell within the categories shown. Shadings distinguish those winters in which the December–January period was wetter (hatching: 590–897 mm) or drier (stipple: 282–590 mm) than the mean. (Drawn from data of Lund, 1978.)

and the timing of floods: those occurring in midwinter, when growth is in any case near its annual minimum, may have a more serious effect upon the capacity of populations to recover subsequently. The performance of species the previous autumn may also be important in that it affects the size of the 'inoculum' carried through the winter months. On the positive side, relatively transparent water, replete with fresh nutrients brought in from the catchment, may eventually favour a rapid rate of growth of the population in the wake of the flood. From a 32-year study of the phytoplankton of the nearby Blelham Tarn (instantaneous retention time: 5–∞ days), Lund (1978) was able to demonstrate that the population of *Asterionella* cells existing at the end of January (i.e. before its rapid vernal increase commences) was typically higher when the previous two months had been relatively dry than when precipitation had been particularly high (see Figure 78).

A final example of the more extreme effects of hydraulic throughput is drawn from Slapton Lower Ley, Devon. This shallow coastal lagoon is situated at the outfall of the River Gara but is isolated from the sea by a long shingle ridge. The lake (area: 0.77 km^2) is uniformly shallow (mean depth: *ca.* 1.6 m). Its storage volume fluctuates between 0.54 and 2.1×10^6 m^3 (mean $\sim 1.2 \times 10^6$ m^3) but is small in relation to the area of its catchment (*ca.* 46 km^2). Some 30 to 40×10^6 m^3 of water is discharged into the lake annually, about two-thirds of it between October and March (Van Vlymen, 1979). Drainage from the lake is culverted through a sluice at the extreme southern end, although it becomes frequently closed by shingle thrown up during storms. At all times, lakewater seeps across the single ridge. When the sluice is blocked, seepage and evaporation together cope with low flows (though without leading to any significant removal of phytoplankton from the lake) but high discharges rapidly fill the lake and necessitate clearance of the culvert. In extremes, the culvert reaches its flow capacity of 0.5×10^6 m^3 d^{-1}, equivalent to between one-quarter and one-third of the lake storage volume when clearance becomes necessary.

At such times, the lake becomes flow-dominated. 'Dilution' of the lake by inflowing river water would indicate a value for k_w of -0.3 to -0.4 d^{-1}, reducing a non-growing phytoplankton stock by 95% in 10 to 7.5 d respectively. In fact, phytoplankton stocks have frequently been observed to have been reduced to this level within three days (unpublished observations of Van Vlymen), irrespective of contemporaneous growth rates, suggesting that direct displacement applies. Van Vlymen (1979) considered that the basin morphometry is conducive to a strong linear flow, likening it to a widened river.

Van Vlymen (1979) also compared the hydrological characteristics of Slapton Ley with 30 or so other lakes, in Europe, North America and Africa, for which appropriate data were available in the literature. In particular, he found a good negative correlation between catchment area: lake volume ratio and mean retention time. The greater is the first and the shorter is the second, then the more likely it becomes that outwash will be a significant component in the losses of suspended material and, hence, a critical factor in the ecology of the phytoplankton. The effects will always be subject to modification by seasonal throughflow patterns and lake morphometry. Whether the lake becomes stratified and whether the temperature of the epilimnion differs from that of the inflowing water (tending either to force the inflow into the hypolimnion and to displace epilimnetic water through the outflow or, alternatively, to flow across it, leaving more of the lake water intact) may further modify the predicted effects of the dilution model. Even so, all throughflow effects act principally on suspended populations; they have little direct effect on benthic propagules, the production of which is clearly advantageous to the survival of those species in lakes subject to markedly seasonal fluctuations in flow.

7.3 Sedimentation losses

It is inevitable that algal particles will sink through water columns if they are both heavier than the water and unable to restore their position by swimming or regulating their buoyancy. So long as this condition holds, the inevitability persists, irrespectively of whether the water is subject to turbulent mixing or not. If the 'algae' do not first decomose or if they are not consumed by animals, they will ultimately settle onto the underlying solid substratum.

In Chapter 2, it was established that the sinking rate of most non-motile, planktonic 'algae' conforms closely to Stokes' Law. If the water were completely static, such particles would sink at an average velocity (v' md^{-1}) predicted by the modified Stokes equation (14); a water column of z_m would be progressively cleared of algal particles, according to a zero-order function, clearance being completed in (t' =) z/v' d. At an intermediate time t, the proportion settled approximates to t/t' or $v't/z$; the unsettled fraction is $(1 - v't/z)$. It was further established that, at fully developed turbulence, when the particles remaining in suspension were continuously mixed throughout the entire column, then the fraction remaining at a time t approximates to $e^{-v't/z}$;

i.e. $N_t = N_0 \, e^{-v't/z}$.

Since, by definition,

$$N_t = N_0 \, e^{k_n t},$$

$$k_n (= -k_s) = -v'/z. \tag{50}$$

The settling rates from still water and from turbulent columns define the lower and upper limits of sinking loss rate. Neither applies to all waters, or even to all parts of the same water body simultaneously. Walsby & Reynolds (1980) considered that sinking losses across metalimnia and into boundary layers adjacent to solid surfaces conformed more closely to the still-water solution but that the exponential reaction applied to water layers frequently integrated by turbulent or convectional motion. The minimum level of turbulence (measured by vertical eddy diffusivity, K_z) at which the reaction applies to particles whose intrinsic (still-water) settling velocities conform with the Stokes equation appears to be in the range 10^{-2} to 10^{-3} $cm^2 \, s^{-1}$ ($Ri \geqslant 0.25$; $d\rho/dz \geqslant 0.02$ kg m^{-4}; see Chapters 2 and 3).

It has also been argued previously that, above this critical level, the actual level of turbulence does not affect the sinking loss rate. Two factors are, however, important. First is the *turbulent intensity* (i.e. frequency of mixing). A natural water column tends to stratify under the influence of solar heating but to 'turn over' at night, as heat loss to the atmosphere may cause the cooled water to drop through the lighter, warmer water beneath. Such convectional motions generate (albeit brief) redistributions of 'algal' particles through the affected layer. The second point is that evaluation of the exponential model of sinking loss is continuously dependent upon the depth of the affected layer. In practical terms, it will correspond to the mean depth of water overlying a persistent stable density gradient. This may correspond to the epilimnion, or that part of it above a secondary metalimnion that survives nocturnal cooling or, if the water body is unstratified, to the bottom boundary layer. This depth may be termed the 'mixed depth'.

In previous attempts to evaluate sinking loss rates, Reynolds (1979a) and Reynolds & Wiseman (1982) adopted a descriptive model that is, at once, less accurate for intensively, continuously mixed columns but, in view of the above points, arguably more practical; this envisaged only one complete mixing per day.

$$N_t = N_0(1 - v'/z_m)^t$$

(where the time scale, t, is measured in days),
whence

$$-k_s = \ln(1 - v'/z_m)/t \qquad (51)$$

Provided that v' is small in relation to z_m, the solution differs little from that of the fully-turbulent model. For instance if $v' = 0.2 \text{ m d}^{-1}$ and $z_m = 10$ m, the sinking loss rate may be evaluated as

$$-k_s = \ln(1 - 0.2/10)$$
$$= -0.020 \text{ d}^{-1}$$

or

$$-k_s = -0.2/10$$
$$= -0.02 \text{ d}^{-1}$$

As the value of (v'/z_m) increases, so their mutual departure increases. When $z_m = 5$ m $(-v'/z_m = -0.040 \text{ d}^{-1})$, the solution to equation (51) is -0.04 d^{-1}; when $z_m = 2$ m $(-v'/z_m = -0.100 \text{ d}^{-1})$, it is -0.105 d^{-1}; when $v' = 1 \text{ m d}^{-1}$ and $z_m = 2$ m $(-v'/z_m = 0.5 \text{ d}^{-1})$, it is -0.693 d^{-1}. Possible discrepancies in the data quoted below, derived according to equation (51), should be borne in mind throughout.

Other factors which influence loss rate from suspension in mixed layers include: the effect of form resistance on entrainment (see §2.5.5.2); the return of once-sedimented material ('resuspension'), especially from shallow deposits affected by wave action; and by changes in the intrinsic settling rate of the particles that are physiologically influenced (§2.6). The latter are especially important in the consideration of the dynamics of specific populations.

The difficulties of obtaining reliable data on the intrinsic sinking rates (v') of 'algae' have been pointed out in reviews by Bienfang, Laws & Johnson (1977) and by Walsby & Reynolds (1980). These stem principally from the variations in individual sinking rates and to the virtual impossibility of excluding convection from settling chambers. Methodologies for estimation may be broadly subdivided among those which (a) follow the movement of a discrete sample layer, or its trailing boundary, through static water, (b) estimate the rate of apparent 'dilution' of an initially homogeneous suspension, or the rate of deposition on the bottom of the chamber, through the early stages of sedimentation (when the concentration curves of straight sinking and exponential loss are most nearly similar), or (c) direct observations of individual particles. In the first two categories, frequent monitoring of cells (e.g. Fritz, 1935; Smayda & Boleyn, 1965; Reynolds, 1973*a*; Smayda, 1974) may be substituted either by direct measurements of optical density in a spectrophotometer (e.g. Steele & Yentsch, 1960), of chlorophyll concentration (Walsby & Xypolyta, 1977) or of *in vivo* fluorescence (Eppley *et al.*, 1967; Titman, 1975), or by using a radioactive label as tracer (Bienfang, 1979). Direct observations (e.g. Lund, 1959; Reynolds, 1973*a,b*) can sometimes be more instructive but they are more liable to inaccuracies caused by convection currents; the use of an electrophoresis microscope perhaps comes nearest to overcoming convectional problems (see Wiseman & Reynolds, 1981). The use of density gradients provides one means of obtaining suitably stable (convection-free) systems (see, for instance, Conway & Trainor, 1972; Booker & Walsby, 1979), provided appropriate corrections for density and viscosity of the medium are incorporated in the final calculations. Nevertheless, most methods can give an approximate measure of the order of typical sinking rates: where intercomparisons of different methods are available, for instance among data pertaining to freshwater *Asterionella* colonies (summarized in Table 18), agreement is generally good. Much of the variation can be attributed to differences in the temperatures at which the various observations were made, in the number of cells per colony, and in the physiological condition of the algae used. Many of the cited works include comparisons of sinking rates in which these characteristics were altered experimentally.

Change in physiological condition is, perhaps, the most perplexing source of variation in specific sinking rates. Differences in the sinking rates of healthy and killed diatoms are well known and have been independently attested by several researches (§2.6; see also Smayda, 1970, for a review). Killing diatoms (for instance, by boiling) or merely introducing sublethal amounts of algicides into suspensions, can increase sinking rate by a factor of three to five; subsequent aggregation of particles may further enhance this effect. Jaworski *et al.* (1981) demonstrated how rapidly live, viable cells of *Asterionella* increase their sinking rate in response to carbon depletion and recover when fresh supplies of carbon dioxide are introduced. Photosynthetic inhibition (by DCMU) achieved a similar effect. Less-exhaustive studies (of the author, unpublished) have suggested that the

Table 18. *Some sinking rates of* Asterionella formosa *colonies (generally 5–8 cells) obtained in the laboratory using healthy, exponentially-growing ('live') material or material which had either been killed or, if live, in stationary phase or narcotized or from CO_2-depleted culture*

θ (°C)	v' (m d^{-1})	Condition	Reference
0	0.12–0.32	live	Lund (1959)
9	0.22–0.26	live	Reynolds (1973*a*)
15	~ 0.25	live	Smayda (1974)
15	~ 0.59	live + propan-1-ol	Smayda (1974)
Apparently 20	0.14–0.26	live	Titman & Kilham (1976)
Apparently 20	0.43–2.53	stationary	Titman & Kilham (1976)
10	0.10–0.24	live	Wiseman & Reynolds (1981)
10	0.58–0.88	stationary	Wiseman & Reynolds (1981)
10	0.49–1.17	killed	Wiseman & Reynolds (1981)
20	0.23–0.40	live	Jaworski *et al.* (1981)
20	0.74–1.20	CO_2-depleted	Jaworski *et al.* (1981)

sinking rates of *Asterionella* and *Fragilaria* are increased in the dark (see also the results of Burns & Rosa, 1980, discussed below). These observations indicate that there may be some inverse relationship between the sinking rate of diatoms and their photosynthetic rates, recalling Lund's (1965) comments on Skabiachevskiĭ's (1960) deduction that diatoms become lighter when actively photosynthesizing. No explanation for these changes can yet be offered; the possibility that the living cell varies the effective viscosity of the medium adjacent to its surface (cf. Margalef, 1957) demands that the electrical properties of that surface are vitally enhanced, whereas no such variation could be found to accompany sinking-rate fluctuations in *Asterionella* (Wiseman & Reynolds, 1981).

In the longer-term, chronic nutrient deficiencies generate physiological stress in cells. Among diatoms, at least, limitation by nitrate, phosphate and (especially) silicate have been shown to increase the sinking rate of several marine species (for a review see Smayda, 1970; see also Bienfang *et al.*, 1982), while those of some freshwater diatoms probably behave similarly (e.g. Lehman, 1979).

Such pronounced and, potentially, rapid changes in *in vivo* sinking rate prevent their precise interpolation into theoretical models of the loss dynamics of natural populations. As is evident from equations 50, 51, calculation of k_s is as sensitive to fluctuations in v' as in z_m. There is thus an obvious attraction to the possibilities of measuring sinking rates *in situ*, despite the inevitable difficulties that are encountered. Burns & Rosa (1980) set up closed settling chambers in the field, two to three metres beneath the surface of stratified lake St George, Ontario. Each chamber included a plate that, when triggered remotely, separated the cylinder into an upper (4.0 l) and a lower (0.35 l) section. Chambers filled with epilimnetic water and the enclosed phytoplankton were allowed two hours or so in which to settle, before the upper and lower sections were isolated. Settling velocities and fluxes of the various species present were calculated from their concentrations in either section. The results successfully distinguished between the behaviour of different species and, intraspecifically, at different times of the day and under different conditions of insolation. The average settling velocities calculated for two species of diatoms (*Asterionella formosa*: 0.40 ± 0.28 m d^{-1}; *Fragilaria crotonensis*: 0.27 ± 0.13 m d^{-1}) are comparable with values obtained by other workers, those obtained during daylight hours being significantly lower (< 0.2 m d^{-1}; < 0.1 m d^{-1} on the brightest days) than those obtained at night (> 0.3 m d^{-1}). Of the other species present, the non-motile forms (*Selenastrum*, *Closterium*, *Scenedesmus* and *Lagerheimia*) showed positive but dielly fluctuating settling velocities, averaging 0.08–0.18 m d^{-1}. Three species of cryptomonad flagellates showed a tendency to move downwards by day but to remain effectively neutrally buoyant in darkness or, in the case of *Rhodomonas*, to show a net upward movement. A gas-vacuolate *Anabaena* was more consistently buoyant throughout (average upward velocity: 0.10 ± 0.11 m d^{-1}).

An alternative approach is to determine the rate of arrival of 'algae' settling into containers ('sediment traps') designed to intercept their movements in open water. Many designs of sediment trap have been used in studies of particulate fluxes in lakes and in the sea (comprehensively reviewed by Bloesch & Burns, 1980); occasionally, their deployment and the analysis of their catches have been directed towards obtaining estimates of the numbers of specific particles trapped per unit area in unit time and, by relating these values to the ambient concentrations of the same particles in water column above the trap, to derive an apparent effective mean settling rate (e.g. Grim, 1950; Järnefelt, 1955; Reynolds, 1976*b*,*c*, 1979*a*; Reynolds & Wiseman, 1982). Of course, it is important to establish the extent to which the catches are affected by the performance characteristics of the trap and to ensure both that the entrapped algae will neither grow nor be consumed by animals or decomposed by bacteria (inclusion of a biological fixative, such as formaldehyde, can meet this requirement). The validity of this technique has been tested against the sedimentation of known quantities of particles (*Lycopodium* spores) of predetermined settling behaviour into traps set in the Lund Enclosures (Reynolds, 1979*a*). It can even be adapted, by suspending inverted traps, to derive net flotation

Table 19. *Some sinking rates of diatoms determined from retrospective measurements in the field, using sediment traps (T) or field chambers (C), or reconstructed from changes in the standing population (R)*

Species	Site	Method	v' (m d^{-1})	Reference
Cyclotella melosiroides	Bodensee (upper 10 m)	T	0.2–2.5	(1)
Cyclotella comta	Bodensee (upper 10 m)	T	0.4 –5.0	(1)
Tabellaria fenestrata	Bodensee (upper 10 m)	T	1.0 –1.5	(1)
	Lohjanjärvi	T	0.3 –6.4	(2)
Fragilaria crotonensis	Bodensee	T	0.5 –1.5	(1)
	Lohjanjärvi	T	0.2 –2.5	(2)
	Crose Mere	R	0.21–0.65	(3)
	Crose Mere	T	0 –0.11[a]	(3)
	St. George	C	0.14–0.40	(4)
	Blelham Enclosures	R	0 –0.59[b]	(5)
	Blelham Enclosures	R	0.1 –1.17[c]	(5)
Asterionella formosa	Lohjanjärvi	T	0.1 –1.1	(2)
	St. George	C	0.12–0.68	(4)
	Blelham Enclosures	R	0.08–0.15[b]	(5)
	Blelham Enclosures	R	0.33–1.02[c]	(5)

(*a*) Traps markedly ‘undertrapped’ (Reynolds, 1976*b*).
(*b*) During population increase phases.
(*c*) During population decline phases.
References: (1), Grim (1950); (2), Järnefelt (1955); (3), Reynolds (1976*b*); (4), Burns & Rosa (1980); (5), Reynolds & Wiseman (1982) and Table 22.

rates of cyanobacteria in the field (Reynolds, 1975*a*). Results obtained by this method, some of which are included in Table 19, support the view that settling behaviours vary with species and with the age and condition of specific populations.

A third possible approach is to deduce sinking losses from the dynamics of natural populations. In most cases, this is rendered virtually impossible because other sources of loss may be difficult to quantify, independent estimates of new cell production can scarcely be obtained and, owing to horizontal variability in distribution, the net change in concentration may be more apparent than real. Reynolds (1976*c*) derived a sinking rate for *Fragilaria* colonies in Crose Mere, during periods of net population decline, assuming zero growth (no silicon uptake), that other sources of loss were negligible and that sinking from the upper 5 m of the lake ($\simeq z_m$) was therefore the only significant source of colony loss (i.e. $k_n = k_s$; $v' = z_m \cdot k_s$). The mean value obtained (0.43 ± 0.22 m d^{-1}) exceeded those calculated from their arrival rate in simultaneously-operating sediment traps, but was in reasonable agreement with those obtained elsewhere (see Table 19).

A more elaborate derivation of the sinking loss rates of diatoms in Lac Hertel, Montreal, was devised by Knoechel & Kalff (1978). Cell growth (k') was estimated directly by an autoradiographic method (Knoechel & Kalff, 1976) and compared with net rates of change (k_n) in their standing populations. The difference was assumed to be equivalent to the combined death and sinking loss (grazing losses were ignored). Their respective rates were separated in a series of trial calculations; death rate was first assumed to be zero (in which case, $k_s = k' - k_n$). Since live and dead cells are associated together in colonies, the sinking rate generated should deplete the dead cells at the same rate and, hence, their concentration at a subsequent time could be predicted. If this figure underestimated the actual number detected, then some further death must have occurred during the interim. Different death rates were then incorporated in turn (each requiring a modified estimate of the sinking loss rate) until the (unique) pair of variables interpolated could simultaneously explain the observed rate of change in both the live and dead cell populations of each species.

Cumbersome though these calculations must have been, Knoechel & Kalff (1978) were eventually able to tabulate k', k_d, k_s and k_n for each of five species of diatom over several 6–10-day periods during spring and summer, spanning the wax and wane of their standing populations. In every instance, specific cell growth (k') continued throughout (range: 0.08 to 0.2 d^{-1} for *Tabellaria*; 0.09 to 0.41 d^{-1} for *Asterionella*), whereas fluctuations in the net rate of change ($\pm k_n$) were more closely influenced by changes in death rates (0 to 0.03 d^{-1} in *Tabellaria*; 0.02–0.15 d^{-1} in *Asterionella*) and sinking-loss rates (0 to 0.27 d^{-1}, 0–0.48 d^{-1} respectively; see Figure 79). Indeed, strong negative correlations were observed between the k_n and the k_s for these two species, and good positive ones between k_s and k_d. Apart from emphasizing, once again, the important link between diatom sinking and physiological condition, Kalff & Knoechel (1978) were able to demonstrate that changes in 'algal' abundance may stem principally from variable loss rates.

Similar findings emerged from the study of loss processes affecting phytoplankton populations husbanded in the Blelham Enclosures (Reynolds *et al.*, 1982*b*; Reynolds & Wiseman, 1982). The work set out to quantify the magnitude of losses to the various 'sinks', in relation to changes in net standing populations. The systems were frequently fertilized with inorganic nutrients with the intent that neither the growth rate nor the crop potential of the dominant organism should be nutrient- limited (Reynolds & Butterwick, 1979). The biomass of populations produced was

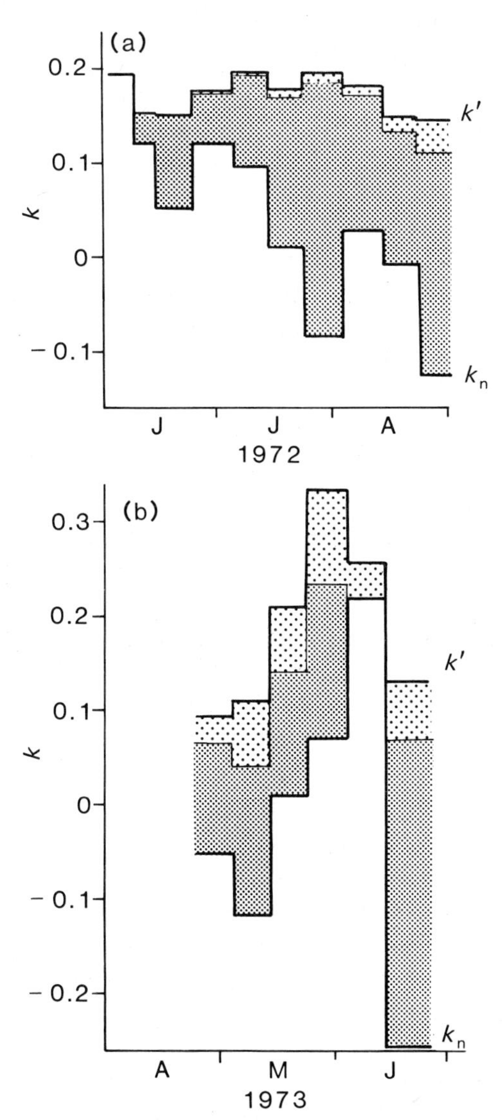

Figure 79. Changes in the rates of growth (k') and net increase (k_n) during the course of development of populations of (a) *Tabellaria fenestrata* (June, July, August) and (b) *Asterionella formosa* (April, May, June) in Lac Hertel. The loss rates are subdivided between death rate (coarse stipple) and sinking rate (fine stipple). (Drawn from data of Knoechel & Kalff, 1978.)

often large but the diversity was generally low (see Figure 44). Nevertheless, conspicuous fluctuations in standing crop were evident, as the population of one species or group of species collapsed and was replaced by another, in a sequence resembling that of small eutrophic lakes (Reynolds, 1980*a*).

Table 20. *Comparison of measured sinking losses of vernally-produced* Asterionella *populations in Blelham Enclosures A and B in 1978 and the total cell production (calculated from uptake of silicon) and the maximum observed standing crops; all data in cells* ($\times 10^{-6}$) cm^{-2}

	Enclosure A	Enclosure B
Si-derived production	23.8–27.1	30.7–36.3
Max. observed standing crop	20.8–22.5	25.3–29.3
Flux into (Tauber) traps	29.8–29.9	32.9–35.3
Flux into (Jar) traps	19.7	34.3
Increment at sediment surface	7.6–54.4	26.1–37.7

The arrival of 'algae' (both live and dead) into sediment traps and onto the sediment surface were independently monitored and the rates of removal by filter-feeding zooplankton were simultaneously assessed. Loss rates were eventually reconstructed from these data, by summation ($k' = k_n + k_d + k_s + k_g$), and, in turn, the cell production and its ultimate fate, partitioned among the various sinks, was hypothesized. In the special case of the diatoms (of which only two species, *Asterionella formosa* and *Fragilaria crotonensis*, were ever abundant), an independent check was reconstructed from the uptake of silicon per unit area of the enclosure (equivalent to the change in the mean concentration in the enclosure plus the effective concentration of the silicate added) and the mean silicon content of the cells of either species collected from the enclosure. It was shown (Reynolds & Wiseman, 1982) that the hypothesized cell production and the eventual flux of intact cells of *Fragilaria* and vernally-produced *Asterionella* into the sediment traps and to the bottom deposits were in good agreement (see Table 20). Observed losses were thus largely attributable to sinking (of live and dead cells) throughout the time-course of each of these populations.

The periods of specific population abundance were subdivided into the component phases of rapid and static exponential increase and of net decline. Values of k_s, calculated from the rate of arrival of cells into the sediment traps during the corresponding periods, were added to k_n ($= k'$) and were found to conform with silicon-uptake derived estimate of growth rate (k'_{Si}: see Table 21). Finally, the sinking rate v', was calculated from k_s and contemporaneous estimates of mean z_m (Table 22).

These data indicate that the reduction in the net rates of increase were initially due to a reduction in k' and not primarily to increases either in mean intrinsic sinking rate (v') or sinking loss rate (k_s). Ultimately, as growth became undetectable, loss rates did increase, owing to simultaneous

Table 21. *Change in the rates of net increase (k_n) over various phases of the development of two vernally-produced*

Asterionella populations in the Lund Enclosures (A and B, 1978), the rates of arrival of cells in (Tauber-type) sediment traps ($\equiv k_s + k_d$) and the range of removal rates attributable to filter feeding zooplankton (see §7.5); the totals $k' = (k_n + k_s + k_d + k_g)$ for each phase are compared with k'_{Si}, the rate derived from the hyperbolic uptake of silica (from the data of Reynolds & Wiseman, 1982, and Reynolds et al., 1982b)

Site	Period	k_n	$k_s + k_d$	k_g	k'	k'_{Si}
Enclosure A	2 Mar–4 Apr	0.097	0.011 to 0.064	0 to 0.006	0.108 to 0.167	0.087
	4 Apr–25 Apr	0.005	0.024 to 0.086	0 to 0.012	0.029 to 0.103	0.025
	25 Apr–10 May	−0.152	0.084 to 0.150	0.001 to 0.014	0 to 0.012	0.007
	10 May–5 Jun	−0.191	0.096 to 0.305	0.002 to 0.795	0 to 0.909	0
Enclosure B	2 Mar–21 Mar	0.147	0.009 to 0.028	0.001 to 0.009	0.157 to 0.184	0.154
	21 Mar–4 Apr	0.065	0.016 to 0.023	0.001 to 0.011	0.082 to 0.099	0.072
	4 Apr–25 Apr	−0.032	0.032 to 0.157	0.002 to 0.134	0.002 to 0.159	0.012
	25 Apr–15 May	−0.242	0.150 to 0.253	0.002 to 0.039	0 to .050	0

Table 22. *Sinking rates of* Asterionella *populations* (v') *calculated from the exponential loss rates during each of the phases shown in Table 21, assuming* $k'_{Si} - k_n = k_s = v'/z_m$ *(from Reynolds & Wiseman, 1982)*

Site	Period	k'_{Si}	k_n	k_s	z_m (m)	v' (md^{-1})
Enclosure A	2 Mar–4 Apr	0.087	0.097	*	11.62	*
	4 Apr–25 Apr	0.025	0.005	0.020	7.53	0.15
	25 Apr–10 May	0.007	−0.152	0.159	5.75	0.91
	10 May–5 Jun	0	−0.191	0.191	3.08	0.59
Enclosure B	2 Mar–21 Mar	0.154	0.147	0.007	11.70	0.08
	21 Mar–4 Apr	0.072	0.065	0.007	11.60	0.08
	4 Apr–25 Apr	0.012	−0.032	0.044	7.50	0.33
	25 Apr–15 May	0	0.242	0.242	4.20	1.02

* negative result defies meaningful solution.

increases in v' and decreases in z_m. These results emphasize the physiological reaction of deep-mixed diatoms to isolation in a well-insolated epilimnion and suggest that increased sinking rate, in this instance, may have been a direct result of impairment to the photosynthetic apparatus, while the decreasing mixed depth and the concomitant reduction in its ability to maintain small particles in suspension contributed to their rapid elimination. Increased sinking rate is arguably a mechanism contributing to the long-term survival of the species in waters liable to periods of microstratification.

Even though the onset of stratification often represents a critical turning point in the maintenance of diatom populations, they are not necessarily prevented from increasing by stratification *per se*. Growth can continue to proceed during sedimentation, provided nutrients are available and light penetration will support it. Populations can continue to increase through much of the summer in large, oligotrophic lakes having relatively deep (say ~ 10 m) epilimnia and persistently high transparency (see §3.4.2 for examples) to which the suspended diatoms can adapt appropriately. The epilimnetic depth in smaller and more sheltered basins, however, can be critical: Reynolds *et al.* (1982*b*) concluded that *Fragilaria* growth probably continued through much of the summer in the well-fertilized epilimnia of the Blelham enclosures although the population in the near-surface (upper 3 m) layer frequently tended to decline. Reynolds *et al.* found a negative correlation between the rates of net change in population and the contemporaneous mixed depth, described by the equation:

$$k_n = 0.292 \log_{10} z_m - 0.028 \quad (52)$$

By interpolating different values of z_m into equation (52) and solving simultaneously, they were able to show that $k' \sim 0.2\ d^{-1}$ and the mean sinking rate was $0.204\ m\ d^{-1}$; while this rate of growth was maintained, a mixed depth (z_m) of > 1.02 m ($v'/z_m < 0.2\ d^{-1}$) was critical to net increase in the upper layer taking place at all. Variance from the predicted behaviour, in this and in later work (Reynolds *et al.*, 1983*b*), was substantially (64%) accounted by changes in the ratio of z_m to the depth of Secchi-disk extinction; when z_m/z_s was high, sinking loss rate (and presumably v') was relatively accelerated, emphasizing once again the implicit role of photosynthetic behaviour in the intrinsic settling properties of planktonic diatoms (see also equation (58) on p. 306).

7.4 Death and decomposition

'Algal' cells dying, for whatever reason, whilst in suspension are effectively lost from the population and contribute nothing to its further development. Cells may die from a variety of causes – deprivation of adequate light to support photosynthesis, nutrient deficiency, exposure to toxic substances (including those produced by other 'algae'), infection by fungi, bacteria and viruses. The susceptibility to these influences and their eventual effects vary with species, with season and with the physiological condition of the 'algae' concerned. Moreover these various influences almost certainly interact: just as a sick person is more vulnerable to infection from other agents, so (for example) a nutrient-deficient 'algal' population may be more susceptible to bacterial or fungal attack than a healthy one.

7.4.1 *'Physiological' death*

Recognition of a dying cell is by no means straightforward: between the extremes of healthy, growing cells and empty walls or heavily bacterized, disorganized contents, there exists a wide 'borderline' spanning an initial loss of physiological vigour (from which cells might potentially recover) and structural disintegration (beyond which recovery is impossible). Any decision as to the condition of individual cells, based on visual microscopic condition alone, tends to be arbitrary and the views of few workers would necessarily coincide. Changes in the ratios of the photosynthetic pigments and, in particular, between the pigments and their derivatives may indicate changes in the physiological state of the populations from which they were extracted (Jensen & Sakshaug, 1973). Interpretations are complicated in natural, mixed populations, however, in that these ratios vary between species while the pigment derivatives may be associated with detrital particles, some of which may be produced by grazing zooplankton: chromatographic separation (for instance) of degradation products of chlorophylls (phaeophytins, phaeophorbides and chlorophyll isomers) may give added resolution (for a review, see Sakshaug, 1980). Other potential indicators of physiological condition of cells include the ratios of *in vivo* fluorescence to chlorophyll *a*, before and after the addition of DCMU (the specific inhibitor of photosynthetic PSII – see §4.1) to the test samples (Sakshaug, 1980). Chlorophyll autofluorescence, excited by suitable light source and observed in cells under an epifluorescence microscope (see Jones, 1979), has also been used in assessing physiological conditions (see, for instance, Sirenko, 1972). 'Track autoradiography', a technique for measuring the post-incubation photosynthetic incorporation of ^{14}C into individual cells (Knoechel & Kalff, 1976), is a sensitive indicator of the potential C-fixing properties of cells. These are apparently retained by diatom cells, at least, containing even a single plastid of normal appearance (Knoechel & Kalff, 1978).

Although the causes of death among natural phytoplankton populations are frequently stated in the ecological literature there have been relatively few attempts to assess the rates of loss therefrom. In many instances, mass mortalities result from specific events, such as those of diatoms when silicon is effectively exhausted (Lund, 1965), of cyanobacteria killed by excessive radiation into the aftermath of bloom formation (Reynolds, 1971) or of specific populations subject to virulent pathogenic infections (see, for instance, Daft & Stewart, 1971; Canter, 1979); were such collapses typical of death among natural populations, the postulation of death rates would be somewhat academic. As growth conditions (e.g. of light, nutrient availability) are modified not least as a consequence of the growth, so some cells inevitably become stressed and the population, as a whole, seems likely to suffer an increasing drain on the stock of live cells. In deriving estimates of combined loss rates, changes in the death rate before, at, and after the maximum are scarcely likely to be unimportant.

One of the most outstanding attempts to quantify death rates in natural populations is that described by Knoechel & Kalff (1978), who, as already mentioned above (§7.3), determined carbon fixation rates and rates of net change in the populations of several diatoms through their wax and wane in Lac Hertel and calculated the contributions of death and sedimentation to the rates of total loss. Data pertaining to two of the populations they studied are represented in Figure 79. Those pertaining to a *T. fenestrata* population (Figure 79a) show a growing discrepancy in the values for k' (which remains remarkably stable) and in k_n, which is apparently due to increasing rates of loss. Consistently, death rate (k_d, range: 0 to $-0.034\ d^{-1}$) is much less than sinking loss rate (k_s: 0 to $-0.269\ d^{-1}$). The

Table 23. *Calculations of* in situ *death rate of* Asterionella (k_d, d^{-1}) *from the proportions of dead cells arriving in sediment traps placed in Blelham Enclosures A and B during the 1978 period of abundance*; *data from Reynolds* et al. (1982*b*)

Enclosure A		Enclosure B	
Period	k_d	Period	k_d
2 Mar–4 Apr	< 0.002	2 Mar–21 Mar	< 0.001
4 Apr–25 Apr	< 0.004	21 Mar–4 Apr	< 0.002
25 Apr–10 May	< 0.033	4 Apr–25 Apr	< 0.031
10 May–5 Jun	< 0.035	25 Apr–15 May	< 0.064
5 Jun–19 Jun	< 0.090		

data for the *Asterionella* population (Figure 79b) show that k' altered significantly during the period considered and that death rate (-0.029 to -0.100 d^{-1}) contributed relatively more to the total rate of loss. Similar results were obtained for the other populations they studied, while those of *Fragilaria crotonensis* and *Synedra radians* behaved similarly to *Asterionella, Melosira italica* to the *Tabellaria.* None of the calculated death rates exceed 0.15 d^{-1} and rarely did any exceed the contemporaneous sinking loss rate.

Reynolds *et al.* (1982*b*) attempted to derive death rates of *Asterionella* from the fraction of dead cells (F_d) entering traps placed below the trophogenic zone of the Blelham enclosure; the use of a preservative in the traps determined that cells described as 'dead' (having disorganized contents; these would have included an unknown proportion which Knoechel & Kalff may well have judged to be 'live') entered the traps in that condition. Since these cells were included in the calculations of sinking loss rate, then k_d corresponds to $F_d \cdot k_s$. If the fraction of dead cells had been disproportionately distributed among the sedimenting colonies, such that colonies with relatively greater dead-cell fractions would sink out faster, then k_d would be liable to exaggeration. Moreover, death may have occurred during settlement through the water column. It is thus better to regard these estimates of k_d as being on the high side. Nevertheless, the derived values (Table 23) were always < 0.09 d^{-1}, even at the end of the populations. Reynolds *et al.* (1982*b*) also commented that corresponding data for *Fragilaria* consistently indicated k_d to be < 0.02 d^{-1}. These findings are in line with those of Knoechel & Kalff (1978).

In contrast, Jassby & Goldman (1974*a*) had to account for daily loss rates of fixed carbon in the order of 0.2 to 0.8 d^{-1} of which, often, only small proportions could be attributed to water transport, sinking or grazing. By elimination, cell mortality was suspected as the major source of loss. The differences between carbon fixation and cell production as indices of k' probably contribute to the different impression of population biology given by their interpretation. It must be pointed out that the limnological characteristics of Castle Lake differ markedly from those of Lac Hertel and from the Blelham enclosures, while the data of both Knoechel & Kalff (1978) and Reynolds *et al.* (1982*b*) apply exclusively to diatoms. Further investigations in other lake systems are needed if any clear picture of specific death rates is to emerge.

Relatively more information is available concerning the fate of carbon in the general metabolism of lakes. It is significant, however, that the conclusion of Jones & Simon (1980), part of whose study was carried out in the same enclosures used by Reynolds *et al.* (1982*b*), was that decomposition of the primarily autochthonous (i.e. produced within the lake) particulate organic carbon occurred at or close to the lake sediments, that is, after sedimentation is more or less complete. This conclusion differs from the consensus of views to be gleaned from the literature (e.g. Kuznetsov, 1968; Kajak, Hillbricht-Ilkowska & Pieczynska, 1972). Perhaps the contradiction is partly a function of the relatively truncated water columns of the Blelham Enclosures (< 12 m) and the prevalence of diatoms in the sedimenting material. Certainly, the examination of material entering sediment traps placed at a depth of about 40 m in Windermere (reported by Reynolds *et al.*, 1982*a*) indicated that vernally-produced diatoms continued to settle for several months after they had disappeared from the plankton but that the proportion of live cells as opposed to empty frustules declined steeply after the onset of thermal stratification; species of Chlorophyceae and cyanobacteria were trapped in relatively very small numbers and, even then, most of those that arrived were already in an advanced state of decomposition.

7.4.2 *Toxicity and allelopathy*

It is now well known that active phytoplankton algae secrete a variety of chemical substances into the water around them. The significance of this has not always been clear, although there is now evidence that, in many instances, the chemicals are photosynthetic intermediates 'funnelled-off' as a consequence of physiological self-regulation (see Chapter 4). Nevertheless, there has long been a good deal of circumstantial evidence that the products secreted by 'algae' include some which are inhibitory or toxic to competing species and that, accordingly, their production may confer some selective advantage. Indeed, some are reputedly toxic to other organisms, including fish, waterfowl and domestic livestock drinking water tainted by these organisms and, in extremes, even to humans. Cyanobacterial 'blooms', especially at lower latitudes, are particularly notorious in this respect (for useful reviews, see Collins, 1978; Carmichael, 1981). There has always been uncertainty whether the toxic substances (which have rarely been identified) are secreted by healthy algae or whether they are produced during the putrescence of algal material (Heaney, 1970). Nevertheless, the secretion by cyanobacteria of dialysable metabolites inhibitory or fatal ('allelopathic': Keating, 1977) to the growth of other 'algal' species in culture has been demonstrated in several recent studies (Keating, 1977; Lam & Silvester, 1979; Vincent & Silvester, 1979), lending support to some earlier observations attributed to direct 'algal' antagonism (Hughes, Gorham & Zehnder, 1958; Vance, 1965). Bishop, Anet & Gorham (1959), Carmichael & Gorham (1978) and others have been able to isolate toxic peptides from cyanobacteria (*Anabaena*, *Aphanizomenon* and *Microcystis*) and possible sites of their production in *Microcystis* have been suggested (Reynolds *et al.*, 1981).

Isolates of *Microcystis* populations studied by Reynolds *et al.* (1981), that eventually became so dominant over all competitors that they were virtually monocultures, were assayed against cultures of common phytoplankton species to see if this commonly-observed phenomenon was attributable to allelopathic exclusion. Apparently toxic effects were observed on *Asterionella*, *Eudorina* and *Chlorella* (but not on *Ceratium*) inoculated into water in which the *Microcystis* had been growing, but not when active cultures were 'diluted' with similar water. The difference in these results may be one of scale: toxicity may depend upon the relative sizes of the producing and affected populations. Whilst it may now be beyond question that 'algae' can produce substances inhibitory to competitors, their effects in the ecology of natural populations are still open to question; it would be wrong to assume that (say) *Microcystis* is always antagonistic to (say) *Eudorina* or *Asterionella* or that dominance of the former was necessarily the result of allelopathy.

7.4.3 *Pathogenic organisms*

Knowledge of the range of pathogenic organisms to which phytoplankton 'algae' are susceptible has been built up over many years, largely through the devoted work of a relatively small number of researchers. Their role in the ecology of phytoplankton 'algae', however, is still poorly understood. This owes partly to the nature of parasitic infections: they may often remain undetected by non-specialists, or their distribution, both in space and time, is genuinely erratic. Epidemics may be apparent enough but their hitherto unpredictable development, combined with their evident host specificity, has made them difficult to study in relation to other factors affecting host populations. Now that many pathogenic organisms can be successfully isolated and maintained in culture on suitable hosts, the opportunities for more detailed study of these aspects seem promising. All that can be attempted here is a brief survey of the types of parasitic attack known to occur in nature and to comment upon their ecological significance, where appropriate.

Viral pathogens ('phycoviruses': Safferman & Morris, 1963; 'cyanophages': Luftig & Haselkorn, 1967) have been isolated from field material and found to actively lyse laboratory strains of cyanobacteria. Goryushin & Chaplinskaya (1968) sampled clear-water 'spots' that developed in natural bloom formations, where evidence was found that lysis had occurred. After filtration through bacteria-filters, the water was dripped on to suspensions of floating *Microcystis* in the laboratory, where it caused local lysis and imitated the spot formation observed in the field; the authors attributed the original clear spots to viral activity. A virus having similar effects on cyanobacteria in several Scottish lochs has also been identified (Daft, Begg & Stewart, 1970); there is no reason to suppose that they do not occur widely and do not also attack other 'algal' groups.

Algal-lysing bacteria have been isolated from a variety of freshwater habitats – lakes, reservoirs and sewage works – and these are known to attack various 'algal' groups, including planktonic genera. Most of the identified species are gram-negative, non-fruiting myxobacteria (Shilo, 1970; Daft & Stewart, 1970; Daft, McCord & Stewart, 1975). Their apparent preference for aerated and neutral or mildly alkaline water means that they are probably potentially abundant almost whenever suitable hosts are to be found, but the available evidence does not suggest that lysis of blooms is a frequent occurrence.

Fungal parasites (and saprophytes) of 'algae' belong mainly to the order Chytridiales or biflagellate groups of Phycomycetes. Many have been identified to species level and their life-histories described (e.g. Sparrow, 1960; Canter & Lund, 1953; Canter, 1972). Chytrids are dispersed as free-swimming, uniflagellate zoospores that seek out suitable hosts on which to grow. A fine mycelial thread penetrates the host cell, through which nourishment is drawn back to the infective zoospore, which itself enlarges into a spherical sporangium. Eventually, liberation of the next generation of zoospores from the ripe sporangium completes the life cycle; the mechanism of dehiscence varies diagnostically between species. The zoospores of biflagellates are typically more elongate and their sporangia are cylindrical or sac-like in shape. The whole fungus alights on or enters the host cell (according to species) but does not produce a mycelium; food

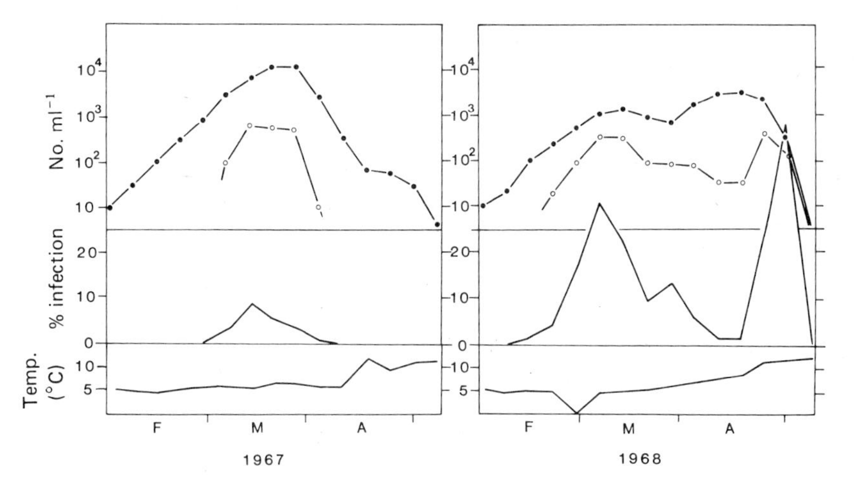

Figure 80. Comparison of the changes in vernal *Asterionella* populations (●) in Crose Mere during February, March, April 1967 and 1968 in relation to the absolute concentration of infecting spores and sporangia of the chytrid parasite, *Zygorhizidium affluens* (○); the percentages of host cells infected and the water temperature are also represented. (Redrawn from Figures 9 and 10 of Reynolds, 1973*a*.)

is absorbed across the contact surface. Zoospores are liberated in open water from a vesicle that is discharged intact (Canter, 1979). Many species, of both groups, produce thick-walled resting spores, which can presumably survive the periods of low host populations.

Infection by fungi usually kills the host cell. If suitable hosts are abundant and fungal increase proceeds rapidly, the parasite population may rapidly reach 'epidemic' proportions. Many factors condition the rapidity of the response. The actual numbers of host cells infected may increase exponentially, which gives a *minimum* estimate of the rate of potential increase in the parasite population ('minimal' because the proportions of zoospores failing to find new hosts are exceedingly difficult to measure accurately). Reynolds (1973*a*) measured rates of increase in the absolute numbers of cells of *Asterionella* infected by the chytrid, *Zygorhizidium affluens*, to be equivalent to 0.3–0.4 d^{-1} (double the rate of *Asterionella* growth); such bursts, however, are short lived (10–14 days). Thus, although the rate of increase exceeds that of the host ($k' < 0.16\ d^{-1}$ at contemporaneous water temperatures), the impact of chytrid infection on the host population is more likely to depend on the relative numbers of hosts and chytrids. An example is illustrated in Figure 80: in 1968, *Zygorhizidium* became numerous relatively earlier in the growth of *Asterionella* than in 1967, so that proportionately more host cells (25% as against 8%) were parasitized. Where the latter was tolerated, the former was evidently not, even though the absolute maximum numbers of cells infected were actually higher in 1967 than in 1968 (see Figure 80); the *Asterionella* population declined, ostensibly as a direct result of the parasitic infection, but recovered during April, when both the numbers of and the percentages of infected hosts increased again. These data suggest that an infection rate exceeding 10% of host cells may be critical to the development of vernal *Asterionella* populations.

The size of the host population when the parasite becomes active conditions its effect in other ways. Low concentrations of hosts mean that infective zoospores must travel relatively further to find suitable cells: Lund's (1957) estimate that the minimum density of *Asterionella* cells at which chytrid epidemics can develop is about 10 cells ml^{-1}.

The question is then prompted 'How do zoospores locate potential hosts?'. In a series of papers, Canter & Jaworski (1978, 1979, 1980) have observed the infection of several strains of diatom (*Asterionella, Fragilaria*) by, and their reaction to, various clones of chytrid, maintained in dual-clone cultures. In *Asterionella* for example, host cells can remain alive for over 24 h following infection, until the fungal sporangium is well enlarged and perhaps fully grown. Normally, released zoospores rapidly take up positions on new host cells; some alight on dead host cells but no evidence that the chytrid could complete its life cycle on dead hosts was found. It was also observed that isolated zoospores of *Rhizophydium planktonicum* were relatively more active in light. There appears to be a threshold level of light below which the zoospores become quiescent, even in the presence of suitable *Asterionella* hosts. In the light, however, the zoospores readily 'swarm' around the host cells, prior to infection. This behaviour suggests an indigenous light/dark activity of the zoospores but that their movements are directed towards suitable hosts in the light, perhaps in response to a chemical stimulus consequential upon photosynthesis.

These observations are supported by the conclusions of ecological studies on chytrid/host relationships, where the development of the parasitic population is most active in the presence of a physiologically-active, growing host population (Masters, 1971; Reynolds, 1973*a*). As already stated, the presence of parasitic chytrids may be 'tolerated' by host populations while they are growing rapidly. Equally, when the hosts are 'inactive', parasites, though present, may also remain relatively static. This makes good ecological sense, since it is the first rule of a parasite that it should not eliminate its hosts. It would be among host populations whose growth rate is light- or nutrient-stressed while remaining photosynthetically active, and so attractive to infective zoospores, that heavy mortalities would be most likely to occur; under such circumstances, parasites can appear to be the factor controlling 'algal' populations.

Host-parasite relationships in the phytoplankton are further complicated by (a) 'hypersensitivity' (cf. Canter & Jaworski, 1979) of hosts to infective spores, such that algal cells die so soon after infection that sporangia fail to develop, and (b) by hyperparasitism – *Zygorhizidium affluens*, for example, is itself frequently parasitized by another chytrid, *Rozella*.

There is ample evidence that many types of Protozoa live on planktonic algae (Canter, 1979). These range between ciliates and rhizopods which devour whole cells and, thus, may be classified as 'grazers' (see §7.5, below) and smaller species of proteomyxids and Monadineae that attack algal cells in the fashion of a parasite; as in other branches of biology, the distinction between predators and parasites is unclear. *Aphelidium* is a chytrid-like protozoan whose zoospores penetrate cells of unicellular or colonial chlorophytes and ingest their contents internally, before a new generation of zoospores are released. Amoeboid vampyrellids attach themselves to the walls of filamentous chlorophytes and 'suck out' the contents of 'host' cells. Other amoeboids (e.g. *Asterocaelum*) ingest whole algal cells and then form special cysts around them, while digestion and, eventually, sporulation take place. Finally, there are larger ameobae (e.g. *Pelomyxa*), ciliates (e.g. *Nassula*, *Ophryoglena*) whose ingestion of large numbers of small 'algal' cells during the course of a single generation, is unmistakably 'grazing'.

7.5 Grazing

The planktonic communities of freshwaters include animals as well as plants (see Table 1). Many of these are herbivores, feeding directly upon 'algae' and bacteria inhabiting the same water masses in which they live. Inevitably, their activities deplete the standing stock of phytoplankton and, hence, may have a significant effect upon their dynamics and population ecology. Equally, the availability of suitable foods is clearly a leading factor in the ecology of planktonic animals and the rate of consumption of the one by the other is relevant in the context of energy flow through freshwater ecosystems. Understanding of this topic has been much slower to evolve than has the appreciation of its importance: progress has been dogged by insufficient basic information on the biologies of many of the animals inhabiting the plankton and by the inadequacies of techniques designed to measure feeding upon natural foods in the laboratory. Although 'grazing' has been the implicit subject of many recent and scientifically rigorous studies in recent years, the overall picture is still far from clear. Even so, the available data sometimes conflict with the widespread preconception that grazing necessarily 'controls' the phytoplankton stock.

7.5.1 *Foods and feeding habits of the zooplankton*

Each of the major groups represented in the freshwater zooplankton – Protozoa, Rotifera and Crustacea – includes species which are believed to feed, partly or wholly, upon planktonic 'algae'. Moreover, the means of obtaining and ingesting food organisms are apparently diverse. It is not the intention here to review these in any detail but simply to give an indication of that diversity. Among the planktonic protozoans, there are relatively small parasitic species which attack algae larger than themselves, ranging through to (see §7.4.3) larger free-living ciliates and rhizopods that ingest small 'algae' whole. The latter include the ciliate genera such as *Nassula* and *Ophryoglena* and the rhizopod, *Pelomyxa*, which have been observed to feed on filamentous cyanobacteria (*Aphanizomenon, Anabaena*) often ingesting sufficient gas-vacuoles to render themselves buoyant (Reynolds 1971, 1975*a*; Canter, 1979). Other rhizopods (e.g. *Vampyrella*) first attach themselves directly to the cells of filamentous diatoms and chlorophytes and proceed, apparently, to suck the contents from the algal cell (Canter, 1979). *Asterocaelum* uses its long tapering pseudopodia to capture centric diatoms which are ingested whole, the body swelling to form a sac-like cyst, which is later vacated, still containing the undigested diatom remains, by the next generation of amoebae (Canter, 1973). When abundant, such organisms (especially the larger rhizopods and ciliates) undoubtedly contribute to severe and rapid reductions of 'algal' populations (Canter & Lund, 1968), though so relatively little is known of other factors governing their distribution, periodicity and population explosions, that their general role in the ecology of phytoplankton is still obscure.

Among the planktonic rotifers, several genera characteristically undergo more or less regular seasonal cycles of abundance that considerably enhance the opportunities for the examination of their gut contents and for testing the correlation between their numbers and those of their food organisms. Many species have been successfully cultured in the laboratory on appropriate foods. In combination, these techniques have been widely applied to elucidate their food and feeding habits, which have been comprehensively reviewed recently by Pourriot (1977). Species having a well-developed ciliary *corona* and crushing *mastax* are generally *filter feeders*. These include the smaller planktonic genera *Keratella, Kellicottia, Filinia* and the colonial *Conochilus*, which can filter out and ingest particles up to about 10–12 μm, and some larger species (of *Brachionus, Notholca*) taking particles of up to 18 μm across. Coronal beating generates a current of water from which the food particles are removed. Within the constraints posed by size, many species apparently show selective preferences for certain foods: *Filinia* and *Conochilus* feed essentially on small detritus particles with associated bacteria and though they ingest small 'algae' like *Chlorella*, they are apparently unable to assimilate them. *Keratella* species are also detritivores, but their food includes small cryptomonads and chrysomonads. *K. cochlearis* is said not to flourish where the nanoplankton is dominated by Chlorococcales, Volvocales or euglenoids, although Ferguson *et al.* (1982) reported that *K. quadrata* responded well to an abundance of *Ankyra*. *Keratella cochlearis* and *Kellicottia longispina* also capture larger *Cryptomonas* by grasping a flagellum, breaking the cell wall and ingesting the pieces.

More specialized feeding mechanisms are characteristics of rotifers in which the corona is reduced or modified but the mastax is specially adapted for particular feeding habits. *Asplanchna* has a grasping mastax and is predatory upon other rotifers, though its diet does include larger 'algae' (Lund, 1965), which are ingested whole. *Ascomorpha* uses its virgate mastax to pierce the walls and suck out the contents of dinoflagellates and, perhaps, other 'algae'. *Trichocerca* spp. are similarly adapted. *Notholca squamula* breaks open the frustules of *Asterionella* cells and ingests the contents (May, 1980). Others have particular preferences for specific 'algal' cells which are ingested whole: *Gastropus* browses on individual cells of *Synura* colonies; *Polyarthra* spp. have an evident preference for *Cryptomonas* cells (cf. also Edmondson, 1965).

When other environmental requirements of rotifers are satisfied (e.g. temperature, oxygen content), a good supply of suitable 'algal' foods can sustain rapid growth and reproductive rates (see Edmondson, 1965). Standing populations of (e.g.) *Keratella*, *Ascomorpha* are known sometimes to exceed 2000 individuals per litre and, at times, could contribute significantly to the rates of loss of the 'algae' on which they feed (see §7.5.2, below).

The third major group of animals represented in the freshwater plankton is the Crustacea, the most numerous and cosmopolitan genera of which are partly or wholly herbivorous. The raptorial ('grasping') feeders – the cyclopoid copepods and the larger cladocerans – are probably typically predatory, but their diets may often include larger filamentous 'algae'. The calanoid copepods (e.g. *Eudiaptomus*, *Diaptomus*, *Limnocalanus*) and many cladocerans (e.g. *Daphnia*, *Ceriodaphnia*, *Bosmina*) are filter feeders.

The essential features of the cladoceran food-gathering apparatus are the four or five pairs of thoracic appendages, the beating action of which generates a current of water through the feeding chamber, formed by the carapace; the long, feathered setae fringing the limbs, which filter out particles from the current; and a further set of 'gnathobase' setae, which carry the filtered particles to the mouth. The size range of particles that can be filtered varies among species and, intraspecifically, with body size (Burns, 1968*a*). The lower size limit is dependent upon the areal density of the setae; the upper limit is probably related to the size of the 'gape', the width of the opening between the carapace valves (Gliwicz, 1980). The abdominal 'claw' is used to clear blockage of the apparatus by larger particles (Burns, 1968*b*).

The precise anatomical extent of the filter-feeding apparatus and its mode of function among the calanoid copepods are still a matter of debate (see, for instance, Kerfoot, 1980). Superficially, the maxillae are the primary strainers, the fringe of bristle-like setae filtering food particles from the current generated by the swimming appendages. Studies on the food intake of *Diaptomus* spp. suggest that the expected filtering rates of animals reliant on this feeding mechanism alone are consistently lower than are observed from the rate of depletion of known concentrations of food particles: the 'feeding envelope' may include the volume enclosed by other outlying appendages that participate in food capture, as proposed by Friedman (1980), perhaps acting in a raptorial fashion, enabling them to grapple and fragment larger 'algae' such as *Cosmarium* and *Stephanodiscus* (Gliwicz, 1977; Richman, Bohon & Robbins, 1980). Adjustments in the orientation of the whole animal may assist it to enhance the prospects of capturing a particle (Alcaraz *et al.*, 1980), indicating that calanoids may be far more selective feeders than cladocerans. Certainly, the size range of the intact food organisms consumed by some *Diaptomus* spp. and the frequency of their ingestion strongly suggest that non-selective filtration of particles in the approximate range 2–12 μm is supplemented by an intake of larger particles (up to or exceeding 30 μm) that are selected, grasped and, frequently, fragmented prior to ingestion (Richman *et al.*, 1980).

However important the various 'algae' may be in the diets of various animals in the plankton, the impact of their feeding activities upon the dynamics of the phytoplankton will be determined by the product of the biomass of their standing populations and their individual feeding rates, and by the capacity of the 'algal' growth rate, net of other sources of loss to withstand the demands of grazers upon it. The relationship also fluctuates in time: while the phytoplankton can satisfy the dietary needs of the zooplankton, the latter may be expected to increase and to thus make increasing demands upon the 'algae' per unit time; if the phytoplankton wanes, the zooplankton is likely, sooner or later, to become severely food-limited and to experience mortalities. 'Recovery' of the phytoplankton would complete the cycle and initiate another. In this context, it must be assumed that the animals do not habitually live close to the capacity of the resources to sustain them, but that such situations do arise frequently in time. Species selection in the zooplankton should be biased towards animals which are able quickly to exploit periods of abundance of a wide range of foods and it is therefore unsurprising that the species-structure and biomass of zooplankton associations should frequently be dominated by filter-feeding crustacea, or that their biomass should fluctuate in response to and, ultimately, influence the availability of foods (see §7.5.2).

7.5.2 *The effects of filter feeding on 'algal' populations*

Several factors contribute to the interrelationship between 'algae' and filter feeders. Of immediate importance to the 'algae' is the *filtration rate* (the volume of water processed by the standing population of filter feeders per unit time), the size- (or other selective-) range of particles that will be ingested and whether distributional differences between the grazers and their potential foods confer any degree of 'immunity' to the 'algae'. The nutritional requirements of the filter feeders and the capacity of the phytoplankton productivity to meet them will influence the ways in which the relationship alters seasonally, as well as between lakes of differing trophic status. These points are considered below.

7.5.2.1 *Filtration rates.* A selection of published values of individual filtration rates, measured either in the laboratory or in lake enclosures, is listed in Table 24. In every case, these are derived by back calculation from the rates of depletion of foods from known concentrations of 'algae' or of radio-labelled 'algae' by known concentrations of the animals under investigation. The rates of filtration are apparent and, of necessity, mean values. Filtration rates are known to be temperature sensitive and they are also sensitive to the size, concentration and, perhaps, 'palatability' of the particles in question. Maximal values have generally been derived when animals are feeding in suspensions of particles well within the size limits of the filtration apparatus and in which their concentration is lower than that which would satisfy their optimum feeding rate. This value, usually known as the 'incipient limiting concentration' can be derived theoretically (see below). Once ILC is exceeded, declines in mean maximal filtration rates of rotifers (see Pourriot, 1977), calanoids (Richman *et al.*, 1980) and Cladocera (Burns, 1968*b*; Jones, Lack & Jones, 1979) may be observed. Reductions in filtration rates of *Diaptomus* spp. (Kibby, 1971; Richman

Table 24. *Individual filtration rates, F, (in* ml d^{-1}*) for a selection of planktonic animals, reported in, or derived from, the literature*

Species	*F*	Reference
Rotifers		
Brachionus calyciflorus	0.08	Halbach & Halbach-Keup (1974)
	0.007–0.017	Starkweather *et al.* (1979)
various	0.02–0.11	Pourriot (1977)
Calanoids		
Eudiaptomus gracilis		
(20°C)	1.32–2.54	Kibby (1971)
(12°C)	0.61–1.76	Kibby (1971)
(adults, 17±3°C)	0.5–10.7	Thompson *et al.* (1982)
(juveniles, 17±3°C)	0.5–6.7	Thompson *et al.* (1982)
Diaptomus oregonensis	< 12.9	McQueen (1970)
(adults, various temperatures)	2.4–21.6	Richman *et al.* (1980)
Cladocerans		
Bosmina longirostris	< 3.0	Sushchenya (1958; quoted by Thompson *et al.*, 1982)
Chydorus sphaericus	0.5–2.6	Thompson *et al.* (1982)
Daphnia middendorffiana		
(~ 2.8 mm, 12°C)	31.0	Peterson *et al.* (1978)
Daphnia hyalina		
(< 1.0 mm, 17±3°C)	1.0–7.6	Thompson *et al.* (1982)
(1.0–1.3, 17±3°C)	3.1–19.3	Thompson *et al.* (1982)
(1.3–1.6, 17±3°C)	3.1–30.7	Thompson *et al.* (1982)
(1.6–1.9, 17±3°C)	14.0–60.0	Thompson *et al.* (1982)
(> 1.9, 17±3°C)	15.7–62.6	Thompson *et al.* (1982)
Daphnia spp.[(a)]		
(< 1.0 mm, 15–20°C)	< 4.99	from Burns (1969)
(1.0–1.3, 15–20°C)	3.67–10.41	from Burns (1969)
(1.3–1.6, 15–20°C)	6.47–18.6	from Burns (1969)
(1.6–1.9, 15–20°C)	10.1–30.1	from Burns (1969)
(> 1.9, 15–20°C)	> 14.7	from Burns (1969)

(*a*) Values given are derived from the appropriate equations in Burns (1969).

et al., 1980) and *Daphnia* spp. (Burns, 1968*b*; Thompson, Ferguson & Reynolds, 1982) also occur when 'large' food organisms are abundant. However, Burns (1969) demonstrated a consistent relationship between the filtration rates of four species of *Daphnia* (*D. magna*, *D. schoedleri*, *D. pulex* and *D. galeata*) and their carapace length; the values of *F* shown in Table 24 are calculated directly from her equations relating filtration rate to body length (L_b) at selected temperatures, namely:

at 15°C,

$$F = 0.153\ L_b^{2.16} \qquad (53)$$

and at 20°C,

$$F = 0.208\ L_b^{2.80} \qquad (54)$$

The observations of Thompson *et al.* (1982) showed that, subject to the availability of suitable food particles, mean individual filtration rates in *D. hyalina* var. *lacustris* varied through less than one order of magnitude but, on average, were well predicted by Burns' (1969) equations.

The total volume of water that can be processed by the filter feeders per unit time (a) (or the 'community filtration rate', 'community grazing index') can be measured directly in the field (e.g. Haney, 1971, 1973) or calculated from the product of the populations of each species/size category and their individual filtration rates (Gliwicz, 1977; Thompson *et al.*, 1982). Thus,

$$a = (F_1 \cdot N_1) + (F_2 \cdot N_2) + \ldots (F_i \cdot N_i) \qquad (55)$$

where F_i is the filtration rate and N_i the standing population of the ith species-size category. Some of their data are reproduced here as (Figure 82) to show the scale and seasonality of fluctuations in the community grazing rate in one of the Lund Enclosures in Blelham Tarn, during which it was frequently fertilized and throughout which predation on the zooplankton was minimal. In the sequence shown, *Daphnia* population developed, by far, the major contribution to the summer community filtration rate, which at times exceeded the equivalent of 1.2 (95% confidence intervals 0.91 to 2.21) times the total enclosed volume per day, whereas during the spring months, it remained below 0.03 and, during summer, it fell several times to < 0.3. These fluctuations, which embrace the range of values measured by Haney (1973) in Heart Lake, Canada and by Gliwicz (1977) in Mikołajske, Poland, are primarily dependent upon fluctuations in the size and species-structures of the zooplankton itself (Figure 81). It will be clear that at the higher ($a > 0.3\ d^{-1}$) community filtration rates, the phytoplankton will experience potential loss rates of a similar order to their potential rates of population growth.

7.5.2.2 *Food selection and grazing-loss rates.* With the likely exception of the calanoids when feeding on larger algae (see above), it is probable that the ranges of particles ingested by filter-feeders are determined largely by size. It would be an oversimplication to regard the animal as simply drawing water across the filtration apparatus and from which suitably-sized

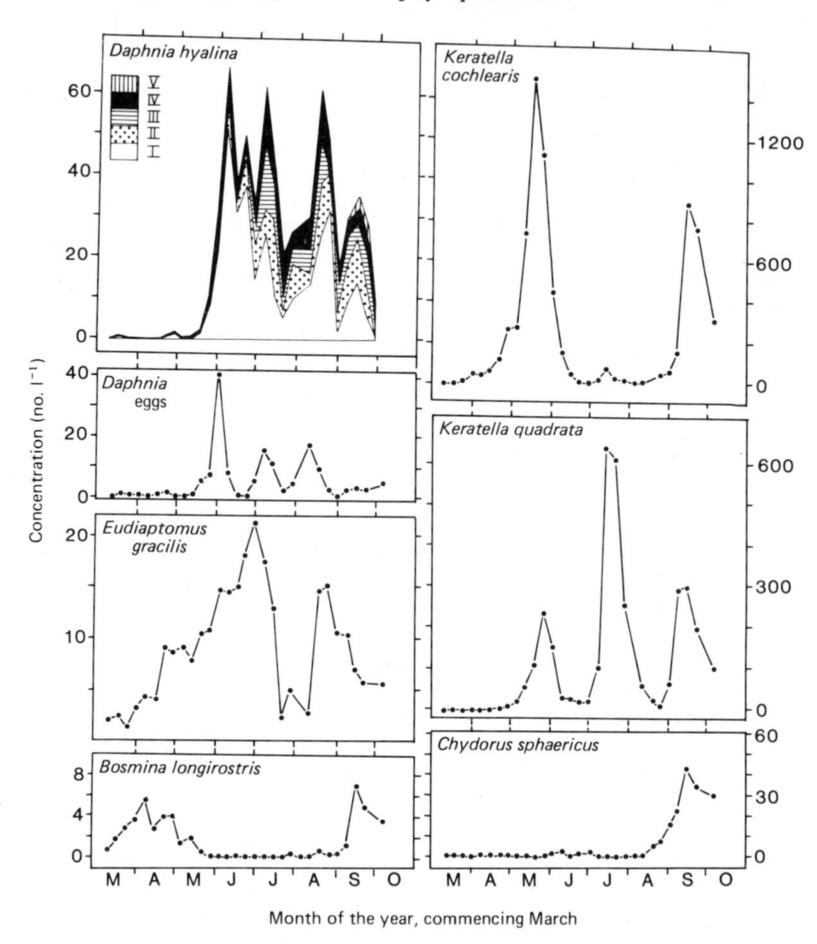

Figure 81. Seasonal variations in the populations of filter feeders in Lund Enclosure A, during 1978. (Redrawn from Figure 3 of Ferguson *et al.*, 1982.)

particles are fortuitously removed; yet this assumption is implicit in the following consideration of the effect of filter-feeding upon 'algal' populations. In practice, the animals exhibit adaptive responses which direct them towards the depth layers wherein their potential foods are likely to be relatively more abundant. Moreover, the individual rates of filtration are presumably subject to 'voluntary' control, such that the pumping actions can be slowed or stopped from time to time. At the same time, the external flows generated by the rhythmic beating of the filtration apparatus are, to some extent, turbulent so that 'algae' morphologically adapted to gain maximal advantage of suspending eddies may often be lost from the feeding current. The simultaneous presence of filter-feeders and suitable

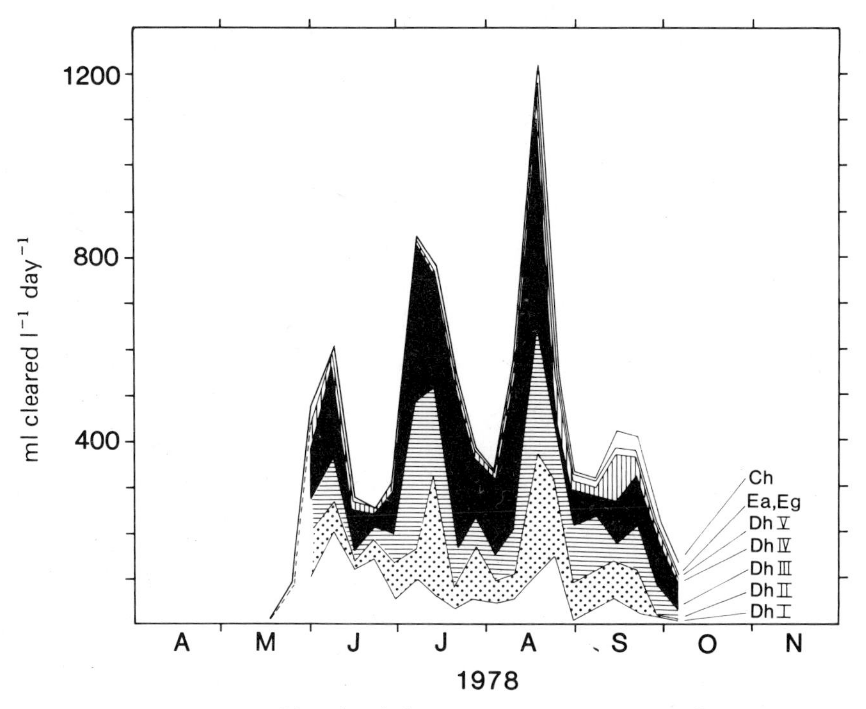

Figure 82. Seasonal fluctuations in the community filtration rate (calculated as equation 55 and expressed as ml l^{-1} d^{-1} of lakewater filtered) in Lund Enclosure A, during 1978. The cumulative contributions of five ontogenetic categories of *Daphnia hyalina* (DhI–V), *Eudiaptomus* adults, fifth copepodids (Ea) and first-fourth copepodids (Eg), and *Chydorus sphaericus* (Ch) are also shown. (Redrawn from Figure 4 of Thompson *et al.*, 1982.)

foods in a water body should never be taken as a guarantee that the latter will become prey to the former. There is, then, a given rate of 'rejection' for each of the various size ranges of particles that may be encountered by given individual animals and this will be complementary to the apparent efficiency of selective removal. It is the latter which will be the relevant statistic in the population dynamics of the 'algal' species, and may be represented by a factor, ϕ. Thus, while the rate of grazing loss (k_g) of a given 'algal' species to filter feeders will bear an overriding relationship to the instantaneous community filtration rate (a; dimension, t^{-1}), it will always be specifically influenced by its size relative to the ranges of particle size 'selected' by the contemporaneous animal populations. Thus,

$$k_g = \phi a \tag{56}$$

The empirical value of ϕ varies between 0 (for complete rejection) and 1 (for complete, non-selective consumption). It can be seen that, for a given community filtration rate, the value of ϕ (and, hence, of k_g) will vary (a) according to the relative dominance of the zooplankton by (say) small rotifers and by (say) large Daphniids, and (b) among different 'algae' *simultaneously* present in the plankton, according to their own mean unit sizes; the latter is of potentially great selective relevance to competing 'algal' species.

Some reported size ranges of particles filtered and ingested by various species are presented in Table 25. Most of the data have been obtained by direct observations of natural foods, though they are apparently well-supported by the results of experiments in which plastic beads of pre-determined sizes have been offered with the suspended food (Burns, 1968*a*; Gliwicz, 1980). There is also good agreement between results independently obtained by different workers. Ferguson *et al.* (1982) observed that *Keratella quadrata* responded to the abundance of *Rhodomonas*, *Chlorella* and *Ankyra* (*GALD* < 18 μm) but not to *Cryptomonas* (~ 21 μm) and this agrees with Pourriot's (1977) statement of the maximum size of particle available to this species; for *K. cochlearis*, a much smaller limiting size of particle is indicated, suggesting that its filtered foods are probably dominated by bacteria. On the other hand, the upper size limit of particles filtered by Daphniids enables them to feed on a much wider range of potential foods and this limit increases linearly with the size of the animal. Burns (1968*a*) expressed this relationship empirically:

$$y = 22x + 4.87 \qquad (57)$$

where $y =$ the size of the largest bead ingested (in μm) and x is the carapace length (in mm). The data of Nadin-Hurley & Duncan (1976) and of Ferguson *et al.* (1982), pertaining to natural foods, suggest slightly higher values than are predicted by Burns' equation, though both comment on the effect of shape. Nadin-Hurley & Duncan (1976) showed that, though the *GALD* of most individual particles ingested (77%) was less than 50 μm, the flexible filaments of *Tribonema* (up to 225 μm in length in *D. magna* and up to 120 μm in the case of *D. hyalina*) were frequently ingested, and they concluded that the critical factor was particle *width*. Though 'available', trichomes of *Oscillatoria* did not figure in the food of *D. hyalina* populations studied by Ferguson *et al.* (1982), unless they were rather shorter than 100 μm in length (Lund & Reynolds, 1982). *Fragilaria* cells were only observed in *Daphnia* guts when colonies of less than 13 or so cells were common in the plankton, that is, when the width of the colonies was less than the length of the individual cells. Colonies of

Table 25. *Examples of the ranges of particle size (diameters or GALD in μm) filtered by selected species of zooplankton reported in the literature; figures in parentheses denote optimum ranges ('peak' selection) where appropriate*

Species	Size	Reference
Keratella cochlearis	(0.5–1)3	Gliwicz (1969)
Keratella quadrata	< 15–18	Pourriot (1977)
		Ferguson *et al.* (1982)
Eudiaptomus gracilis	1(5–10)16	Gliwicz (1969)
Diaptomus oregonensis	2.5(12–15)30	Richman *et al.* (1980)
Bosmina corregoni	0.5(12–23)35	Gliwicz (1969)
Chydorus sphaericus	0.5(12–23)35	Gliwicz (1969, 1980)
Daphnia cuculata	2(11–23)35	Gliwicz (1969, 1980)
Daphnia middendorffiana	> 2.6[a]	Peterson *et al.* (1978)
Daphnia hyalina	(1–20)100	Nadin-Hurley & Duncan (1976)
	< 50–60	Ferguson *et al.* (1982)
Daphnia spp.[b]	0.9 mm < 24.7	Burns (1968*a*)
	1.3 mm < 33.5	
	1.6 mm < 40.1	
	1.9 mm < 46.7	

(*a*) Peterson *et al.* quoted 10 μm^3. The diameter of a sphere of this volume is ~ 2.6 μm.
(*b*) Stated values are derived from the equation of Burns (1968*a*).

Eudorina and *Microcystis* extruded from *Daphnia* guts were always < 50 μm in diameter (Ferguson *et al.*, 1982). On the other hand, the lower size limits of freely-suspended particles filtered by *Daphnia* spp. (0.5–2 μm) enable them to feed on bacteria, which are widely reported to figure in their diets (McMahon & Rigler, 1965; Haney, 1973; Peterson, Hobbie & Haney, 1978; Ferguson *et al.*, 1982). Indeed, the range of potential foods available to *Daphnia* is such that their gut contents often give a good qualitative representation of the species composition of the phytoplankton, though quantitatively, the proportions may be biased in favour of particles of 1–30 μm (Nadin-Hurley & Duncan, 1976; Ferguson *et al.*, 1982). This dietary 'flexibility' is of considerable survival value to *Daphnia* and may be compared with that afforded to calanoids by their alternative adaptations for capturing food. From the point of view of the phytoplankton, however, feeding by *Daphnia* constitutes a serious threat of mortality, especially when the scale of its contribution to the community filtration rate is taken into account (see Figure 82).

While the value of the selectivity coefficient (ϕ) for some algal populations in the presence of known populations of filter feeders may be readily approximated from a knowledge of the filterable-size limits of the dominant

Table 26. *Coefficients for selection* (ϕ) *of various species of phytoplankton by mixed zooplankton communities, dominated by* Daphnia hyalina *var.* lacustris, *calculated by Reynolds* et al. (*1982b*).

Species	Range of ϕ
Asterionella formosa	0.28–1.0
Fragilaria crotonensis	0–0.35
Cryptomonas ovata	$\rightarrow$ 1.0
Ankyra judayi	$\rightarrow$ 1.0
Eudorina unicocca	
(< 50 μm)	$\rightarrow$ 1.0?
(> 50 μm)	0
Microcystis aeruginosa	generally 0 but perhaps $\rightarrow$ 1 for small (< 60 μm) colonies

zooplankton ($\phi \rightarrow 1$ for 'algae' whose mean size is < 20 μm; $\phi \rightarrow 0$ for species > 50–60 μm), many species will fall into an intermediate category, either because their sizes are variable or their shape and orientation in the feeding current enable them to escape it, or again they may be actively rejected by the filter feeders themselves. These processes have yet to be examined in any detail. Reynolds *et al.* (1982*b*) used indirect methods to approximate ϕ for several species of netplankton grazed by mixed communities of animals (rotifers, calanoids and cladocera). Some of these are quoted in Table 26 as ranges. These would have applied only to the conditions then obtaining and should not be regarded as being necessarily typical for those species in other lakes or at other times. They simply illustrate the principle that the dynamics of some species are relatively more sensitive to filter-feeding grazers than are others. Using these data Reynolds *et al.* (1982*b*) were able to approximate simultaneous and seasonal fluctuations in the values of k_g affecting each of the populations concerned and, thus, to account for their role in influencing their seasonal wax and wane.

Interpolating such values of ϕ, Reynolds *et al.* (1982*b*) were able to estimate k_g for the populations they studied in Lund Enclosures. Several species of small algae (nanoplankton: *Ankyra*, *Chromulina* spp.) were observed to eventually 'disappear' from the plankton community at rates explicable largely in terms of the (*Daphnia*-dominated) filtration pressure. For example, the rapid net decline ($k_n = -0.647$ d^{-1}) in a large, probably light-limited, standing population of *Ankyra*, numbering over 0.3×10^6 cells ml^{-1}, was attributable to the contemporaneous community filtration pressure of between 0.595 and 1.341 d^{-1}. Although the latter range is sufficiently imprecise to permit other explanations for the demise, other evidence (the quantified sinking and death rates) did not accord with the alternatives to the contention that grazers could have been responsible for at least 65%, and perhaps > 97% of the observed loss of cells, i.e. that $-k_n \simeq k_g \simeq a$ and, therefore, $\phi \rightarrow 1$. Assuming the deduction to be valid, it is also required that a similar value of ϕ applied during the increase of the population when k_g would have also been commensurate with the value of a then obtaining. Thus, at an observed rate of net increase of approximately 0.5 d^{-1} and a community filtration rate of > 0.2 d^{-1}, the true rate of growth would have been > 0.7 d^{-1} (equivalent to every cell in the population dividing once per day). As the grazing population increased in response to the presumed increase in the availability of a suitable food source, a increased to > 0.6 d^{-1}, while the net rate of increase in *Ankyra* dropped to about 0.12 d^{-1}. These figures suggest that the true growth rate of the *Ankyra* had not declined and that the reduction was due entirely to grazing. Potentially, the increased grazing rates could soon have led to a net decrease in the *Ankyra* population whilst still maintaining the same rate of growth, as was indeed deduced in a later (smaller) population of *Cryptomonas*, but this failed to occur before the *Ankyra* population had reached light-limited proportions. Once growth was resource-limited, however, the zooplankton did not take long to clear the water of *Ankyra* cells, almost to the limits of detection. This particular case illustrates particularly well (i) that k_n and k_g are not necessarily inversely correlated, (ii) that net increases of nanoplankton species are not incompatible with the presence of grazing animals, as has often been implicitly assumed in the past, but that (iii) the comparative dynamics of growth (which are regulated by factors other than grazing – light, temperature, thermal stability, nutrients, etc.) and of community filtration ultimately determine the dynamics of net change, and (iv) that grazing can nevertheless contribute to the elimination of an 'algal' population and so influence both the composition of the phytoplankton and its periodicity (see also below and Chapter 8).

Reynolds *et al.* considered the effects of crustacean filter feeding on other, slower growing 'algae'. The stellate colonial form of *Asterionella*, measuring 120 μm or more across, is apparently readily ingested by *Daphnia* (which may engulf exclusively the smaller colonies and, hence, within the theoretical compass of the Daphniid feeding gape, or first fragment the larger colonies by physical damage); periods of *Asterionella* abundance and sustained population increase coincided with low community filtration rates. Thus, its spring increase ($k_n \sim 0.065$–0.150 d^{-1}) and decrease ($k_n = - > 0.03$ d^{-1}), could be satisfactorily related to changes in growth rate ($k' = 0.15$, decreasing to 0.01 d^{-1}) and sinking loss rate (k_s, 0.01 d^{-1} increasing to $> 0.2\ d^{-1}$) without invoking crustacean grazing ($< 0.02\ d^{-1}$). In a later population that flourished briefly and declined ($k_n = -0.180\ d^{-1}$) in the summer months, sedimentation (0.01–0.06 d^{-1}) could not account for the observed loss, which was attributed to grazing ($a = 0.18$–$0.42\ d^{-1}$).

The dynamics of *Fragilaria* and *Microcystis* populations were found to be independent of a, except when small mean colony size permitted them to be ingested by *Daphnia*. A special case is presented by *Eudorina* whose mode of growth determines that daughter colonies are first produced in a series of divisions of mother cells within the colony and subsequently released into the medium, where they rapidly gain in size and mass. Only the released daughters are briefly of filterable size (30–40 μm) before they enlarge to achieve effective immunity from grazers (within about a day). Reynolds & Rodgers (1983) showed that a grazing pressure of $> 0.9\ d^{-1}$ is necessary before sufficient daughters can be ingested to prevent a net increase in the population. Thus, *Eudorina* populations may continue to increase at moderate filtration pressures. The same is not necessarily true of all colonial algae: for instance the smaller mean size and longer maturation periods of actively growing *Sphaerocystis schroeteri* colonies makes them relatively more susceptible than *Eudorina* to *Daphnia* grazing (author's unpublished observations). Nevertheless, the gelatinous–colonial habit must generally be considered to constitute, however fortuitously, a pre-adaptive deterrent to grazers in freshwaters.

7.5.3 *Phytoplankton–zooplankton interactions*

Lest the impression conveyed by this section should be that grazing by zooplankton is to be regarded merely as a constraint upon the potential growth of phytoplankton, a few concluding paragraphs will serve to emphasize that those same grazers are dependent upon a fluctuating food resource, that their own dynamics respond to these fluctuations, and that these interactions play an essential part in moulding the composition and structure of the pelagic community. These topics have been considered in some detail in a long series of admirable papers by Z.M. Gliwicz and his co-workers in Poland (see especially Gliwicz, 1970, 1975, 1977, 1980; Gliwicz & Hillbricht-Ilkowska, 1972; Gliwicz, Ghilarov & Pijanowska, 1981) and in the excellent review by Porter (1977). Here, the intention is simply to illustrate these points by reference to calculations highlighting the capacity of the phytoplankton to meet the needs of the zooplankton, to review one set of field observations and finally to offer one or two concluding statements.

How much food does zooplankton require? Several authors have attempted to determine the food requirements of *Daphnia* but the most exhaustive published study is probably that of Lampert (1977*a*,*b*,*c*). Working with *D. pulex*, Lampert (1977*a*) derived equations to describe the maximum hourly assimilation rates, A, in the form $a(L_b)^b$, $a(W_b)^b$ at various temperatures, (where L_b and W_b are respectively the length and weight of the individual). At 15°C, a 0.8-mm individual can assimilate approximately 2.4 μg C d^{-1}, a 2.1-mm one ~ 15.7 μg C d^{-1}. The assimilation efficiency varied with the food source offered, being up to 60% on *Synechococcus*, *Scenedesmus* and *Asterionella* but relatively much poorer on *Anabaena*, *Microcystis* and *Staurastrum*. Thus the ingestion requirement for the above individuals will be > 4.0 and > 26.2 μg C d^{-1}, respectively. Metabolic carbon losses are in the order of one-sixth of the maximum assimilation rates and these therefore represent the minimum food requirement for maintenance (i.e. about 0.6 and 4.3 μg C would be the minimum daily food intake in the examples; assimilation efficiency, however, probably increases at low food intake). Below this, the animals will starve and die. The capacity to grow and eventually to reproduce is therefore related to the range between the minimum maintenance and the maximum food requirements.

From the maximum (food-limited) filtration rates observed for *Daphnia* of similar sizes at similar temperatures (7.6, 62.6 ml d^{-1}) by Thompson *et al.* (1982; see Table 24), the concentration of 'suitable' foods needed to 'saturate' the requirements of the large and the small individuals may be respectively approximated as 0.4 and 0.5 μg C ml^{-1}; equally, those required to meet the basic maintenance requirements can be shown to be 0.07 and 0.09 μg C ml^{-1}. These values may be compared with Lampert's (1977*b*) calculation of the food concentration saturating the assimilation requirements of *Daphnia* ~ 0.3 μg C ml^{-1} and with the incipient limiting concentrations (ILC) below which the individual filtration rate is maximal (0.2–0.75 μg C ml^{-1} of filterable foods; Jones *et al.*, 1979; Thompson *et al.*, 1982). It should be stressed that these various concentrations are independent of the population density of the *Daphnia* and, virtually, of individual size. The net production efficiency is strongly influenced by generally-limiting food concentrations and not by variations at high concentrations (Lampert, 1977*b*). While bearing in mind differing food preferences and the means of obtaining them, it may be presumed that parallel constraints operate for other filter feeders (Gliwicz, 1970). The cell concentrations of some 'suitable' organisms representing minimum and saturating food sources for *Daphnia* are given in Table 27. The concentrations cited should be interpreted as the C-requirement were the food

Table 27. *Equivalent concentrations of selected algal and bacterial foods representing the minimum maintenance* (~ 0.08 μg C ml^{-1}) *and saturating* (~ 0.5 μg C ml^{-1}) *requirements of* Daphnia *at about 15°C; contents calculated as* 0.21 pg μm^{-3} *(see Chapter* 1), *except for* Asterionella (C = 50% *ash-free dry weight) and free-living bacteria, which are here assumed to have conformed to Thompson's mean value (quoted by Thompson* et al. *1982) of* 0.013 pg C *cell*$^{-1}$)

Species	cell vol. (μm³)	cell C content (pg)	Populations equivalent to 0.08 μg C ml^{-1}	Populations equivalent to 0.5 μg C ml^{-1}
Cryptomonas ovata	2710	569	140	880
Asterionella formosa	645	60	1330	8330
Chromulina sp.	440	92	870	5430
Rhodomonas minuta	72	15	5300	33330
Ankyra judayi	24	5	16000	100000
Free-living bacteria	—	0.013	6.15×10^6	3.85×10^7

organism present as a monoculture. For example, were *Cryptomonas* the sole food source available, a population of 140 ml^{-1} represents the minimum that would sustain *Daphnia* individuals; whilst this concentration obtained, an 0.8-mm *Daphnia*, filtering 7.6 ml d^{-1} could expect to encounter 1064 cryptomonads per day or, on average, one every 81 s; a 2.1-mm animal filtering 62.6 ml^{-1} would ingest one every 10 s. Below this concentration, neither animal would ingest *Cryptomonas* cells sufficiently often to satisfy its minimum needs. In practice, of course, several potential foods may be present simultaneously, but if their combined food values is much less than 0.08 μg C ml^{-1}, the *Daphnia* is likely to starve and die, regardless of its own concentration. Equally, *Daphnia* is unlikely to remain abundant in lakes when edible food concentrations fall below 0.08 μg C ml^{-1}, or at all in lakes where nutrient limitation determines that the food concentration is frequently or always below this level (see below).

Though this limiting condition remains absolute (for a given temperature range) its occurrence is influenced by dynamic changes in the phytoplankton and in the zooplankton. The effect of a given level of filtration on the net rate of increase of phytoplankton has already been considered (§7.5.2.2, above). Thus, we may envisage that a population of *Daphnia* filtering 50% of the water in which a population of (say) 5×10^3 actively-growing *Rhodmonas* ml^{-1}, each undergoing one cell division each day will, superficially at least, preserve a steady state in the food supply (i.e. the remaining 2.5×10^3 *Rhodomonas* produce a further 2.5×10^3 in each ml of water).

Above the minimum maintenance level, however, the *Daphnia* will be able to direct a fraction of the assimilated carbon to growth and reproduction, such that, first, bigger, and then, more *Daphnia* contribute to an increasing community filtration rate. Without a commensurate increase in the rate of production of the foods, the developing population will, sooner or later, become food-limited. Data obtained by Ferguson *et al.* (1982), working with artificially contrived, predation-free *Daphnia* populations, initially dominated by small animals, in plastic enclosures containing abundant food ($> 0.15 \times 10^6$ *Ankyra* ml^{-1}), indicated that the mean length of the *Daphnia* cohort increased from 0.8 to 1.4 mm in 6 days and to 1.7 mm after 13 days by which time a recruitment of juveniles, five times as many as the number originally enclosed, had taken place. The completion of this generation cycle had led to a population capable of generating an aggregate filtration rate 12 times that of the originally-enclosed population. Extrapolating to natural populations, it is easy to see how the community filtration rate can increase over two to three orders of magnitude, when foods and temperatures permit, within a month or so (Thompson *et al.* 1982; see also Figure 82).

The impact of such gross changes in community grazing rate upon the phytoplankton depends upon the relative dynamics of phytoplankton growth. On the one hand, it is reasonable to suppose that if a Daphniid population is already abundant because of the recent presence of a large population of grazable 'algae', they may be able to keep the populations of other 'suitable' algae at low levels until starvation reduces their numbers. Equally, a small but growing Daphniid population, initially filtering 5 $ml\ l^{-1}\ d^{-1}$ of a medium in which algae (equivalent to $> 0.5\ \mu g\ C\ ml^{-1}$) are suspended and dividing once every day, would require 26 days before it was capable of reducing the algal population. The algae, however, have the potential to reach a light-limited maximum of a 3-m mixed layer of $\sim 5\ \mu g\ C\ ml^{-1}$ or of a 1-m layer containing $\sim 15\ \mu g\ C\ ml^{-1}$ within, respectively, 3 and 5 days; grazing on the 'algae' does not prevent them from achieving their theoretical maximum populations. However, as has already been shown 'algal' growth rate declines as light or other factors become limiting, such that grazing rate overtakes algal growth rate relatively earlier than these figures would indicate; indeed at high community filtration rates, the water may be cleared so rapidly, that the 'algae' are unable to adapt to the improving light conditions or recommence their growth before most of them have been removed (Reynolds *et al.*, 1982*b*).

Examples of these effects on the composition of the phytoplankton are illustrated in Figure 83, which shows the distribution of the phytoplankton

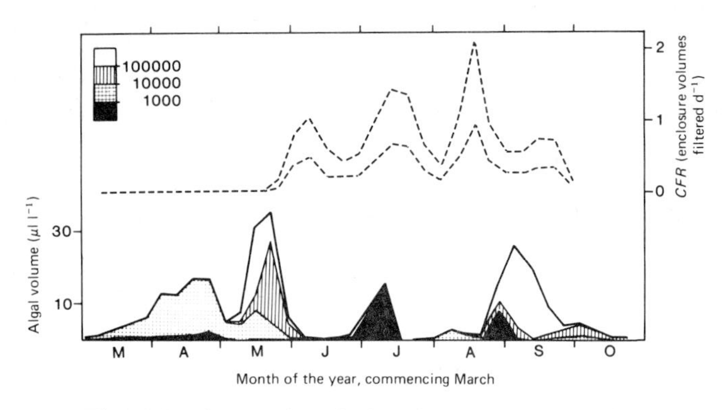

Figure 83. Seasonal variations in the abundance and size distribution of phytoplankton in Lund Enclosure A, 1978, in relation to the zooplankton community filtration rate (*CFR*). Generally only those particles < 10000 μm^3 unit size would be available to the filter feeders as food. (Redrawn from Figure 10 of Reynolds *et al.*, 1982*b*.)

unit volumes among size categories: < 1000 μm^3 (which includes bacteria, *Rhodomonas*, *Chlorella* and *Ankyra*); 1000–10000 μm^3 (*Cryptomonas*, *Asterionella* and small colonies of *Eudorina*, *Fragilaria* and *Microcystis*); > 10000 μm^3 (inedible forms – e.g. most *Eudorina* and *Microcystis* colonies); this is the same period to which the zooplankton composition and community filtration rates – shown in Figures 81 and 82 – apply. In the early part of the year, potentially edible foods dominated the plankton and were abundant (~ 1.0 μg C ml^{-1}); at the lower water temperatures obtaining, the abundance, grazing activity and growth rates of *Daphnia* were too low to have been considered 'food-limited'. During May and June, however, the foods were severely depleted during the first pulse of *Daphnia* growth, but subsequently recovered as the *Daphnia* declined.

It is of interest to note that mortalities through starvation were relatively more prevalent among the smaller size categories than the larger ones (see Figure 81); presumably, the greater filtration volumes and wider selection limits of large *Daphnia* 'buffered' them against these extremes. They also enhance the recovery capability when food supplies increase, as indeed happened in the wake of the late June and August populations of *Ankyra* and *Cryptomonas* and, in turn, contributed to their rapid elimination (Figure 83). The overall impression is that the growth cycles of the nanoplankton and of the *Daphnia*-dominated zooplankton alternate. The zooplankton may 'control' the abundance of the phytoplankton over short periods, but the opposite (i.e. that the phytoplankton 'controls' the zooplankton) seems more generally valid (see also Gliwicz & Hillbricht-Ilkowska, 1972; Gliwicz, 1975).

The maintenance of an artificially-high nutrient availability in the Blelham Enclosure during the year to which these observations apply (see Reynolds & Wiseman, 1982) biases the system towards supporting large populations of fast-growing 'algae' (i.e. nanoplankton), so these responses of the community may be exaggerated. Later observations (as yet unpublished) on phytoplankton–zooplankton interactions in the Lund Enclosures maintained at lower fertilities more faithfully imitate the cycles of most eutrophic lakes, where larger algae (*Eudorina*, *Sphaerocystis*, *Anabaena*, *Microcystis*) come to dominate the plankton after the first *Daphnia* pulse. These seem to be to due to the nanoplankton being unable to sustain or recover maximal growth rates, so their net increase rates are comparable with those of the ungrazed species (0.2–0.5 d^{-1}). The latter are then able to develop relatively large populations simultaneously with those of the nanoplankton and to continue to maintain or increase them (so far as nutrients will allow) during periods of high grazing (Gliwicz, 1975; Porter, 1977; see also §7.6 and Chapter 8). Thereafter, neither the nanoplankton, nor therefore the Daphniid or filter-feeding rotifer populations, can recover until more nutrients become available later in the summer or early autumn. These different responses do not alter the conclusion regarding the mutual 'controls' exerted by either the phytoplankton or the zooplankton on the other.

Calanoids are less exclusively dependent upon the concentrations of smaller food particles to supplement their energetic requirements, gained by actively seeking out larger algae (see §7.5.2.2); they are better suited for survival in oligo- and meso-trophic lakes that do not support dense populations ($> 0.08\ \mu g\ C\ ml^{-1}$) of nanoplankton and in those eutrophic lakes where nanoplankton falls to low levels in summer. This is a powerful selective factor influencing both the typical species composition in less-productive lakes and the succession of species dominance, often tending from rotifers and Daphniids to calanoids or chydorids, observed in mildly eutrophic lakes (see also Lampert, 1977*c*; Gliwicz, 1977, 1980; Hillbricht-Ilkowska, Spodniewska & Węgleńska, 1979; Gliwicz *et al.*, 1981).

Indeed, the availability of foods and the means of obtaining them are probably as important in regulating the abundance, size- and species-distribution of zooplankton as is the widely accorded role of predation by planktivorous predators (especially fish): for a full discussion, see Gliwicz & Preis (1977), Gliwicz (1980), Gliwicz *et al.* (1981). The well-known size-efficiency hypothesis for the regulation of structure in zooplankton communities (of Brooks & Dodson, 1965; see also Hall, Threlkeld, Burns & Crowley, 1976) states that large planktonic herbivores (large Cladocera, calanoids) compete more efficiently for a wider range of potential food particles than smaller ones (small Cladocera, rotifers) and so dominate the plankton, provided that predation is low; intense size-dependent predation, however, selects against the larger animals, enabling the smaller forms to dominate. Community structure responds to predation. Work reviewed here suggests that fluctuations in the supply of specific food requirements develop similar effects, quite independently of changes in predation: *Daphnia* can grow actively and become dominant provided foods are plentiful, whereas many small species and small *Daphnia* are more vulnerable to starvation. Equally, an abundance of non-filterable foods tends to favour dominance by small chydorids and calanoids. While predation undoubtedly selects against larger zooplankton, its impact upon reproductive capacity may be exaggerated compared with that of a fluctuating food supply. Gliwicz & Preis (1977) considered that *Daphnia* reproduction is not seriously constrained until the fresh biomass of planktivorous fish exceeds the range 600–900 kg ha^{-1}. This is in considerable excess of natural fish stocks even in naturally eutrophic lakes (typically < 300 kg ha^{-1}) and in artificially-stocked Trout fisheries in eutrophic UK waters (Reynolds, 1979*c*) and is probably approached or exceeded only in continental fish ponds. It must be doubted that lake fish populations necessarily control zooplankton to the extent that their grazing potential is seriously inhibited. This view conditions the widespread belief that to remove the fish from water bodies will improve water quality dramatically because an unconstrained zooplankton would control the phytoplankton more effectively. Even if such a move did (temporarily) increase grazing on the phytoplankton, it would be more likely to alter the balance against nanoplankton and in favour of larger standing crops of larger inedible species, including acknowledged 'nuisance' species like *Microcystis* and *Fragilaria.* The impressive demonstrations of phytoplankton responses to the level of fish predation given by Andersson, Berggren, Cronberg & Gelin (1978) and by Leah, Moss & Forrest (1980) apparently support the original contention and run counter to the alternative view advanced here; it should be pointed out that the stocking density imposed by Andersson *et al.* (1978) was equivalent to between 900 and 2200 kg ha^{-1}, while the small impoundments studied by Leah *et al.* (1980) were extremely fertile, habitually supporting relatively very large crops of small algae (to the virtual exclusion of larger forms), and apparently able to sustain the development of several generations of filter feeders; a small shift in the relative dynamics of 'algae' and feeders would, indeed, be expected to eliminate the 'algal' biomass rapidly (cf. the case of *Ankyra* in the Lund enclosure, above).

Some other facets of the interactions between phytoplankton and zooplankton may be mentioned briefly. One conditioning factor is that the vertical distribution and migrations of animals and 'algae' may permit the former access to the latter; theoretical aspects of this problem have been discussed by Bowers (1979). In a recent study of seasonal changes in the vertical distribution of planktonic Crustacea in the Lund Enclosures, George (1983) has confirmed that the animals are very sensitive to light, the underwater penetration of which is often a correlative of phytoplankton concentration (see §4.3.1.2). He showed that, by day, the animals were preferentially distributed around given depths in the light gradient; at high phytoplankton concentrations (steep attenuation gradients) they became tightly bunched but, generally, were more dispersed (within similar light ranges) when 'algal' concentration was low and attenuation gradients were more 'flattened'. This behaviour optimizes the food-gathering potential of the animals, placing them well within the layers of maximum algal density (even when these are located well below the surface) but extending the feeding range when the foods are sparse.

Another facet of the interaction is that excretion and defaecation by zooplankton return ammonia and phosphates to the water, where they can be recycled by the 'algae'. This may be extremely important when the rate of 'algal' growth is severely nutrient-limited (Lehman, 1980). The efficiency of the recycling will be less than complete and, in any case, both grazed and ungrazed species will be competing for them. Uptake by the ungrazed species is less likely to be recycled *in situ*, either sustaining their continued net increase or being transported to the hypolimnion in sedimenting cells. Again, this accumulation of nutrients in ungrazed species must contribute towards the eventual dominance of larger planktonic species, like *Microcystis*, *Ceratium* and *Peridinium*, over smaller ones and is therefore a contributory factor in the seasonal succession of the phytoplankton in nutrient-limited epilimnia (Porter, 1977). Nutrients transported to the hypolimina are unlikely to become available to epilimnetic production until thermal stratification breaks down, or some other whole-lake recycling process occurs, later in the season (see also Chapter 8).

Finally, it is appropriate to point out that grazing by herbivores is not an entirely negative process so far as individual algae are concerned: Porter (1973) has reported that some chlorophytes that form mucilaginous colonies (e.g. *Sphaerocystis*) may be ingested whole by *Daphnia* but, not only do they remain undigested and survive passage through the gut, they may absorb nutrients that enhance their renewed growth on return to the external medium.

7.6 Loss processes and phytoplankton composition

The rates of loss of cells, variously attributable to either washout, sinking, *in situ* death (for whatever cause), decomposition and grazing, combine to detract from the potential rates of population increase, sometimes to the point where the populations decline. The relative magnitudes of these losses vary from lake to lake, under the influence of such characteristics as basin morphometry, hydrology and trophic status, and from season to season, as changes in temperature influence the stability of the water column and the dynamics of zooplankton growth and reproduction. Whether the 'algae' are able to withstand these losses is also influenced by their potential for growth that is regulated mainly by the interactions of light and temperature and by the availability of limiting nutrients: when $k' > (k_w + k_s + k_d + k_g)$, the population is increasing; when it is not, vegetative population is in decline. The combinations of these components are endless, every solution virtually unique. Two important principles, however, can be emphasized. The first is that, while the values of some components (e.g. k_w) will be instantaneously nearly identical for all species present, others (k', k_s, k_g) definitely will not; that is, the net rates of change ($\pm k_n$) in each of the populations of species simultaneously present may well be quite different: some will increase while others are decreasing, because they are relatively less susceptible to contemporaneous rates of (say) sinking loss or removal by filter feeders. The second principle is that, though differential simultaneous specific growth and loss rates may well determine which species will be contemporaneously selected as potential dominants and which are likely to be eliminated, they act only on the specific populations present at the time; thus, compositional changes in the phytoplankton are still dependent upon specific growth responses to environmental fluctuations; loss processes mediate the net responses and the competitive outcome (Reynolds *et al.*, 1982*b*). Put another way, loss processes operate like the rules of a board-game; they do not choose the players.

That the losses sustained by various algal populations partition in different ways may be judged from Figure 84, which presents the differences between k' and k_n during the course of each of several algal populations husbanded in the Lund Enclosures (A and B), as reconstructed by Reynolds *et al.* (1982*b*). In each case, the vertical bar represents the total loss rate and the shading shows the contributions attributable to grazing and to sedimentation. The bars convey an overall impression that summer populations of *Asterionella*, *Ankyra* and *Cryptomonas* were eliminated largely or wholly by grazers but that vernal populations of *Asterionella*,

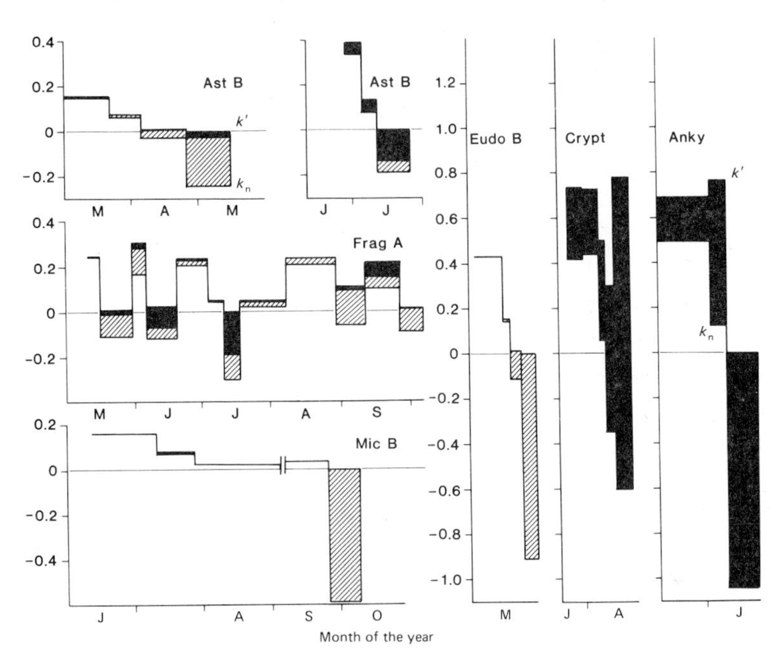

Figure 84. Simultaneous plots of net increase rate (k_n, lower lines) and true growth rate (k', upper lines) for selected populations of 'algae' husbanded in the Lund Enclosures, during 1978. (Ast, *Asterionella formosa*, March–July; Frag, *Fragilaria crotonensis*, May–October; Eudo, *Eudorina* cf. *unicocca*, May; Crypt, *Cryptomonas ovata*, July–August; Anky, *Ankyra judayi*, June–July; Mic, *Microcystis aeruginosa*, June–October). The areas between k' and k_n represent Σ_{k_L}, which is shaded in the proportion of k_g (black):k_s (hatched). (Original, drawn from data in Reynolds *et al.*, 1982*b*.)

as well as of *Fragilaria*, were principally subject to sedimentation losses. Much of the final elimination of *Microcystis* is through sedimentation: colonies remain viable on the sediments over winter and potentially provide an inoculum for the following season's growth. It is debatable, then, whether this sedimentation should be labelled as 'loss'.

That differential rates of loss may, at times, select between species is also illustrated by reference to the data of Reynolds *et al.* (1982*b*). They envisaged a simple 'community' in which three species, *Cryptomonas ovata*, *Fragilaria crotonensis* and *Microcystis aeruginosa* were mutual competitors. It was assumed that all three species would be growing at the near-maximal rates observed during their study and that none was nutrient- or light-limited; the *Cryptomonas*, however, was assumed to be sensitive to grazing pressure; similarly, *Fragilaria* was assumed to be sensitive to the depth of column mixing; *Microcystis* was assumed to be insensitive to either. Quantitative expressions for their respective rates of net population change, derived from the field data, were then solved to apply to four sets of hypothetical environmental conditions combining two levels 'grazing pressure' ($a = 0.2$ or $0.8\ d^{-1}$) and two contrasted conditions of column mixing ($z_m = 1.0$ m or 6.0 m). The equations were:

Cryptomonas $k_n = 0.622 - 0.870\ a$

Fragilaria $k_n = 0.292 \log_{10} z_m - 0.028$
(where z_m = the mixed depth in m)

Microcystis $k_n = k' = 0.160$

At the lower grazing pressure, the net growth rate of *Cryptomonas* ($k_n = 0.448\ d^{-1}$) is adequate to allow it to increase more rapidly than either of its competitors. When the higher grazing pressure is imposed, *Cryptomonas* declines ($k_n = -0.074\ d^{-1}$), leaving *Fragilaria* with the fastest rate of increase (when $z_m = 6.0$ m, $k_n = 0.199\ d^{-1}$) or *Microcystis* ($k_n = 0.160\ d^{-1}$) when the mixing depth is 1.0 m. Indeed, *Microcystis* should increase faster than either of its competitors whenever $a > 0.531\ d^{-1}$ and $z_m < 4.4$ m.

These hypothetical simulations were intended to illustrate the concept of the selective importance of simultaneous differential loss rates, rather than to explain a particular sequence of events. Nevertheless, it was applied qualitatively to explain observed fluctuations in the standing biomass of each of several species dominating the Lund Enclosures during the course of the year. Moreover, it was argued (Reynolds *et al.*, 1982*b*; Lund & Reynolds, 1982) that because the magnitudes of loss rates were often comparable with the rates of growth and that the selective action distinguishes between the net rates of population change among potentially competing species, loss processes play a fundamental role in shaping the communities of natural water bodies.

8

Periodicity and change in phytoplankton composition

'The flower of their age'
I Samuel II, v. 33

The composition of phytoplankton communities and the relative abundances of component species undergo continuous changes and on varying scales. These range between frequent reorganizations of existing community structures, in response to advective mixing processes, through annually recurrent cycles of compositional change that accompany underlying cyclical fluctuations in insolation, temperature and vertical differentiation (stratification) of the environment, to longer-term floristic changes, where one recognizably recurrent cycle is supplanted by another, in response to sustained limnological shifts in morphometry, hydraulic throughput and nutrient loading. The responses are observed relatively easily, either directly or, as in the case of long-term florisitic changes, in the retrospective synopses that can be gained from systematic analyses of the fossilized remains of species representative of past assemblages, recovered from intact cores of lacustrine sediments (Round, 1971).

Why do phytoplankton communities respond in these ways? Do the responses conform to any well-defined patterns? And, if so, how predictable are they? The answers to these questions are fundamental to any professed understanding of phytoplankton ecology. Implicitly, they have been sought for over half-a-century by researchers investigating the spatial and temporal distributions of phytoplankton in relation to measured environmental variables in a wide range of water-bodies of differing morphometry, hydrology and trophic status. Yet it is humbling to have to concur with the view expressed by Kalff & Knoechel (1978) that 'although this [research] activity has resulted in an enormous and growing literature, progress in understanding and prediction has been very slow'. Certainly, there are still no widely-accepted explanations either of the mechanisms that drive the seasonal waxing and waning of phytoplankton, or of the factors that condition long-term floristic changes. Nevertheless, most of the points to be made in this chapter have been stated previously by others.

The problem may be less one of finding new and holistic approaches to the functioning of planktonic ecosystems, for which Kalff & Knoechel (1978) called, but more one of establishing a framework into which existing knowledge may be more readily assimilated.

It is this 'target' which I have pursued in the preparation of this book. The sequence in which the various themes have been arranged reflects a somewhat anthropomorphic view of the hierarchy of survival problems that must be overcome if an 'alga' is to be a successful component of the phytoplankton: first, it must maintain a place in the fluid environment; it must then be capable of assembling the materials necessary for growth and be able to translate them into increased biomass represented by succeeding generations (growth), in the face of physical, chemical and biotic environmental constraints. Throughout, every effort has been made to emphasize the diversity of phylogenetic affinities, morphological and physiological adaptations, growth strategies and susceptibilities to loss processes that is encompassed by phytoplankton organisms. This has been done with the deeply-held conviction that these do not represent differing specific solutions to the *same* problems of pelagic life (where one might expect a much greater degree of evolutionary convergence of form and function than is apparent) so much as differing strategies evolved to meet perceptible differences of emphasis in the interactions among the potentially constraining environmental factors that characterize pelagic systems. In short, they offer far more 'niches', in terms of stability, temperature, underwater light climate, nutrient availability and biological interactions, than is commonly appreciated. The evolutionary adaptations which are required to exploit effectively one range of those niches may well prove to be less suitable in another; probably none is adequate under all environmental conditions that may be encountered in open waters. One set of conditions will tend to select for ('best-fitted') species with the most appropriate adaptations; as conditions fluctuate, so does the identity of the 'best-fitted species'. Analogues can be found among terrestrial plants: just as there is parallelism in a common need to obtain light and nutrients (in the broadest sense), so there is obvious differentiation in structual organization (exemplified by lichens, mosses, herbaceous and arboreal phanerophytes); none is continuously dominant, local selection being influenced by climate, aspect, soil type and depth, water relations, cropping and so on. The principal difference in the ecology of phytoplankton algae is that selection attributable to environmental variability generally (but not exclusively) operates over much shorter time scales than occur in terrestrial environments.

In this concluding chapter I shall try to establish those selectively significant differences in the growth and survival strategies that are discernible among planktonic 'algae' and to show how these influence some well-documented seasonal and long-term floristic changes.

8.1 Seasonal periodicity of phytoplankton

8.1.1 *Patterns of periodic change*

In answer to the second of the questions posed above, the variations in the abundance and species composition of phytoplankton that occur in a given lake through a given twelve-month period are, as a generalization, often broadly repeated the following year, thus constituting one of the most striking examples of pattern conformation in freshwater communities. It is often the case that the phytoplankton biomass achieves high levels and falls to low ones at approximately the same times in successive years (Figure 42); frequently the same species will be prominent at the corresponding stages (Figure 43) while individually, each will tend to increase or decrease on a specific annual cycle. The precise timing of these phenomena may vary from year to year and the performances of individual species may fluctuate such that the dominant species at any given stage will not always be the same. A single example will suffice to illustrate this point: the phytoplankton biomass in Windermere increases during the spring of each year, typically reaching maxima equivalent to between 15 and 25 mg chlorophyll m^{-3} in May or June (Lund, 1966*b*, and unpublished). In every one of over 30 consecutive years, this maximum has been attributable to the abundance of several species of diatoms – *Asterionella formosa*, *Melosira italica* subsp. *subarctica* and *Cyclotella comensis*. Each has been present in nearly every year; each has undergone sustained increase during April and May, but the starting population (or 'inoculum'), the duration and rate of increase and the maximum population achieved by each species in each year have been more conspicuously variable. Usually *Asterionella* has dominated over the other two species with a maximum standing crop of between 6×10^3 and 10×10^3 cells ml^{-1}. Occasionally, the final population of *Asterionella* has been relatively smaller ($> 2 \times 10^3$ cells ml^{-1}), while *Melosira* or *Cyclotella* has been relatively more numerous and so constituted the largest fraction of the biomass. At the species level, we might find considerable difficulties in establishing a pattern of dominance, still more in predicting the dominant species in advance. At a higher level of categorization, however, it would be fair to state that the group of species (or the species assemblage), *Asterionella–Melosira–Cyclotella*, regularly dominates the vernal plankton of Windermere, and it would have been a 'safe bet' to have predicted that the assemblage would dominate during the following spring. In a sense, part of the variability has been ignored. This is not simply a matter of expediency or convenience; it is a means of classifying and defining seasonally-observable phenomena of planktonic compositional changes for the purpose of establishing pattern.

This concept of seasonal change in biomass and community composition was introduced in Chapter 3 (see Figures 42 and 43). By grouping together species which share similar annual cycles of abundance, or which frequently increase contemporaneously, such that they are mutually alternative dominants, sub-dominants or co-dominants, the characteristic sequences of community compositional changes became more readily apparent. Of far greater significance, however, is that the annual cycles of biomass fluctuation and the sequence of representative assemblages are often broadly repeated in geographically remote lakes; that is, that the sequential patterns are common to whole series of water-bodies sharing similar properties of morphometry and trophic status. Equally, other distinctive and repeatable patterns are to be found characteristically among other types of water bodies (see §3.4.1, 3.4.2).

The existence of several such characteristic patterns of seasonal change were firmly established in Hutchinson's (1967) review of published descriptions of the phytoplankton periodicities of various (mostly temperate) lakes. Since that time several dozens of descriptive accounts of the phytoplankton cycles in individual lakes have been added to the literature, many with appropriate quantifications, and a relatively small number in which explanations for the observed transitions of dominance have been advanced (see, for instance, Nauwerck, 1963; Pavoni, 1963; Gliwicz, 1975; Knoechel & Kalff, 1975; Reynolds, 1976*a*; Lewis, 1978*b*; Lehman, 1979; Sommer, 1981; Reynolds *et al.*, 1982*b*). Several good general reviews of periodic mechanisms in freshwaters are also available (Lund, 1965; Round, 1971; Fogg, 1975; Porter, 1977; Kalff & Knoechel, 1978); parallel discussions of the factors operating in marine environments can be found in Margalef (1958, 1978) and Smayda (1980). In general, the observed patterns conform, to a greater or lesser extent, with one or other of the main types recognized by Hutchinson (1967), although some expansion of these has become necessary in the light of later information, particularly where this applies to small, highly enriched (hypertrophic) ponds and lakes, to tropical lakes (neither were covered in Hutchinson's treatment) and to deep, oligotrophic lakes that have been subject to anthropogenic eutrophication during recent years. Neither the information summarized in Figures 42 and 43 nor the brief descriptions of periodic sequences given in §3.4 can be expected to accommodate precisely every observable sequence that may have been or may be shown to occur naturally.

Nevertheless, they do apply to elements of a wide range of known periodicities. That the same basic patterns occur so frequently and that some of them, at least, are imitable in experimental enclosures (see Lund & Reynolds, 1982) strongly suggests that the motivating mechanisms of change are common to all of them.

The driving forces and mechanisms of seasonal change are acknowledged to be related to variations in the physical, chemical and biotic environment and to the many possibilities brought about by their mutual interactions, which together effect differential specific growth and loss rates among the 'algae'. To attempt to match known general patterns of compositional change against likely seasonal changes in the interaction of environmental factors constitutes one possible approach to the elucidation of periodicity. There are so many possibilities to be considered that the exercise would be speculative and, probably, unfruitful. As an alternative approach, I propose to examine the course of seasonal changes in the interactions among broad environmental variables (§8.1.2), identifying at each stage those factors that are most likely to limit 'algal' production critically. Based on the synthesis of previous chapters, I will attempt to identify those physiological and morphological characteristics of plankton 'algae' which will be selectively favoured or disfavoured and to compare these with the documented patterns of assemblage dominance. The section on periodicity concludes with discussions on the rapidity and frequency of interassemblage transitions (§8.1.3) and the directions of the changes they mediate (§8.1.4).

8.1.2 *Environmental selectivity*

8.1.2.1 *The winter and vernal periods.* A convenient starting point for this discussion is to consider temperate lakes in mid-winter. Away from oceanic seaboards, they will usually be liable to periods of ice-cover, the frequency and duration of which generally increase with latitude and altitude of the locality. The ice may also be subject to a substantial covering of snow. The principal characteristics of the water column beneath the ice include its low temperature (between 0 and 4°C), low radiation income (short day length, low angle of incident sunlight, compounded by high reflectance of the ice surface or, if present, extinction by snow) and by a generally low level of eddy diffusivity. In relation to the latter, it should be stressed that it is rarely the case that ice-covered water is hydraulically static. Immediately beneath the ice there will be an inverse temperature gradient (and hence a density gradient) which flattens with increasing depth to a point where there is almost zero resistance to mixing (see, for instance, Stewart, 1972). Light passing through the ice will tend to warm the immediate surface layers,

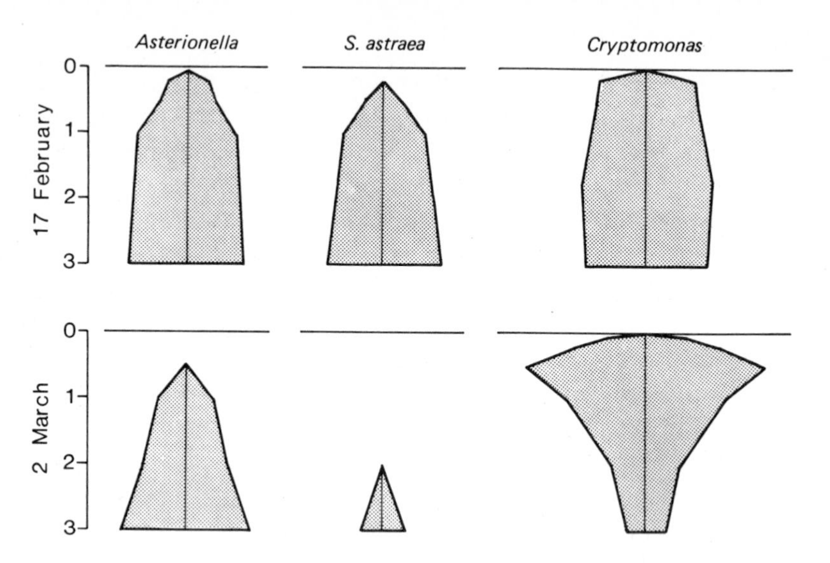

Figure 85. Vertical distributions (plotted as 'cylindrical curves') of two diatoms and *Cryptomonas ovata* in Crose Mere on two occasions during a period of continuous ice cover in 1969. (Original, from author's unpublished data.)

rendering them more dense, causing surface water to sink and generate convectional rotations. Corioli's forces, the movements of the ice itself and inflowing seepages, especially towards the end of the period of ice cover, also contribute to internal water movements. Temperature gradients may thus become compressed to occupy thin sub-ice layers, overlaying water in which the mean eddy diffusivity may be in the order of 10^{-5} to 10^{-4} cm^2 s^{-1} or greater. This is weak enough for many inert particles to sink directly to the sediments, with little prospect of resuspension even from shallow deposits. The convection may, however, delay the settling of smaller diatoms at their intrinsic (still-water) rates sufficiently to maintain a growing population (see Figure 85). These conditions would tend to select for those 'algae' whose growth is less severely restricted by low temperatures, that is, those species whose high surface area:volume ratio determines that the Q_{10} value of growth is relatively low (~ 2; see Chapter 6) but are nevertheless sufficiently large that they can survive the long periods of darkness that inevitably occur. 'Algae' with appropriate adaptations include the smaller pennate diatoms, some *Oscillatoria* species and larger flagellates. The greater is the under-ice convection, the more the diatoms would be selected; the less, likewise the flagellates. *Oscillatoria* spp. might be expected to be tolerant of both sets of conditions. Several published studies indeed make reference to the increase of cryptomonads and/or small diatoms under ice (e.g. Rodhe, 1955; Nauwerck, 1963; Cronberg, Gelin & Larsson, 1975; Jónasson & Adalsteinsson, 1979) and, remarkably, to the intitiation of vernal increase of centric diatoms, such as *Stephanodiscus*, during the later stages of ice cover (Coveney *et al.*, 1977); *Oscillatoria* spp. are able to maintain relatively higher biomass levels, subject to flushing rates, in the lakes where they were already high during the previous autumn (e.g. Zürichsee: Pavoni, 1963). The phytoplankton cycles in Char and Meretta Lakes, Northern Canada (74° 42′ N, 94° 57′ W), both of which are ice-covered for 10 months out of every 12, are dominated by flagellates, first by *Gymnodinium* spp., then towards the end of the ice-covered period, by chrysophytes (*Chrysococcus* spp.), and by diatoms (*Cyclotella* spp.) during the ice-free periods (Kalff, Kling, Holmgren & Welch, 1975). In most of these instances, the average biomass of the sub-ice populations (usually $< 5\ \mu l\ l^{-1}$) is apparently regulated by light availability; where higher levels are observed (e.g. of *Oscillatoria* in Zürichsee), these owe to larger populations existing before, and persisting after, the onset of surface freezing.

Although the events accompanying final melting or breakage of the ice cover may be protracted and, especially in very large water bodies, may be accompanied by striking horizontal differences in physical characteristics (see for instance the description of 'thermal bar' formations in Lake Ontario and the associated inshore production of *Melosira binderana*: Munawar & Munawar, 1975), the hydrological conditions in erstwhile ice-covered lakes, ponds and rivers eventually come to resemble those in their ice-free counterparts subject to less severe winters. Their temperatures will still tend to be uniformly low and, lacking stable thermally-determined density gradients, their waters are exposed to full isothermal wind-driven, mixing and, to a greater or lesser extent, by increased inflow discharges. At the same time, the rates of gaseous interchange at the surface will be enhanced, but so will those of relative dilution by outwash (k_w), while the concentrations of dissolved, potentially limiting, nutrients will have reached, or be rapidly approaching, their annual maxima. One or other of the latter may be characteristically low for the system in question, such that both the rate of production and the final yield of phytoplankton biomass may be nutrient-regulated; the relative concentrations of limiting nutrients may also be expected to influence the outcome of nutrient-based competition, as predicted (for instance) by the Tilman–Kilham model (see Figure 61). That vernal increase is often initiated in ice-free lakes without significant increases in temperature or nutrient concentration suggests that increasing insolation is frequently the overriding controlling factor.

However, the unrestricted movement of water through virtually the entire depth of the water column and its relationship with the duration, intensity and penetration of insolation represents the most striking departure from the characteristics identifiable with ice-covered lakes and is the major environmental factor regulating phytoplankton production and species selection in well-mixed columns. The greater is the relative proportion of the 'dark' water, whether this owes to the overall mean depth of the circulating water mass, *per se*, or to its apparent ('optical') depth attributable to low incoming irradiance intensities or to high vertical extinction coefficients, then the greater is the constraint on 'algal' production generally. In this context, it will be recalled from Chapter 4 (§4.3, see above) that the physiological attributes favoured in optically deep water columns at low water temperatures and low insolation are high photosynthetic efficiency (P v. I curves, Figure 45), high photosynthetic capacity (i.e. a high chlorophyll content per cell) and lowered thresholds of light saturation (i.e. low I_k). Moreover, the photosynthetic apparatus must be well adapted to contend with frequent alternations between exposure to light and darkness. Indeed, the balance between the mean durations of the light and dark periods determines the point where photosynthetic gains exceed respirational losses and, hence, when net photosynthetic production can occur. Assuming constant mixing and turbidity, the timing of this point will itself be determined relative to increasing day length and intensity of solar radiation. The greater is the optical depth of the water column then the relatively later is the onset of a substantial vernal increase in biomass (compare, for instance, the timing of the initiation of net vernal growth in Crose Mere [February], Windermere [April] and Rostherne Mere [May], three mainly ice-free lakes at similar latitudes, of ascending optical depth: see Figures 55 and 57).

Growth at low temperatures and low average insolation also requires the efficient assimilation of photosynthetically-fixed carbon and its 'translation' into 'new' biomass and the minimization of respirational losses during dark periods. In §6.1.2, it was argued that the former constraint selects against the growth of species with low (< 0.5–0.6) surface area: volume ratios (L_i-limited species), the latter against species of low individual cell size ($< 500\ \mu m^3$). Moreover, the vertical extent of mixing, provided this is not unduly constrained by extreme shallowness (say, < 1–2 m), is conducive to prolonged suspension of non-motile phytoplankton, which, conversely, will experience proportionately low rates of sinking losses (see §7.3); the relative inactivity of the filter feeders, combined with their slow rates of growth at low temperatures, virtually ensures that grazing loss rates will be minimal (§7.5). Losses by outwash may well be significant in depleting suspended stocks and delaying the onset of vernal increase though these will be generally non-selective (§7.2).

These considerations would suggest that diatoms, especially the colonial pennate forms (*Asterionella, Diatoma, Fragilaria, Tabellaria*), by virtue of their photosynthetic behaviour, cell size and surface area:volume ratios coupled with an ability to maintain overwintering vegetative populations in suspension in ice-free lakes, are strong potential candidates to dominate the vernal increase. To varying extents, some centric diatoms (*Melosira, Cyclotella, Stephanodiscus* spp.), some solitary (e.g. *Chrysococcus, Mallomonas*) and smaller colonial chrysophytes (e.g. *Dinobryon, Synura*), desmids (smaller *Staurastra*, attenuate *Closteria*), colonial chlorococcales such as *Scenedesmus* and *Pediastrum* spp., solitary filamentous cyanobacteria (*Oscillatoria*) and larger *Cryptomonas* spp. might meet similar criteria. The relative competitive positions of these 'algae' may be resolved by the relative sensitivities of their photosynthetic behaviour and growth rates to deep-mixing and total insolation. They will also be influenced by ambient nutrient concentrations, sinking rates and the capacity for recruitment from vegetative or overwintering stocks. As at all other times, it is the species, *of those 'available' at the time*, that can maintain the highest average rate of net increase under the given environmental conditions, or provide the largest initial inoculum, that is most likely to dominate.

In reality, all but the most oligotrophic lakes (Figure 42) eventually experience a rapid vernal increase in phytoplankton biomass, of which a large, if not dominant, proportion is indeed contributed by diatoms (see Figure 43). One of the most familiar and ubiquitous species in meso- and eutrophic lakes is *Asterionella formosa*. Its evident success may be attributed principally to a high photosynthetic efficiency (~ 5 mg C (mg chl $a)^{-1}$ E^{-1} m^2; §4.2.1), to a facultative ability to increase its chlorophyll content (to up to 2.2 pg $cell^{-1}$) relative to ash-free dry weight, which confers a high photosynthetic capacity – up to 6 mg O_2 $(10^9 \text{ cells})^{-1}$ h^{-1} – and a low threshold of photosynthetic saturation (I_k: 30–60 μE m^{-2} s^{-1}), and to the efficient (light-dependent) assimilation of carbon at low total irradiance levels. The net photosynthetic productivity of (*Asterionella*-dominated) plankton in Crose Mere in early February in 1971 (see Figure 56: $LDH = 28$; $P_{max}/R \simeq 9$; $I/0.5I_k \sim 14$; $\varepsilon_{min} \sim 0.7$ ln units d^{-1}) rose to some 0.10 mg C (mg C biomass)$^{-1}$ d^{-1}, which if directed entirely to population increase corresponds to a growth rate of ($k' =$) 0.10 d^{-1}. Comparison with the maximum rates of vernal increase observed in nature (0.15–0.16 d^{-1}: Table 17; see also Reynolds *et al.*, 1982*b*) suggests that such efficient conversion occurs. As a further gauge of this efficiency, the maintenance of a growth rate of 0.10 d^{-1}, during a perceived 'day' totalling only 6–7 h (k' 0.014 to 0.017 per light-hour), may be compared with the extrapolated (continuously light-saturated, temperature-limited) growth rate of *Asterionella* at 5°C, shown in Figure 68 ($\sim$ 0.85 divisions d^{-1}, equivalent to a growth rate of $\sim$0.6 d^{-1} or 0.025 h^{-1}). These properties are selectively advantageous in three ways: (1) coupled with low sinking loss rates, they are likely to yield a faster rate of net growth than can be maintained by many potential competitors; (2) rapid growth can be initiated relatively earlier in the year; (3) extension of the period of light-limited growth deeper into the midwinter months presumably goes some way to countering winter losses through outflow, or to enhancing the size of the surviving population present at the inception of more rapid net increase.

Other diatom species that can increase their numbers during winter or early spring in ice-free systems in temperate latitudes are, potentially, sub-dominant, co-dominant or even dominant members of vernal assemblages. Examples among the pennate forms include *Fragilaria crotonensis*, *Tabellaria flocculosa*, *Diatoma* spp. and *Synedra* spp. and, among the centrics, species of *Melosira*, *Cyclotella*, *Stephanodiscus* and *Rhizosolenia*. Present knowledge of their precise specific environmental requirements, adaptabilities and dynamics is still too poorly developed to permit a predictive appraisal of likely interactions, but the literature does provide some pointers to the critical mechanisms. It is possible, for instance, to cite the analysis of seasonal changes in the dynamics of five diatoms in Lac Hertel, presented by Knoechel & Kalff (1978), which demonstrated that species with a higher I_k than *Asterionella* (*Synedra radians*, *Melosira italica*) remain more severely light-limited in well-mixed columns, but sustain a higher P_{max} as transparency and/or day length increases and when the growth of low-light adapted forms is saturated or on the verge of inhibition. *Tabellaria fenestrata* had a lower overall P_{max} (and k) than any of the other species but, characteristically, its photosynthetic rate stabilized at the higher light intensities apparently inhibitory to the low-light adapted species (although *Asterionella* and *Fragilaria* populations developing later in the year were also evidently adapted to maintain a flattened curve of cell division rate over a wide range of incident daily radiation). These observations fitted the observed sequence of fastest-growing species, which moved from *Asterionella*/*Fragilaria* to *Synedra*/*Melosira* and, eventually, to *Tabellaria*.

That Knoechel & Kalff (1978) observed growth rates of *Tabellaria* to be sensitive to nutrient supply is consistent with the wide consensus of many studies of nutrient-limitation, reviewed in Chapter 5. In particular, application of Tilman's (1977) resource-based competition theory to limiting resource-ratios (§5.6.1) provides a strong basis for selection between potential alternative dominants. Thus, other factors being equal, *Asterionella* would be expected to out-compete *Cyclotella meneghiniana* at high (> 100) Si:P molecular ratios but to fare less well when the ratio is low. Certainly, in Windermere (where Si:P at the initiation of vernal increase is frequently in the range 100–1000) *Asterionella* usually dominates over *Cyclotella* spp. even though both species are usually found to increase simultaneously through the same vernal period (March–June), during which the dissolved silica level is severely depleted (Si:P reduces). On the other hand, there is no clear evidence that *Cyclotella* continues to increase significantly after the *Asterionella* maximum, so that the relationship is much less clear than that detected in Lake Michigan (Kilham & Titman, 1976): this may be because, whatever the extent of P-limitation on the growth rate of either species, the removal of silicic acid from solution is absolutely limiting *Asterionella* and insufficient remains to sustain any perceptible growth of the *Cyclotella* before it, too, becomes limited. Attempts to regulate artificially the Si:P ratio in the Lund Enclosures in Blelham Tarn (a lake in which *Cyclotella meneghiniana*, though present, is however, relatively, unabundant) have not greatly affected vernal dominance by *Asterionella* (Lund & Reynolds, 1982): a low phosphorus content restricted *Asterionella* growth rate but the final yield could still be large if the silicon was there to support it (see also Reynolds & Butterwick, 1979), while growth of *C. meneghiniana* was insignificant.

On the other hand, fertilization with phosphorus (Si unchanged; P increased; Si:P low) has not altered the basic pattern of *Asterionella* dominance, though enhanced growth rates of *Stephanodiscus* species, belonging to the *S. hantzschii* complex (see note in Lund, 1981) were observed. This seems consistent with observed diatom dominance in small, shallow, phosphorus-rich lakes where closely-related species, together with *Diatoma* and *S. astraea* tend to be seasonally most abundant. The performance of *S. astraea* in Crose Mere (see Reynolds 1973*a*, 1978*b*) suggests that its net growth rate, though rarely exceeding those of contemporary *Asterionella* populations (k_n 0.11 – 0.12 d^{-1}), is maintained over a similarly wide range of (winter – spring) temperatures and at similar average radiation inputs; *S. hantzschii* increased at an average rate of 0.28 d^{-1} in the same lake during February 1968, coincident with unusually high P levels (> 150 μg P l^{-1}). The ability of both species to contribute to and even dominate the vernal biomass of such eutrophic waters may be related to the higher nutrient levels.

When Lund Enclosures have been continuously isolated from Blelham Tarn through the preceeding summer and winter and left unfertilized throughout (giving low Si, low P and low Si:P, similar to waters of nearby P-deficient oligotrophic lakes), the (small) vernal diatom maximum continued to be dominated by the same species present in the Tarn, viz: *Asterionella formosa*, *Melosira italica* and *Cyclotella pseudostelligera*, rather than by the *Cyclotella comensis/Rhizosolenia eriensis* assemblage represented in the more oligotrophic lakes and, indeed, in the Blelham Tarn of 25 years beforehand (Lund, 1978). The 'failure' to mimic the 'oligotrophic' diatom succession in the Lund Enclosure was attributed to the insufficiency of suitable inocula relative to those of the more recent populations in the Tarn and the Enclosure (Lund & Reynolds, 1982). Apart from suggesting that the *Asterionella–Melosira italica–Cyclotella pseudostelligera* assemblage of species has now largely eclipsed the *Cyclotella comensis/Rhizosolenia* groups as the eutrophication (P-enrichment) of the Tarn has proceeded (Lund, 1978), the observation also emphasizes the importance of the overwintering stocks in influencing the outcome of interspecific competitiveness, through their individual abilities either to maintain growth under physically-limiting winter conditions of light and water movement or to produce benthic overwintering stocks of 'resting' cells (e.g. *Melosira* spp.: Lund, 1966*a*; *Stephanodiscus astraea*: Lund & Reynolds, 1982). The availability of 'inocula' may thus be at least as important a factor conditioning short-term compositional responses of phytoplankton communities as those attributable to light, temperature or nutrient ratios.

A further factor which must be taken into account is interspecific differences in specific loss rates. Although grazing will be of minor importance and outwash will affect most species by a similar quantity, death rates (especially where these are mediated by species-specific parasites) may differ interspecifically and, largely, unpredictably. Sinking loss rates should be small in relatively well-mixed columns but they nevertheless differ between species having different sinking rates. 'Heavy' diatoms (*Melosira*, *Stephanodiscus astraea*) will always suffer greater relative losses than those of lighter species; the latter may be increased under conditions of physiological stress, which, moreover, will not be simultaneously experienced by all species present. Here may be instanced the comparative reactions to exposure to high light intensities of low-light adapted *Asterionella* and *Tabellaria* populations discussed by Knoechel & Kalff (1978; see above) or the evidently more acute sensitivity of *Fragilaria crotonensis* sinking rates to high $z_m:z_{eu}$ ratios, than (say) those of *Asterionella* (Reynolds *et al.*, 1983*b*), which perhaps go some way to explaining the relatively poor performance of the former in vernal eutrophic assemblages.

Several other 'algal' groups are conspicuously represented in vernal assemblages, although, as a generalization, instances of their habitual dominance are rare. In many lakes, assemblages are complemented by small chlorophyte genera such as *Chlorella* and *Ankistrodesmus* (in mesotrophic or mildly eutrophic lakes) or *Scenedesmus*, *Oocystis*, *Tetrastrum* and *Elakatothrix* (in eutrophic lakes) and, more ubiquitously, by various *Cryptomonas* spp. To varying extents, 'algae' abundant later in the year may also make their initial appearance during the vernal period: *Eudorina*, *Pandorina*, *Coelastrum*, and even *Ceratium* were regularly observed to increase during the period of *Asterionella*/*Stephanodiscus* dominance in Crose Mere and several similar nearby lakes (Reynolds, 1973*c*,*d*, 1978*b*); *Anabaena* and *Aphanizomenon* also often began to increase during March or early April.

In more oligotrophic lakes, colonial chrysophytes (notably *Dinobryon* and *Synura* spp.) sometimes occupy a corresponding niche. *In situ* rates of growth (as opposed to increases attributable to direct recruitment from overwintering benthic stocks) among these species may be slower than those of contemporaneous diatom populations. The hypothesized constraints (see above) or low total insolation and exposure to long periods of darkness upon the smaller species and low temperature on the larger, low *SA*/*V* species, could be the critically controlling factors.

In waters which are either unamenable to diatom growth, owing to (for instance) lack of dissolved silica, or are so shallow and turbid that diatoms sediment rapidly into aphotic sediments but in which relatively high nitrogen and phosphorus concentrations still favour the maintenance of potentially large algal biomass, or again, in lakes where typical diatom growth is for some reason impeded in certain years, these 'other algae' may well become dominant during the vernal period. Examples of vernal dominance by cryptomonads in small eutrophic meres in Shropshire (including facultative dominance of Crose Mere in 1966) and by dense populations of *Scenedesmus* in very (N-, P-) rich ponds (Folly Pool, Shropshire; Priest Pot, Cumbria) have been given by Reynolds (1973*c*,*d*, 1978*b*, 1979*b*) and Vincent (1980*a*). Apart from being favoured by high concentrations of inorganic nutrients, *Scenedesmus* spp. (and perhaps some other colonial representatives of the Chlorococcales) have been shown to be capable of assimilating organic solutes and may be facultative heterotrophs (Algéus, 1950; Berman *et al.*, 1977; Vincent, 1980*b*). Other relatively P-rich ponds are known to be dominated by *Dinobryon* (Lehman, 1976) or *Synura* spp. (Reynolds, unpublished thesis); this is also remarkable in that these algae are traditionally associated with unproductive, P-deficient lakes and *Dinobryon* spp. have been supposed to be restricted to such habitats (Pearsall, 1932; Rodhe, 1948).

So far as it is possible to judge, the rates of specific increase in these cases are not significantly greater than those observed in populations sub-dominant to diatoms, under similar light, temperature and nutrient conditions: comparative inocula and the relative performance of diatoms under these conditions again appear to be critical. The later in the year that the vernal growth is initiated, when day length, radiation intensity and water temperatures will likely all be higher, the more the competitive balance is likely to move away from diatoms. As the increase gets under way, however, and the phytoplankton biomass increases, so light penetration decreases relative to the (assumed) constant depth of mixing. The increased turbidity will offset the increasing insolation at the surface to the extent that the mean photoperiod of suspended algae is not increased and light limitation may continue to apply, yet the extremes of radiation intensity perceived between the surface and the aphotic reaches of the water column will increase. Such conditions would be expected to apply most rigorously in richer waters supporting potentially large 'algal' biomasses, and will presumably favour the continued dominance of low-light adapted species. An interpretation of this sort may be advanced to account for the vernal dominance of lakes by *Oscillatoria* spp.

Planktonic *Oscillatoria* spp. have a wide distribution among temperate lakes but gas-vacuolate species of the *O. agardhii–O. rubescens–O. prolifica* group are especially prominent in (and frequently dominate) two quite distinct types of lakes (Reynolds & Walsby, 1975). One is the mildly eutrophied large, deep, usually alpine basin, exemplified by Zürichsee (Thomas, 1949, 1950; Pavoni, 1963), and Vierwaldstättersee, Switzerland (Zimmerman, 1969), Lago Maggiore, Italy (Ravera & Vollenweider, 1968), Gjersjøen, Norway (Faafeng & Nilssen, 1981) and Lake Washington, North-west USA (Edmondson, e.g. 1970). The other is the shallow, enriched, unstratified basin, examples of which cover a wide range of basin areas: Lough Neagh, Northern Ireland (Gibson *et al.*, 1971; Foy *et al.*, 1976), Sjön Norrviken, Sweden (Ahlgren, 1971), various small Dutch lakes (Drontermeer, Veluwemeer, Wolderwijd: Berger, 1975) and the ornamental lake in St James' Park, London (Whitton & Peat, 1969). The species dominating the latter group include *O. agardhii* and/or *O. redekei*; *O. agardhii* (together with its variety *isothrix*) and *O. rubescens* are characteristic of the first group. In addition, *O. agardhii* and *O. rubescens* may be abundant during parts of the year in small, seasonally-stratified basins in alpine (e.g. Wörthersee, Klopeinersee, Austria: Findenegg, 1947, 1966*b*) and lowland districts (e.g. lakes of Holstein, W. Germany: Overbeck, 1968; Indiana: Eberley, 1959, 1964; and Minnesota: Brook, Baker & Klemer, 1971). Characteristically, the *Oscillatoria* populations in these lakes tend to form stable depth-maxima in the summer metalimnion, provided that this is located within the euphotic zone, where the populations are able to maintain themselves or even grow slowly (see Reynolds & Walsby, 1975; Klemer, 1976). Growth may be more rapid when the lakes are isothermally mixed in the autumn (Lund & Reynolds, 1982; Reynolds *et al.*, 1983*b*).

To understand this apparently discontinuous pattern of distribution, it is necessary to take account of the extensive investigations of the physiological characteristics and growth conditions of *Oscillatorias* that are available (Meffert, 1971; Foy *et al.*, 1976; Ahlgren, 1978; van Liere, 1979). *Oscillatoria* spp. (of the *agardhii–rubescens* group) evidently possess an extremely effective light-capturing mechanism across a wide band of the visible spectrum. Chlorophyll *a* content is extremely variable (see Table 4), ranging in *O. agardhii* photostat-cultures (at 20°C) between 0.3 and 0.4% of dry weight, when grown at $> 110\ \mu E\ m^{-2}\ s^{-1}$, and 1.3–1.7% when grown at $< 28\ \mu E\ m^{-2}\ s^{-1}$ (see p. 76 of van Liere, 1979). The content of the accessory pigment, C-phycocyanin (2–11% dry weight) is similarly variable; moreover, some *O. agardhii* strains and *O. rubescens* also contain quantities of the red pigment C-phycoerythrin. Alterations in the relative proportions in which the pigments are synthesized occur in response to changes, with time and with depth, in the spectral composition and intensity of irradiance (*chromatic adaptation*). Secondly, the transfer of photosynthetically-fixed carbon to growth is particularly efficient at low-irradiance levels, while the energy requirements for maintenance seem remarkably low (equivalent to about $5\ \mu E\ m^{-1}\ s^{-1}$); at 20°C and under continuous illumination in photostat cultures, the growth rate (k') of *O. agardhii* increases linearly to $\sim 0.7\ d^{-1}$ at $\sim 47\ \mu E\ m^{-2}\ s^{-1}$, becoming rapidly saturated (light-independent: cf. §6.1.2) at $0.85\ d^{-1}$ at $60\ \mu E\ m^{-2}\ s^{-1}$ and sharply photo-inhibited above $\sim 185\ \mu E\ m^{-2}\ s^{-1}$ (p. 36 of van Liere, 1979). At lower temperatures, the growth rate becomes increasingly L_i-limited (Q_{10}: 3–4) and saturates at progressively lower irradiances (at 5°C, $k' = 0.05$, $I_k \sim 17\ \mu E\ m^{-1}\ s^{-1}$: see Figure 68).

These observations indicate that the *Oscillatoria* spp. in question are essentially adapted as 'shade-plants' that are particularly favoured by conditions of low irradiance and short day length. That they are also able to achieve their maximal growth rates at relatively low nutrient availabilities (K_s [P]: 0.03 μM; K_s [$(NO_3)^{-1}$]: 1.2 μM; van Liere, 1979) potentially enables them to inhabit waters over a wide range of trophic status. Maintenance of approximately neutral buoyancy and a virtual immunity of trichomes exceeding 100 μm or so in length to ingestion renders them liable to generally low rates of loss.

It is arguable that it is the coupling of these physiological and morphological characteristics that contributes to their simultaneous success in both large, deep, mildly-eutrophied alpine lakes and highly-enriched shallow lakes. In both instances, the depth of mixing (z_m) relative to light penetration (z_{eu}) generates a low average perceived insolation owing to relatively long dark periods. Thus, vernal *Oscillatoria* growth is potentially initiated before that of any other major competitor, while their increasing biomass (increased turbidity) may so offset the effects of increasing day length and incident irradiance on the 'light climate' that the increase of faster-growing competitors continues to be severely light-limited.

In the shallow lakes, this delay can be indefinite. For although the near-surface irradiances well exceed 1000 μE m^{-2} s^{-1}, the duration of exposure may be so short that neither photosynthesis nor growth is necessarily inhibited. At a similar level of vertical diffusivity in the mixed layers of deeper lakes, a lower vertical attenuation of light does lead to increasing risk of photo-inhibition. Once the lake stratifies ($z_m \sim$ 10–20 m), the risk is directly increased. A compensatory reduction of pigment content may improve the prospects for survival, but *Oscillatoria* spp. will normally invoke a much more effective mechanism of buoyancy regulation: by first reducing buoyancy, they will sink through the epilimnion until they reach a depth to which perhaps only 0.5–3% (< 35 μE m^{-2} s^{-1}) of the immediate sub-surface irradiance (I'_0) penetrates. Provided that this depth falls within the range of the metalimnion (where density gradients and hence stability, are greatest), the *Oscillatoria* trichomes will increase their relative gas-vacuole content to the point of approximate neutral buoyancy, and remain stratified there (Reynolds & Walsby, 1975). They may make further compensatory adjustments in buoyancy, depending upon changes in light penetration and nutrient levels (Walsby & Klemer, 1974; Klemer, 1976). There may be continued growth but the principal function of *Oscillatoria* statification appears to be as a mechanism for survival of the summer period ('*aestivation*') until more suitable hydraulic conditions obtain in autumn. Similar behaviour is observed in many smaller, thermally-stratifying and moderately eutrophic continental-type lakes (for references, see Eberley, 1959; Reynolds & Walsby, 1975; Klemer, 1976).

Two major ecological factors may interfere with the seasonal biology of *Oscillatoria* spp. and exclude their dominance from many more stratifying-lake systems. One is that, in many smaller lakes (especially those near oceanic seaboards, which may become alternately microstratified to within a few centimetres of the surface for days on end, and then be wind-mixed to several metres) the formation of stable metalimnetic gradients within the euphotic zone may be too erratic for the *Oscillatoria* trichomes to stratify for the summer. The situation is excerbated in more eutrophic waters, wherein larger epilimnetic 'algal' populations habitually wax and wane, adding another dimension to environmental variability perceived by the stratified *Oscillatoria*, namely that of changing light penetration. The response of the *Oscillatoria* is to adjust its buoyancy appropriately moving upwards or downwards in the water column; the combination of low light penetration and epilimnetic deepening may well introduce the *Oscillatoria* population into the upper euphotic zone (where its growth may also be temporarily stimulated) and where it may become subsequently stranded, overbuoyant and exposed to damaging irradiances, as the water column stabilized and/or clears once more. Examples of populations behaving in these ways have been given by Reynolds *et al.* (1983*b*). The surviving stock may eventually be so heavily depleted that it fails to recruit a competitively effective stock to the autumn or winter (mixed) water column. Such an explanation may account for the progressive elimination of characteristic autumn *Oscillatoria* maxima from Esthwaite Water, Cumbria, as the eutrophication of the lake has advanced (see Lund, 1972*a*).

The second factor operates in small lakes subject to periodically high winter discharges, which inflict high rates of outwash on winter populations of *Oscillatoria*. Such floods may deplete the overwintering stock to a greater extent than can be compensated by the slow growth rate before the competitive advantage moves to (say) diatoms. The effect of winter closure of Lund Tubes on the size of the overwintering *Oscillatoria* populations and the eventual vernal maxima developing therefrom, compared with their equivalents in Blelham Tarn, has been clearly demonstrated by Lund & Reynolds (1982).

In conclusion, it may be appreciated how the interactions of the basic biologies of diatoms, *Oscillatoria* and many other species of 'algae' with the physical (photic, thermal and hydraulic) properties of individual lake basins, not only during the winter–spring period but throughout the year, exert an overriding influence on the phasing, size and species composition of vernal phytoplankton maxima in freshwaters.

8.1.2.2. *The summer period.* The advancing season in temperate regions, with its attendant increases in duration and intensity of solar radiation and net heat input, heralds changes in aquatic ecosystems that, to a greater or lesser extent, come to resemble those already obtaining in low-latitude counterparts. The most immediate effect, that of temperature rise, is such as to increase the relative anabolic rates of almost all 'algae', especially those of the larger, low *SA*/*V* species; not only is the potential rate of increase in the total biomass stimulated but the range of competing species is also potentially expanded. Corresponding effects may be directly attributed to increased incident insolation. In theory, however, increases in the standing biomass might 'keep pace' to the extent that, despite the increases in *LDH*, the vertical extinction coefficient (ε_{min}) and hence the optical depth of the mixed water column (see equation 27) continue to ensure that light remains a limiting factor. Otherwise, any increase in the ratio $\Sigma\Sigma NP/24\Sigma NR$ will indeed raise the potential productivity of the 'algal' community as a whole and of the nanoplankton in particular.

This seasonal alleviation of the severe light- and temperature-limitation upon the biomass, productivity and species composition need not, by itself, alter the dominance of later vernal phytoplankton assemblages by already large-pre-existing populations. To the examples of persistence of *Oscillatoria* spp. in optically-deep waters and of *Scenedesmus* in Folly Pool (§8.1.2.1) may be added those of large or dominant diatom populations maintained throughout much of the year in shallow, well mixed lakes such as Loch Leven (Bailey-Watts, 1976) and the Neusiedlersee (Dokulil, 1979) or in the deeper mixed systems of Windermere (Lund, 1950), Millstättersee (Findenegg, 1943), Michigan (Schelske & Stoermer, 1971) and in such tropical lakes as Lake Victoria (Talling, 1965, 1976*b*) and Represa de Broa, Brasil (Tundisi *et al.*, 1977). Although diatoms generally have high intrinsic sinking rates (v': Table 19) the rates of sinking loss (k_s) are kept low in well- or deep-mixed columns (v'/z_m low: see §7.3).

Elsewhere, two or possibly three, major factors are eventually responsible for the net loss of dominant vernal diatom populations and their replacement by species representative of other algal groups. One is the exhaustion of nutrients, especially of dissolved silicon; the second is the onset of thermal stratification; the third may be excessive losses, separately attributable to grazing by zooplankton (whose activity, growth and reproductive rate respond to increasing temperature and food availability) or to simultaneous parasitic attack. These factors interact with each other, and not always to the immediate detriment of diatoms. The isolation of the epilimnion from the deep waters below leads directly to the localization of further insolation and temperature increases within the relatively restricted confines of the upper part of the water column: within it, the stimulatory effect upon the growth of phytoplankton able to remain there is correspondingly magnified. This applies equally to diatoms, so long as the absolute depth of the epilimnion and the intrinsic sinking rate of the diatoms are such to ensure that $k_s \ll k'$. However, the 'physiological shock' to which hitherto deep-mixed, low-light adapted diatoms may be subjected in high-insolated, shallow epilimnion (low z_m) can lead simultaneously to a 3–4-fold increase in intrinsic sinking rate (increasing k_s) and virtual collapse of k'. The more abrupt the transition in limnological conditions, the more severe is the effect and the more rapid is the loss of diatoms from suspension (cf. Round, 1971). At the same time, the confinement of an actively-growing phytoplankton (with or without the participation of diatoms) to an epilimnion in which the growth of herbivores and pathogens is also stimulated and to which the upward vertical hydraulic transport of dissolved nutrients fails to overcome the rate of nutrient uptake therein and removal of sedimenting biogenic particles therefrom determines that the medium becomes increasingly 'segregated' in the vertical plane. At depth, light and, to some extent, temperature continue to effect the principal constraints on algal growth; in the epilimnion, where these factors potentially favour the maintenance of growth and high biomass levels, sinking and grazing loss rates and, sooner or later, nutrient depletion now assume more critical roles in controlling community composition.

Among the species (Figures 43, 74, 75) that respond to increasing structural stability in temperate lakes are many nanoplanktonic Chlorococcales (e.g. *Ankyra*, *Chlorella*, *Crucigenea*, *Monoraphidium*, *Tetrastrum*), chrysophytes (e.g. *Chrysochromulina*), and cryptophytes (*Cryptomonas*, *Rhodmonas*), larger colonial Volvocales (*Volvox*, *Eudorina*, *Pandorina*), and other chlorophytes (*Sphaerocystis*, *Gemellicystis*), colonial (*Dinobryon*, *Uroglena*) and unicellular chrysophytes (*Mallomonas*), filamentous (Nostocales: *Anabaena*, *Aphanizomenon*, *Gloeotrichia*) and colonial cyanobacteria (Chroococcales: *Microcystis*, *Gomphosphaeria*, *Coelosphaerium*) and 'large', unicellular dinoflagellates (*Peridinium*, *Ceratium*). In different ways, their morphological and behavioural adaptations counter large and irretrievable sinking losses through microstratified layers: small size, density 'reduction' (mucilaginous colonies) and, especially, motility (flagellates) or the control of buoyancy should be seen primarily as mechanisms prolonging the residence times of 'algal' populations in the upper illuminated layers. In addition, circumstances may well arise ($z_{eu} > z_m$) in which the flagellates and buoyancy-regulating cyanobacteria are able to maintain approximate vertical stations offering optimal combinations of light and limiting-nutrient concentrations (§3.3), or even to migrate up or down through stable layers on diel cycles.

A rapid transition from full mixing to stable thermal stratification and the elimination of diatom-dominated assemblages, such as occurs in smaller temperate lakes ($z_m \rightarrow < 4$ m: see Figures 24, 74), in theory, opens the habitat to the development of representatives of all the above assemblages. To some extent, this may occur: in the Lund Enclosures, for example, it is frequently the case that the growth rates of cryptomonads, *Eudorina*, *Sphaerocystis*, *Uroglena* or *Dinobryon*, *Anabaena*, and *Ceratium* all increase approximately simultaneously (e.g. Reynolds *et al.*, 1983*b* and unpublished; see also Figures 70–72). However, their attainment of large or dominant populations (if at all) is subject to the same pronounced and generally predictable periodic sequences observed in natural lakes (cf. Figure 43).

Several factors contribute to this sequencing. The most obvious of these is the comparative rates of increase, net of losses (k_n): given that light levels and nutrients are initially saturating and the 'algae' are able to grow at their maximal (temperature-, photoperiod-limited) rates, we may confidently expect that nanoplanktonic forms will build up their numbers relatively more rapidly ($k' \geqslant 0.6$ d^{-1}) than (say) *Eudorina* ($k' \sim 0.4$–0.5 d^{-1}) and, in turn, (say) *Sphaerocystis*, *Dinobryon*, *Anabaena* and *Ceratium*. That this sequence resembles that with which these same genera are abundant in lakes is probably not coincidental but the former cannot be held to be the cause of the latter. It might equally be that the growth rates of the later dominants are lower because they are, by then, limited by other factors. Moreover, it would be quite wrong to give the impression that species B continues to increase at a steady but slower rate throughout the period in which first A grows faster, then stops and disappears: through each stage, changes in relative light penetration and nutrient availability would be expected to influence *in situ* rates of growth of all the species present.

Nevertheless, the effects of different specific rates of loss do influence the relative abilities of the 'algae' to compete for dominance. In the same way as intolerably high sinking losses are primarily responsible for the elimination of diatoms (see above and §7.3), so removal by filter-feeding zooplankton might be expected to limit the net rates of increase of the smaller algae and, perhaps, overcome them. Thus, it is conceivable that the increase of 'species A' (in this instance, an edible 'alga') is reversed before it becomes abundant and before the increase of 'species B' (in this case, an ungrazed 'alga') is significantly impaired by deteriorating photic conditions, so that 'B' soon becomes the dominant species. Even this relationship is complicated by the fact that the zooplankton feeding is not a passive regulator; filter-feeding scarcely becomes a significant regulator of nanoplankton growth until the nanoplankton has first been sufficiently abundant (equivalent to > 0.08 μg C ml^{-1}) to sustain the growth of the filter feeders (§7.5.2; 7.5.3). The relationship is also influenced by the relative magnitudes of k_g and k'. If k' is high, then k_g has further to go to 'catch up', i.e. a larger standing stock of filter feeders has to be established, whereas in the interim, the magnitude of k_n may allow the 'alga' to produce a large, light-limited maximum. In this way, the phase of potential dominance by nanoplankton may vary between a prominent biomass maximum (supported by a high k_n) and a long early post-stratification phase of low biomass (k_g rapidly overtaking a lower k') during which the biomass of larger, ungrazed but slower-growing ($k_n \simeq k'$) species progressively develops.

A second (but no less important) factor influencing the sequence of dominance by summer algae is the relative sizes of inocula recruited at the time of stratification. The generalized equation (42) of population dynamics will never permit N_t to be large if N_0 is initially very small, regardless of the value of k, although the constraint is diminished the longer or the more frequent are the periods, t, during which positive k_n can be maintained. 'Colonization' of the newly-formed epilimnion by species that rely mainly or exclusively on maintaining dilute, vegetative populations in open water is likely to be subject to chance events (survival of winter floods, arrival in the lake) and, therefore, to be less predictable in occurrence. Equally, the investment in rapid growth rates will facilitate the rapid colonization of suitable environments by the 'available' species.

In contrast, colonization by species that overwinter as morphologically or physiologically distinct propagules should be relatively more assured and, hence, more predictable. Events subsequent to the establishment of suitable epilimnetic growth conditions will depend upon the proportions of viable propagules (itself dependent upon the successful completion of the life-cycle in previous summers), the nature of the stimulus to 'germinate' (remembering that the majority of propagules are not exposed to the same set of supposedly favourable conditions obtaining in the epilimnion) and upon the relative numbers of propagules which can respond in this way. In many instances, the full annual cycles of 'algae' in lakes are known too incompletely to permit thorough evaluation of the role of overwintering. It does appear that *Ceratium* spores excyst during the vernal, mixed period (Heaney *et al.*, 1981) and *Anabaena* akinetes germinate in open water, having been resuspended from the sediments during full isothermal circulation, provided that certain minimal temperature and photoperiod conditions are satisfied (Reynolds, 1972). In both cases, suspended stocks of vegetative cells probably already exist at the time of stratification and, superficially, are alone sufficient to account for the base (N_0) of summer growths.

In *Microcystis*, however, vegetative colonies are recruited to the epilimnion after growth has been initiated in benthic overwintering colonies, apparently in response to the penetration of light to micro-aerophilic or even completely anoxic deep-water sediments (Reynolds *et al.*, 1981). In many stratifying lakes that support *Microcystis* populations, these stimulatory conditions will occur only after the lake has already become thermally and chemically stratified. This must mean that *Microcystis* is always likely to be a late entrant in the summer sequence; as its maximal growth rate is typically low ($k' < 0.25\ d^{-1}$), in part compensated by potentially substantial recruitment, its population maxima are also likely to occur in the later part of the summer.

In all three cases, however, the 'algae' are likely to be prominent in given lakes if they have been successful over a number of previous seasons in building up appropriate resources of propagules and seasonal variations in the characteristics of the ambient growth medium continue to furnish suitable growth conditions during part of the subsequent year. Equally, a number of 'poor years', perhaps due to long-term environmental changes, lessens the prospects of large populations in the future, simply because the fund of propagules is weakened.

The differences in performance between fast-growing species sensitive to high rates of loss and slow-growing, loss-resistant species producing specialized overwintering propagules emphasize the contrasting roles of *r*- and *K*-selection in planktonic community composition (see §6.4). A general statement describing the species-sequence of summer phytoplankton is that because biomass of 'algae' able to utilize the resources of light and nutrients available at the onset of stratification is low, the ratio of demand to supply favours an initial colonization by populations of ephemeral, opportunist *r*-selected species. Sooner or later, as biomass and demand increase, resources (be they light or nutrients) decrease and exploitive loss processes become more prevalent, so the more conservative, slower growing *K*-strategists are progressively selected. The lower is the D/S ratio, then the more prolonged (in terms of standing biomass attained) is the period of *r*-selection.

Resource gradients also exert a powerful selective role among the *K*-strategists and must be considered as the third main component conditioning summer compositional sequences. However, neither the gradients themselves nor the comparative performances of individual species along those gradients have yet been adequately researched to permit the interpretation of either seasonal changes in dominance in particular lakes or the differences between characteristic sequences that occur in different types of lakes. Nevertheless, sufficient pointers exist to permit speculation about the possible mechanisms. The broad differences between the meso- and eutrophic sequences imply that the distinctions are dependent upon different gradients of nutrient availability in either kind of lake. Moreover, since phosphorus is often the critical regulator of biomass in the less eutrophic waters, it follows that species dominance is influenced by P. Thus, the chrysophyte-, colonial Chlorococcales- and dinoflagellate-dominated assemblages might be more tolerant of low ambient P-levels than the Volvocales- and Nostocales-dominated groupings of eutrophic, high-P lakes. Even without experimental evidence of the half-saturation constants of P-limited uptake and growth, it seems unlikely that these assemblages should apparently be so rigorously partitioned into one or other of the categories. Yet when the Lund Enclosure B was deliberately manipulated so that, after stratification, it would furnish a declining gradient of soluble reactive P, representatives of all these assemblages appeared successively in its plankton, in descending order of their k'_{max} values and even culminated in a small *Ceratium* maximum (see Figure 72).

Similarly, if nitrogen becomes the major growth-limiting factor, a clear response is predictable: that, of the species confined to the upper reaches of the water column, only the nitrogen-fixing cyanobacteria (Nostocales) will be expected to maintain significant positive growth. In many naturally-eutrophic lowland lakes, especially in regions of glacial deposition, species of *Anabaena*, *Aphanizomenon*, and *Gloeotrichia* do tend, sooner or later, to become dominant. In Crose Mere, the change from dominance by Volvocales to dominance by Nostocales follows in the wake of N-depletion to below 0.3–0.4 mg N l^{-1} (Reynolds, 1978*b*). The nitrogen-fixing forms commence their seasonal growth approximately simultaneously with *Eudorina*, *Pandorina* and (occasionally) *Volvox* spp. but the latter dominate first, presumably because they maintain relatively high rates of growth (k': 0.4–0.5; cf. 0.2–0.3 d^{-1}). This appears, then, to be a distinctive response along a critical gradient of declining nitrogen availability, to which phosphorus is saturating. It may be noted that in lakes severely deficient in P and N (oligotrophic lakes) nitrogen-fixers are usually conspicuously absent, as are the majority of the other summer assemblages, save for the sparse representation by dinoflagellates. As P is depleted from moderately eutrophic, N-deficient epilimnia, it is the dinoflagellates (especially *Ceratium* in temperate regions, *Peridinium* spp. at lower latitudes) which often become dominant. Alternatively, the dominance of nitrogen-fixers may be supplemented or replaced by *Microcystis*, whose cell quotas of P and N, imported during the epilimnetic recruitment phase, sustain prolonged growth. Equally, if neither N nor P ever becomes limiting the continued dominance of colonial Volvocales and Chlorococcales might be anticipated. Some formerly N-deficient, P-rich lakes in Cheshire are now more continuously dominated by chlorophytes in this way (Reynolds, 1978*b*).

To these examples of 'sub-sequences' apparently regulated by interacting gradients of P and N, may be added the responses along gradients of other 'nutrients'. One calling for special mention is carbon (as carbon dioxide, in its various forms). Talling's (1976*a*) experiments ranked five species of phytoplankton (*Melosira italica*, *Asterionella formosa*, *Fragilaria crotonensis*, *Microcystis aeruginosa*/*Ceratium hirundinella*) in order of photosynthetic- and growth-tolerances of high pH and low CO_2 concentrations. This principle might be applied further to the species assemblages presently under consideration, although the only evidence to hand comes from (largely unpublished) data from the experiments in the low-alkalinity water of the Lund Enclosures. The chrysophytes generally seem most sensitive in this respect, ceasing growth and forming resting cysts at pH values > 8.5 or so. The growth rates of volvocalean species of *Eudorina* and *Volvox* apparently become inhibited and sexually-reproduced resting zygotes are often induced when pH exceeds 9.5 but both *Microcystis* and *Anabaenas* (of the *flos-aquae* species-complex) can continue to grow above even this level. These observations are in line with those of Shapiro (1973), whose experiments showed that diminished supply of CO_2 selected against chlorophytes and in favour of cyanobacteria. However, we have occasionally observed the growth of *Sphaerocystis schroeteri* (unpublished data) and *Staurastrum pingue* (Reynolds & Butterwick, 1979; see also the comments of Brook, 1981) at pH values in the range 9.5–10.5. Limiting carbon gradients might therefore be expected to exert some influence on specific growth and dominance sequence in enriched soft-water lakes, though less so in hard-water lakes that maintain more abundant reserves of bicarbonate and impart an important buffering effect against high pH (range 8.2–8.8). If the observation (above) that pH > 8.5 adversely affects the growth of chrysophytes is correct, bicarbonate buffering of hard-water lakes, regardless of the availabilities of other nutrients, might be responsible for their apparent absence from many such lakes, hitherto sometimes attributed to P-toxicity (see also Lehman, 1976).

Finally, one other gradient potentially selects among summer *K*-strategists: light. The depth to which a water column is regularly mixed (z_m) is influenced mainly by heat income and exposure to wind and other mixing processes and is largely independent of the phytoplankton density, to which light penetration is so strongly coupled. Thus, the increase in biomass of *K*-selected species, resistant to sinking and grazing removal, leads directly to an overall diminution in the euphotic depth (z_{eu}) and, on average, to a deterioration of their light-climate, as defined by the z_m/z_{eu} ratio. The higher are the external concentrations of limiting nutrients, then the possibility of populations becoming 'self-shaded' is correspondingly increased. The high-frequency fluctuations in perceived light that accompany a z_m/z_{eu} ratio much exceeding unity are ultimately counter-productive to all 'algae', though some species appear more sensitive than others in this respect. Indeed, in suitably enriched lakes, a persistently high z_m/z_{eu}

ratio favours the dominance of either *Oscillatorias* or certain non-mucilaginous colonial Chlorococcales (*Scenedesmus*, *Pediastrum*, *Coelastrum*; see §8.1.2.1). At the other extreme, the growth of *Eudorina* and *Sphaerocystis* appears to be particularly sensitive in increases in z_m/z_{eu} (Figures 73, 74). The buoyancy-regulating cyanobacteria and the large motile dinoflagellates, on the other hand, are, in theory, well-equipped to contend with a fluctuating mixed depth, being able to make rapid compensatory movements. This self-regulating behaviour is of considerable advantage, once cells are already abundant, and will work against their potential competitors. Migratory movements are of additional advantage in maintaining favourable nutrient-uptake gradients at the cell surface and if, as is often supposed, they also give access to nutrient resources located at depth that may be largely unavailable to their competitors (Reynolds & Walsby, 1975; Reynolds, 1976*c*). These considerations go some way to explaining the large biomasses that these 'algae' achieve towards the end of summer in temperate lakes (sometimes the largest achieved during the entire year) when many of their competitors have ceased to grow.

The provisional conclusion to be drawn is that the several resource gradients that exist, both in time and in space, are each likely to favour the growth of particular algae in 'sub-sequences' that contribute to the overall sequencing of dominance through the summer period. The observable sequence in given lakes may be that determined by one overriding resource gradient (say extreme P-deficiency) or by a series of sub-sequences imposed when another limiting resource becomes more critical (say carbon or light limitation). These possibilities are summarized by reference to the 'tolerances' of critical factors by the principal 'algae' considered in this account (Table 28). The latter are listed in descending order of maximal *in situ* growth rates, which, if other factors are subcritical, might also approximate to the anticipated periodic sequence. To the right are represented their tolerances of given potential limiting resources. Should one or other of these resources be severely limiting, then the original algal sequence should still be followed but the species less tolerant of limitation by that factor (indicated by a diminished number of + symbols) will tend to be omitted. This illustration is a way of representing the various observations and is in no way intended to explain any particular series of responses. It does, however, fit the observations into a framework which might form the basis of some future matrix predicting the events in the summer phytoplankton of a wide range of temperate lakes. Temperature and grazing effects and entries for nanoplankton and cryptomonads are included for completeness.

Table 28. *The tolerance of various 'algae', representative of summer assemblages and arranged in descending order of* k'_{max}, *to critical limiting factors. Species with high tolerances (+ + +) may be expected to participate in modified periodic sequences (reading downwards) and from which 'intolerant' species may be omitted. For further explanation, see text. Based on data shown in Figures 70 to 75, Reynolds* et al. (1983*b and unpublished*).

	Growth rate tolerance of:					
Species	Low temperature (< 10 °C)	High grazing (a > 0.6 d^{-1})	Low [*P*] (< 0.4 $\mu g\ l^{-1}$)	Low [*N*] (< 300 $\mu g\ N\ l^{-1}$)	Low [*C*] pH > 9.5	High z_m/z_{eu} (> 1.0)
r-selected nanoplankton	++	·	·	·	++	+
Cryptomonas	++	·	++	+	+	+++
Eudorina	+	+++	·	+	+	·
Sphaerocystis	·	+	+++	·	++	·
Dinobryon, Uroglena	++	++	+++	+	·	++
Anabaena	+	+++	·	+++	++	+
Microcystis	·	+++	+	++	+++	++
Ceratium	·	+++	++	++	+++	++

8.1.2.3 *The autumn period.* The characteristic diminution in day length and in radiation intensity after the summer solstice eventually leads to a net seasonal cooling of the atmosphere in temperate latitudes. Concomitantly, the frequency of penetration of polar air masses into the dominant atmospheric flow patterns at lower latitudes, and of the pressure changes and frontal activity that this penetration generates, also begins to increase; near oceanic seaboards, at least, the likelihood of episodes of strong winds increases proportionately.

In turn, these seasonal climatic variations affect the environmental characteristics of limnetic systems, through a number of interrelated processes. The drop in mean air temperature leads to net heat losses from lake surfaces and, eventually, to thermal instability: cooled water near the surface must sink and be replaced by now-warmer water from below, by *convection.* As convectional cooling proceeds, so water from increasingly greater depths in the column becomes involved in the surface flow: in deeper, hitherto-stratified lakes, the volume of the epilimnion is increased and the metalimnion is depressed. Moreover, as the mean epilimnetic temperature approaches that of the hypolimnion, metalimnetic gradients are weakened, until they disappear altogether. Equally, any increase in wind stress accelerates the rate of heat interchange at the surface, while the increased mean velocity of the circulation increases shear at the metalimnetic surface, leading to internal wave formation, (Kelvin-Helmholtz) instability, 'erosion', and entrainment of the water of the erstwhile metalimnion. Even when the wind drops, internal *seiche* movements, compounded by horizontal rotation (attributable to Corioli's forces), continue to de-stabilize the limnetic thermal structure.

The comparative contributions of convectional cooling and increased wind-generated water movements vary according to geographical location, size and mean depth of the lake itself and its exposure to wind. In some deep, well-sheltered lakes, the vertical transfer of wind-induced turbulent energy may never be sufficient to be wholly responsible for complete isothermal mixing (*holomixis*) of the water-column which is ultimately due to the convectional collapse of the metalimnetic density gradient. Accordingly, the point at which the water body becomes isothermally mixed usually occurs as late as November or December in Rostherne Mere and Windermere but in late October in Blelham Tarn and mid-September in Crose Mere, despite the similarity of latitude in the respective locations (see Table 29).

So far as the environments of the phytoplankton are concerned, the most important effect of the progressive breakdown in stratification is the

Table 29. *Approximate dates of the onset of column holomixis in some British lakes of differing depth and surface area, located between 53 and 55°N*

	Windermere (North Basin)	Rostherne Mere	Blelham Tarn	Crose Mere	Lough Neagh
Area (km^2):	8.05	0.49	0.10	0.15	367
Max. depth (m):	64	27.6	14.5	9.1	34
Mean depth (m):	25.1	13.3	6.7	4.8	8.9
Isothermally mixed:	December	November	October	September	all-year[(a)]

(*a*) According to Gibson *et al.* (1971). In the summer of 1976, an unusually hot one in Britain, Lough Neagh stratified briefly.

increased optical depth of the water column. Even with no change in turbidity, both the photoperiods and the total insolation of suspended 'algae' become reduced. These factors are exacerbated by the shortening day length and the reduction in radiation intensity falling on the lake surface. Other factors being equal, the standing populations become diluted in the increased circulation, so that the coefficients of light attenuation due to the presence of 'algae' (ε_a) are reduced and, ignoring the effect of the extinction owing to the (increased) depth of water (ε_w), the euphotic depth (z_{eu}) is correspondingly increased. However, the resuspension of fine sediments, the *in situ* precipitation of dissolved metal ions entrained from the anaerobic hypolimnia of eutrophic lakes (especially of $Fe^{2+} \rightarrow Fe(OH_3)\downarrow$) and, most importantly, the additional production of phytoplankton that may be stimulated by the abrupt reintroduction of limiting nutrients (e.g. SRP, NH_4^+) into surface waters all contribute to an increase in turbidity and a further deterioration in the light climate.

Owing to the weather-related occurrence of the mixing episodes, these various changes are experienced more in a stepwise sequence than as a gradual process. However, their impact increases as the summer progresses: at each stage, the planktonic environment increasingly comes to resemble that of the vernal mixed period, with the very important difference that the temperature is several degrees warmer. This means that, although the vernal assemblages might be expected to benefit at the expense of summer ones, selection against the larger, low *SA*/*V* forms should be generally weaker than against smaller, nanoplanktonic species. Moreover, the growth-rate responses should be expected to be more rapid and to be limited by light, rather than by temperature.

Equally, the resistance of the warmer, less-dense water to integration with colder water masses below, that was primarily responsible for the maintenance of the summer stratification, may initially be sufficiently strong to check the early stages of destratification to the extent that the early episodes of post-solstice mixing are contained within a deepened epilimnion and this itself may develop renewed microstratification during subsequent mid- or late-summer anticyclonic phases of reduced wind stress and solar warming. In this way, the selective pressures favouring the growth and, perhaps, dominance of autumn assemblages may become manifest at first spasmodically, alternating with those favouring the summer ones, and finally eclipse them only in the later stages of destratification.

Examples of phytoplankton whose net increase and potential dominance is evidently stimulated by late-summer and early-autumn mixing episodes in temperate lakes are furnished by the diatoms. Generally, the species correspond with those that were abundant in the vernal, prestratified period, that is, those best-placed to provide suitable inocula either as residual suspended populations, or as vital survivors entrained from depth or resuspended from shallow sediments. The floristic distinctions between oligotrophic and eutrophic lakes are really or apparently maintained: *Asterionella formosa*, *Tabellaria flocculosa*, *Melosira italica*, and *Cyclotella* spp. tend to be more prominent among less-rich lakes; *Fragilaria crotonensis*, *Melosira granulata* and *Stephanodiscus astraea* are the more so in richer waters.

Within individual lakes, the relative competitive positions of co-existing diatom species are influenced, as always, by the comparative sizes of their inocula and by their comparative net increase rates. The maintenance of viable benthic stocks of physiologically-resting propagules affords a potential advantage to those species of *Melosira* and *Stephanodiscus* that have been observed to produce them in abundance; provided that these can be returned to suspension for renewed vegetative growth to commence, such recruitment may outweigh the advantages (of superior light-adaptation, nutrient-limited growth kinetics and, hence, of faster rates of growth) apparently conferred upon the pennate genera. Comparative loss rates, determined by specific sinking rates and susceptibility to grazing, further influence the outcome of interspecific competition. Comparisons of net increase rates, the size of standing crops and persistence of summer maxima of *A. formosa*, *F. crotonensis* and *M. granulata* in Crose Mere, shown in Reynolds (1973*a*), suggest that *Asterionella* and *Fragilaria* grow more rapidly and sink more slowly than the *Melosira* and that *Asterionella* and *Melosira* are less sensitive to high turbidity than is *Fragilaria*. Later work (Reynolds *et al.*, 1982*b*) has supported the contention that *Asterionella* is more sensitive to high crustacean filtering-rates than is *Fragilaria* (or, probably, *Melosira*).

Assuming silicon and other potentially limiting nutrients (including carbon) to be readily available, the onset of summer and autumn diatom pulses is regulated primarily by the extent of convective mixing. There is an absolute dependence of net increase rate upon the depth of mixing, that is determined by the relative magnitudes of k' and k_s. Reynolds *et al.* (1982*b*) developed an equation in terms of z_m to describe the net change in near-surface populations of *Fragilaria crotonensis* in the Lund Enclosures, that showed the critical depth of mixing, where $k' = k_s$ ($k_n = 0$), to be 1.25 m. This condition was satisfied by mean values of $k' = 0.20\ d^{-1}$ and $v' = 0.204\ m\ d^{-1}$, although both quantities were considered to be subject to seasonal variations in light and temperature. Indeed, the responses of *Fragilaria* populations to high turbidities (high z_m/z_s) generated in later experiments were shown to detract from observed k_n. A modified equation, which accounted for 64% of the variance in the original, was subsequently developed (Reynolds *et al.*, 1983*b*):

$$k_n = 0.292 \log_{10} z_m - [1/\{1.048 - 0.039(z_m/z_s)^2\}] + 0.972 \qquad (58)$$

The equation may also be invoked to explain the observed poor performances of *Fragilaria* during the later stages of destratification ($z_m > 6$ m) when z_m/z_s exceeded ~ 2.0.

A second group of 'algae' which can respond positively to increased mixing in summer is represented by planktonic desmids, especially species of *Cosmarium*, *Staurastrum*, *Staurodesmus* and *Closterium*. This behaviour may be related to a lack of motility and relatively large size, which, according to the Stokes Equation, confers a fast rate of sinking; the presence of a mucilaginous envelope nevertheless reduces average density, but also lowers form resistance (cf. §2.5). Large size also increases (L_i-) dependence upon higher water temperatures, which possibly accounts for their apparent confinement to summer assemblages. Desmids form a significant part of the plankton in the well-mixed epilimnia of oligotrophic, soft-water lakes, where z_{eu} values are generally high (yielding favourable z_m/z_{eu} ratios), although some species of *Staurastrum* (e.g. *S. pingue*, *S. cingulum*) and *Closterium* (*C. aciculare*) can grow well in eutrophic waters and *S. pingue*, at least, has been shown to be tolerant of high pH values (> 9.5; Reynolds & Butterwick, 1979). Elsewhere, where z_m/z_{eu} is high, owing either to extreme depth or to low light penetration, mixing is conducive to growth by *Oscillatoria* spp. (§8.1.2.1).

Mixing *per se* is not necessarily deleterious to the growth of species associated with microstratified columns. When mixing extends much beyond the euphotic zone ($z_m/z_{eu} > 1$) their growth rates, however, are liable to be depressed or halted. Field experiments, designed to simulate the effect of the sudden imposition of 'deep mixing' (i.e. beyond the euphotic depth), indicated that whereas growing *Sphaerocystis* populations rapidly fell moribund and decreased in numbers, those of *Anabaena*, *Microcystis* and *Ceratium* remained more constant (Reynolds *et al.*, 1983*b*).

The gas-vacuolate cyanobacteria characteristically became more buoyant and more resistant to turbulent circulation, as has been observed in nature (Reynolds, 1973*b*; Reynolds *et al.*, 1981); cessation of artificial mixing was followed by an abrupt rise of the population towards the surface to constitute a surface bloom, also in the classic manner (cf. Reynolds & Walsby, 1975; see also Figure 27). It may be noted that late-summer fluctuations in stability of water-columns supporting populations of the bloom-forming cyanobacteria remain the most likely cause for widespread bloom occurrence in temperate lakes at that time of the year. Exposure to potentially damaging irradiance and wavelengths, while the cells are lodged at the surface often leads, directly or indirectly, to mass cell mortality. Non-growing populations of *Microcystis* surviving deep-mixing episodes and bloom formation often persist well into the autumn. Ultimately, however, there is a change in buoyant behaviour: cells lose buoyancy (through a combination of gas-vacuole collapse, perhaps aided by enzymatic weakening of the component membranes, and a failure to assemble new vacuoles) and are rapidly recruited to the bottom sediments, where they overwinter (Reynolds & Rogers, 1976; Reynolds *et al.*, 1981).

Large flagellates are also potentially able to regain station rapidly after deep, turbulent mixing episodes (see, e.g., Figure 31) but prolonged exposure to mixing can also result in the formation of overwintering propagules. Cyst formation in *Ceratium* has been so induced experimentally (Reynolds *et al.*, 1983*b*), though the precise mechanisms and the pre-conditioning of cells for encystment are not, as yet, clear (see Chapman *et al.*, 1980).

Such specific responses of relatively large, dominant summer populations of *K*-selected 'algae' combine with the uneven progress of destratification to 'blur' the transition in community composition to dominance by autumn and winter forms. In all but the most oligotrophic temperate lakes (see Figure 42), this long transition is often characterized by a biomass maximum, of a magnitude approaching and sometimes surpassing that of the vernal period. Eventually, however, declining insolation and temperature fall to levels that will neither sustain the existing biomass of 'summer' species nor facilitate the continued growth of autumn or winter forms; the total biomass generally falls to a midwinter minimum, whence the annual cycle recommences the following spring.

8.1.2.4 *Seasonal sequences in other freshwater systems.* Before re-examining the frequency and direction of seasonal compositional changes from a more strictly ecological standpoint, it is appropriate to digress briefly to consider the annual sequences that occur in other types of standing water and in rivers. Lakes at lower latitudes generally behave in a similar way to their temperate counterparts, with the exception that irradiance intensities may be higher, surface water temperatures are correspondingly higher (instances of winter ice-cover are rare, save at high altitudes) and temperature-controlled density gradients will be potentially more stable. Owing to the greater density differences per unit temperature rise in warm water, gradients may develop rapidly and approach more nearly to the surface; equally, a small drop in surface temperature (at night, for instance) generates instability; and tropical climates do not exclude seasonal rises and falls in temperature, wind activity and rainfall. Moreover, tropical lakes come in a variety of sizes, depths and trophic status and exhibit a corresponding diversity of annual environmental variabilities. Thus, their communities will be influenced by broadly similar constraints to those that operate in temperate waters. Exposed or well-mixed lakes, or the epilimnia of deeper ones, may be dominated typically by diatoms, desmids or larger colonial chlorococcalean genera (e.g. *Pediastrum*, *Sorastrum*); cyanobacteria and dinoflagellates tend to dominate more stable columns in which the euphotic depth is frequently greater than the mixed depth. Lewis' (e.g. 1978*b*) work on Lake Lanao, Philippines, shows particularly well how tropical lake communities can respond to seasonal changes in mixing and relative light penetration. Meromixis (incomplete seasonal destratification) of (e.g.) the deep African rift-valley lakes probably presents similar opportunities for community variability, although nutrient poverty (especially with respect to nitrogen) may condition the compositional responses critically. More frequent (diel) alternations between stable microstratification and holomixis, such as that described for the equatorial Lake George, present an extreme case of environmental selectivity that apparently favours one of the strongest *K*-strategists, *Microcystis* (see Ganf, 1974*a*, and §3.2.3).

Although the freshwater 'algae' are generally ubiquitous, there are evident differences among the commoner representatives of the major taxonomic groups in tropical and temperate waters, not all of which are necessarily attributable solely to temperature. Centric diatoms (especially *Melosira* and *Cyclotella* spp.) are relatively more abundant than pennate species in the tropics; among the dinoflagellates *Peridinium* spp. tend to replace *Ceratium* nearer to the equator; and 8-, 12- and 20-armed species of *Staurastrum* tend to replace the 4- to 6- armed forms of temperate waters. Cyanobacteria are generally absolutely and/or relatively more abundant in tropical lakes, where several genera of spasmodic occurrence at higher latitudes are more common in planktonic assemblages (*Anabaena*, *Lyngbya*). *Spirulina* is another common genus in shallow, rich (and often saline) lakes, where its ecology shows some parallelism with temperate *Oscillatoria* spp.

Despite their absolute abundance throughout the world, too few small water bodies ('ponds') have been studied to permit any generalized assessment of the ecology of their planktons. Such small water bodies are typically shallow as well, so that there is little expanse of open water in relation to the containing basin area or to its perimeter. Potentially, then, ponds are more directly affected by 'run off', having short hydraulic retention times and, potentially, high nutrient loadings. They will rarely be subject to extreme fetch and wave action. Neither will they be expected to become stably stratified for weeks or months on end but to be fully mixed at irregular intervals, alternating with short periods of microstratification, under the influence of solar warming. The upper limits of biomass that can be supported will be subject to similar constraints as those operating in larger lakes (see §4.3.2) but the truncated depth enables a proportionately higher concentration of organisms to be maintained, with correspondingly more severe attenuation of light. Column-mixing and light penetration can therefore interact in much the same ways as they do in stratified lakes. At the same time, sedimenting particles will encounter the bottom boundary layers relatively quickly, although they may experience more frequent episodes of wind-induced resuspension.

Thus, the selective pressures acting upon the floral composition of phytoplankton are not dissimilar from those of lake epilimnia, but they operate on truncated temporal and spatial scales and are subject to a much stronger element of chance (cf. Talling, 1951). The flora of ponds show many affinities with those of lakes, especially among the more *r*-selected species among the Chlorococcales, Volvocales and Chrysophyceae. Larger non-motile diatoms, dinoflagellates and colonial cyanobacteria are less well-adapted to life in truncated water columns. Where nutrients severely limit the development of large populations, the competitive bias may frequently shift to epipelic 'algae' or to aquatic macrophytes rooted on the bottom vegetation. The growth of floating vegetation (e.g. *Lemna*), plants with floating parts (e.g. water-lilies, Nympheaceae) or even overhanging trees, generally select against phytoplankton development by shading. Frequent flooding, seasonal drying-up, extreme acidity, dystrophy (peat- and iron-staining) and metal pollution each introduce special characteristics regulating plankton development.

Development of phytoplankton communities in rivers depends largely upon the physical characteristics of turbulent flow and the turbidity due to suspended particulate loads (Belcher & Swale, 1979). Structural instability and high-frequency fluctuations in the perceived light represent extreme cases of well-mixed, turbid lakes and indeed, many of the algae of such environments predominate in rivers – centric diatoms (*Cyclotella*, *Stephanodiscus* spp.), certain pennate genera (small *Nitzschia*, *Synedra* spp.) and small chlorophytes (*Scenedesmus*, *Chlorella* and *Chlamydomonas*); larger or more-slowly growing species generally do not compete well, except during low flow regimes in large rivers (see, for instance, Szemes, 1967; Talling, 1980). In many temperate lowland rivers, at least, the supplies of nitrogen and phosphorus are sufficient to saturate the requirements of plankton regulated by physical factors although deficiencies of silicon may limit vigorous fluvial populations of diatoms in spring or summer and contribute to dominance by Chlorophyceae (Belcher & Swale, 1979). Standing crops are generally lower than in lakes of similar trophic status, although they are inevitably influenced by the presence of natural and artificial impoundments and backwaters along the water course, wherein substantial growths of limnetic phytoplankton may take place (for examples, see Talling, 1976*b*).

8.1.3 *Measurement of community change*

It has been shown in the preceding section (§8.1.2) that seasonal changes in the abundance and composition of phytoplankton communities owe primarily to the growth- and loss-rate responses of individual component species, representing distinctive assemblages, to broad, large-scale fluctuations in environmental characteristics. In order to formulate generalized ecological explanations of the relationships between the mechanisms and directions of community change and the major environmental driving variables it is helpful to invoke a quantitative measure of the process of change itself that will distinguish between periods both of relative stability and of rapid restructuring of the community. To be effective, the expression must be sensitive to changes in the relative proportions of the species present, even when no gain or loss of species takes place.

The 'succession rate' parameter (σ) derived by Jassby & Goldman (1974*b*) largely satisfies that requirement:

$$\sigma = \Sigma_i \, (\mathrm{d}c_i/\mathrm{d}t)^2 \qquad (59)$$

where the community composition at time t is

$$c_i(t) = b_i(t)/[\Sigma_i \, b_i \, (t)^2] \qquad (60)$$

and where $b_i(t)$ is the biomass of the ith species. The instantaneous succession rate, σ, has the dimension of $(\text{time})^{-1}$. Jassby & Goldman (1974*b*) used σ to describe patterns in the seasonal changes in phytoplankton of Castle Lake, California, over several summers. Values were maximal ($\sigma > 0.1 \, \mathrm{d}^{-1}$) in May or June but fell to low levels ($< 0.02 \, \mathrm{d}^{-1}$) towards the end of summer, corresponding to a 'stabilization' of community composition after the initial ice-melt, vernal overturn and onset of stratification, all of which usually occur within a short period in April or May and which each constitute a major reorganization of the limnetic environment inhabited by phytoplankton.

In a carefully argued critique of the assumptions implicit in the Jassby–Goldman index, Lewis (1978*a*) pointed out that since change was a phenomenon of the entire community, the changes in the abundances of each individual species should be related to the abundance of all the species present and that all individuals should be equally weighted. Thus, he formulated a modified index that obviated the species weightings accorded by Jassby & Goldman (1974*b*). This index, σ_s, he called the summed-difference (SD) index, which was derived as:

$$\sigma_s = \Sigma_i \, \{\mathrm{d}[ib_i \, (t)/B(t)]\}/\mathrm{d}t \qquad (61)$$

Over a short time interval, $t_2 - t_1$, this is estimated as

$$\sigma_s = \Sigma_i \{[b_i \, (t_1)/B(t_1)] - [b_i \, (t_2)/B(t_2)]\}/(t_2 - t_1) \qquad (62)$$

where $b_i \, (t)$ is the abundance (concentration) of the ith species and $B(t)$ is the size of the community $\Sigma_i \, b_i(t)$. In parallel empirical tests of the SD and JG indices against his own data (e.g. Lewis, 1978*b*) pertaining to the phytoplankton periodicity of Lake Lanao, Philippines, Lewis was able to demonstrate that both could be positively correlated with absolute rate of change in autotroph abundance, the absolute rate of change in primary production and with the grazer biomass, regardless of whether numbers (concentration) or biomass was used as a measure of abundance. In nearly every instance, the SD index proved to be more sensitive to community change than did the JG index. Individual values of SD index applied to concentration in Lake Lanao ranged between 0.02 and 0.13 d^{-1}.

In attempting to apply Lewis' SD index to describe community stability in periodic sequences observed in several British Lakes, I experienced some difficulty in adhering rigidly to Lewis' stipulation that cell concentrations alone would suffice (Reynolds, 1980*a*); this arises because cells differing greatly in size make differential contributions to the community as a whole. As an illustration, a hypothetical community comprising *Ceratium* and *Rhodomonas* cells is envisaged: at time t_1, their standing populations are each 100 cells ml^{-1}; after 5 days (t_2) of typical growth, *Ceratium* reaches 200 ml^{-1}, *Rhodomonas* 800 ml^{-1}. Interpolating in equation (61), σ_s is resolved to be 0.12 d^{-1}, indicative of a relatively rapid rate of community change. Yet in reality, *Ceratium* overwhelmingly dominates the hypothesized community in both instances, so the change in community structure is a small one. By according to each *Ceratium* cell an arbitrary score of 500 cell units, the disparity in cell sizes is overcome: the solution of σ_s then approximates to ~ 0.002 d^{-1}. Parallel weightings were accorded to other 'large'-celled species in order to give unitary estimates comparable with those of potentially alternative dominants of different cell size, before change and stability in the communities under investigation were evaluated (plotted in Figure 86). Sensitivity to changes in dominance (between the codified species) is preserved, whereas the intervening periods of continuing dominance are represented by a low index of change. In retrospect, these arbitrary score values could have been represented better by expressing b_i and B as cell volumes, though the overall result would differ little. Interpolation of the products of the hypothetical cell concentrations and the respective mean volumes of *Ceratium* and *Rhodomonas* cells (Table 4) into the SD equation (62) yields a value of $\sigma_s \sim 0.003$ d^{-1}. For comparison, the same biomass values inserted into the JG equations (59, 60) give an insignificantly different result ($\sigma \sim 0.003$ d^{-1}). Whichever index is adopted, empirical description of the transitions in community dominance is afforded better resolution the shorter the time intervals between individual standing crop estimates (i.e. sampling occasions) are kept (< 7 days; Reynolds, 1980*a*).

The representations of community change in Figure 86 apply to phytoplankton periodicities in several natural lakes (of differing size, morphometry and trophic status) and to Lund enclosures maintained at differing fertilities. They also serve to illustrate a number of general points about the nature of compositional changes. The first is that the onset of the transitional phases ('peaks' exceeding 0.1 d^{-1}) is typically abrupt, but rate of change weakens progressively as the 'new' assemblages becomes dominant (hence the certain 'skewness' of the plots through the transitions). The second is that during the transition, when the 'old' assemblage is being replaced by the 'new' one, there is often an increase in the number of species, relative to the total number of individuals, present per unit volume

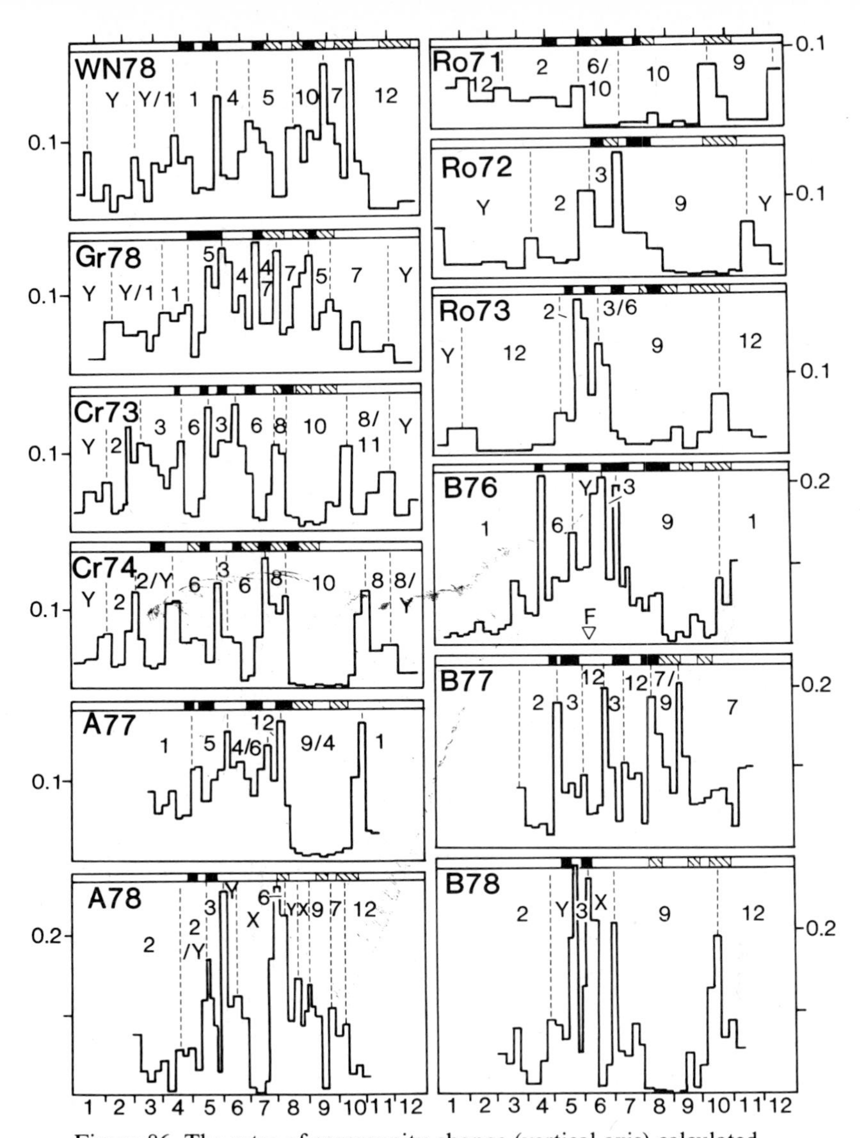

Figure 86. The rates of community change (vertical axis) calculated according to a weighted version of Lewis' (1978*a*) S. D. index through a number of annual sequences in natural lakes (WN, Windermere, North Basin; Gr, Grasmere; Cr, Crose Mere; Ro, Rostherne Mere) and experimental enclosures (Blelham Tubes A,B) tracing the progression of dominant species assemblages recognized by Reynolds (1980*a*): (1, *Asterionella*/*Melosira italica*; 2, *Asterionella*/*Stephanodiscus astraea*; 3, *Eudorina*/*Volvox*; 4, *Sphaerocystis*; 5, *Uroglena*/*Dinobryon*; 6, *Anabaena*/*Aphanizomenon*; 7, *Tabellaria*/*Fragilaria*/*Staurastrum*; 8, *Melosira granulata*/*Fragilaria*/*Closterium*; 9, *Microcystis*; 10, *Ceratium*; 11, *Pediastrum*/*Coelastrum*; 12, *Oscillatoria agardhii*; X, nanoplanktonic opportunists; Y, cryptomonads. Solid bars at the top of each diagram represent phases of microstratification (increasing stability); hatching denotes episodes of increased column mixing. (Redrawn from Figure 4 of Reynolds 1980*a*.)

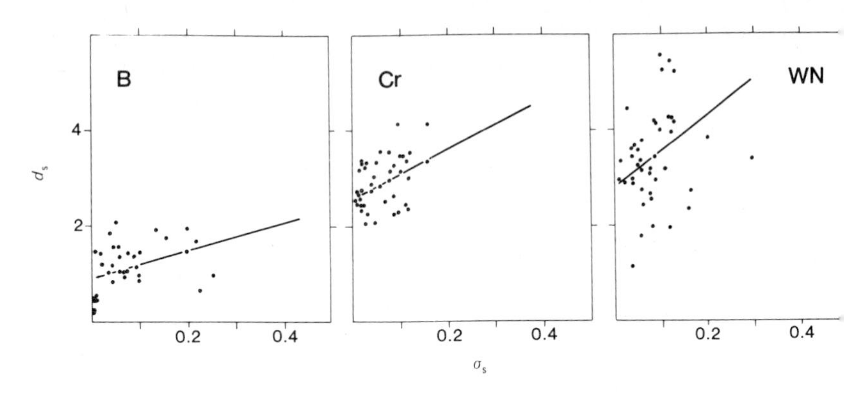

Figure 87. Values of species diversity (d_s), from Figure 44, plotted against corresponding values of community change rate (σ_s), from Figure 86, for three data sets applying to (B) Blelham Enclosure B in 1978, Crose Mere (Cr) in 1973 and (WN) Windermere (North Basin) in 1978. (Original.)

of water, i.e. in the species diversity (cf. §3.5). Comparison of the three examples of seasonal changes in Margalef's index of species diversity (d_s, representing in Figure 44) with the corresponding seasonal fluctuations in the rate of community change (Figure 86; Lund Enclosure B, 1978; Crose Mere, 1973; Windermere, 1978) suggests that, for each basin, the peaks in either plot occur coincidentally. There is, indeed, a significant correlation ($P = 0.05$) in each case, although there is a wide scatter about the fitted regressions (Figure 87), which explain only 9–17% of the variance. They are, of course, independent variables: diversity is not dependent upon dominance as is the index of community structure; in any case, species diversity is influenced by other considerations.

Perhaps the most important point about compositional changes that is underlined by (e.g.) Figure 86 is that community change occurs *episodically*. In broad terms, an increased rate of change is directly due to a 'new' species or group of species ascending to dominance. This alteration may be brought about by several contributing mechanisms, the extent of whose contribution is moreover variable. Among these are an accelerated rate of net increase in the 'new' species, which as shown in the previous section (§8.1.2) can be related to changed environmental selectivity. Equally, a decline in net increase and/or an increase in rate of loss (whether through sinking, grazing or death), is also related to the same environmental change. However, the eclipse of the 'old' dominant by the 'new' one will generally involve a reduction in the standing population of the first. If this is not the case, because a dominant but non-growing population resists attrition (e.g. *Microcystis*), then it will continue to influence both the composition of the 'new' community and, to a greater or lesser extent, the prospects for further net increase in the biomass of the 'new' assemblage. The weaker is the tendency for the existing dominant to be replaced by another, then the less is the perceivable change in community organization and the less will be the instantaneous value of σ. Thus, though it cannot distinguish their relative contributions, the derivation of the index of community change ensures that it is sensitive to the *mechanisms* of change, i.e. to changes in the specific rates of growth and attrition of the species present in a given system at a given time.

8.1.4 *Directionality of community change*

It is clear that, although the sequences of alternating periods of relative community stability (low σ) and the transitions of rapid reorganization (high σ), shown in Figure 86, provide a convenient representation of the seasonal periodicity of phytoplankton dominance in the named basins and years, they do not conform to a single pattern. The use of the index quantifies the *extent* of change but is unprejudiced as to its direction. If periodic sequences are to become fully understood and predictable, it is important to resolve the directional aspects of change. Part of the problem is terminological. Many authors equate 'seasonal succession' with 'seasonal periodicity'. Though loosely defined, 'succession' is a widely-accepted biological concept implying a strongly predetermined bias in the sequence in which species or species-groups will dominate a community. This is certainly true in the context of terrestrial plant ecology, where the successional concept has perhaps reached its highest development (see, for instance, Pielou, 1966; Odum, 1971). An area of the earth's surface (for instance, freshly-bared land, a sandy foreshore or an estuarine mud flat) is initially colonized by '*pioneer*' plant species that, having once arrived there (as seeds or other propagules), are suitably adapted to survive and grow there. By modifying the nature of the habitat (binding the substratum and their own organic deposition) the 'pioneers' open the habitat to further invaders that in turn alter the character of the environment further. Eventually, through a series of distinctive stages, the original bare land gradually evolves to support a '*climax*' vegetation (usually forest) that is then virtually closed to the species representing earlier successional stages. The advancing succession is typically characterized throughout by increasing species diversity and increasing production per unit area but declining productivity (i.e. production:biomass ratio). Nutrients become increasingly partitioned within the biomass of the plants. In terms of the component species, there is an overriding tendency in favour of increasingly *K*-selected species. In terms of community organization, each stage in the successional 'sere' alters the environment in such a way that it is prepared for the next. Thus, the main successional sequence owes principally to *autogenic* processes, emanating from the adaptabilities and responses of the plant species themselves.

However, classical ecological interpretations of plant community organization recognize that other types of change may occur. Several successions are liable to sudden reversals to more primitive stages by the external imposition of ('*allogenic*') perturbations that occur independently of the established biota. A forest may be clear-felled; grassland may be ploughed up; windstroms or high water levels may destroy pioneer waterside communities. In each case the habitat is rapidly returned to the equivalent of an earlier successional stage, whence the succession may subsequently resume. It should be noted that there is in no sense a reversed successional sequence. Thus, in a main sequence

pioneer → B → C → D → climax

we may expect to see modifications of the type

pioneer → B → C → D → climax
↙
perturbation
↙
pioneer

or, less severely,

pioneer → B → C → D
↙
perturbation
↙
B → C → D → climax

but not

climax → D → C → B → pioneer.

Equally, the imposition of cropping (in the agricultural sense) or the grazing of livestock, may arrest the succession at a sub-climactic stage or *plagioclimax*. In this way, land potentially bearing a natural forest vegetation can be maintained indefinitely as grassland or arable fields.

Following this reasoning, the inadequacies of referring all observable periodicities of phytoplankton to 'succession' and of seeking illuminating analogies with strict seral succession in terrestrial communities become readily apparent. What is required is a less-restricted terminology which accommodates the basic kinds of periodic sequences of phytoplankton that may occur.

It is necessary first to recognize the distinguishing characteristics of autogenic and allogenic alteration of the limnetic environment, and the responses of phytoplankton communities thereto. We may envisage an initially clear water column, charged with dissolved nutrients: that is, the analogue of the bare land stage of terrestrial successions. Autotrophic phytoplankton responds, within the limits of light and temperature, by increasing its biomass rapidly; as has been pointed out previously, the community composition will be biased in favour of faster-growing species which may become dominant. This high productivity (low biomass, high production) may continue for some time but progressively, light penetration is reduced, production becomes restricted to a declining volume of water and, though those organisms placed well within the illuminated zone may still maintain a high rate of growth, the productivity of the entire biomass is bound to fall. Moreover, as nutrients are depleted from the productive zone, the competitive advantage is likely to move in favour of slower-growing, conservative species that are better adapted to nutrient scarcity (cf. Pearsall, 1932; Margalef, 1958, 1968; Nauwerck, 1963; Lewis, 1978*a*) or to low average insolation. So long as the physical stability of the water column remains relatively constant, the environment becomes increasingly segregated into an upper, nutrient-deficient layer and a lower, light-deficient one (see also §5.6), eventually selecting in favour of 'algae' with suitably specialized adaptive strategies. Sequential responses of this kind may vary, according to which nutrient or other factor is the more critical, but they share the common property of being self-imposed consequences of community development (i.e. they are autogenic). Indeed, in the context of phytoplankton, it is to such sequences that the use of the term 'succession' should be confined (Reynolds, 1980*a*).

The successional trend may be interrupted at any time by externally-imposed (allogenic) perturbations which significantly alter the physical structure of the environment. The most important of these are wind-induced mixing episodes, especially where these both randomize the vertical distributions of the existing species and recharge the surface waters with nutrients. Round (1971) aptly described such perturbations 'shock periods'. Potentially, they are conducive to the eclipse of existing populations by species better adapted to well-mixed columns in which light is likely to be the most serious limiting factor. These may be regarded as being the equivalent of an earlier stage in the former succession or as an alternative. If the perturbation is sustained, however, the earlier succession cannot be followed and is supplanted by a new line of development. Its 'progress' is likely to be soon arrested, as either nutrients or light become limiting, at a stage equivalent to a plagioclimax, or what I have termed '*shift*' (Reynolds, 1980*a*).

Alternatively, the perturbation may be temporary, the water column quickly regaining its former stability. Depending upon the severity of the perturbation, the water will have been partially cleared and recharged with nutrients. The habitat is now set for *reversion* to the earlier successional sequence (or something like it) but, in all probability, recommencing at a more primitive stage. In this way, periodic changes in dominance may be viewed as community responses not to one but to three categories – true succession, shift and reversion.

The second step is to accommodate parallel levels (successional 'stages') of community organization that are not always dominated by the same planktonic species or even by representatives of the same assemblages. Again, precedents can be found in terrestrial plant ecology: there is little difficulty in recognizing 'grassland', regardless of whether it is dominated by *Festuca* or *Agrostis* or for that matter, by *Lolium* or *Cynosurus*. I would contend that the concept of mutually-alternative assemblage dominants (see Figure 86 and Reynolds, 1980*a*) goes some way to meeting this requirement. Thus, the *Eudorina–Pandorina–Volvox* [3] and the *Uroglena–Mallomonas–Dinobryon* [5] groupings may be viewed as being representative of a given successional stage (the early summer, post-stratification phase), the former being associated with P- and N-rich lakes, the latter with P-deficient ones. In any given year, the identity of the species dominating either assemblage may vary; it is the representation of that assemblage that may be confidently predicted. In this way, it becomes possible to resolve a common pattern to sequences of dominant algae (say) *Stephanodiscus* → *Eudorina* → *Anabaena* → *Ceratium* and *Asterionella* → *Volvox* → *Aphanizomenon* → *Microcystis*, either simultaneously in neighbouring lakes or in the same lake in successive years, by describing the sequence of assemblages; in the alphanumeric codifications of Reynolds (1980*a*) both can be represented [2] → [3] → [6] → [9/10].

The final step is then to determine the mutual relationships between the various assemblages. Fourteen assemblages [1–12, X, Y] were deemed to be represented in the twelve periodic sequences shown in Figure 86, mediated by 82 transitions. Twenty-one of these were directly related to the onset of stable stratification and a further 19 follow phases of increased mixing and epilimnetic deepening. The most consistent response throughout the series is the replacement of vernal diatom-dominated assemblages [1, 2] soon after the onset of stratification by the *Eudorina–Volvox* assemblage [3], in the more eutrophic waters, or by the *Sphaerocystis*–chrysophyte groupings [4, 5], in the more oligotrophic examples. Occasionally, as was the case in Crose Mere (1974), Rostherne Mere (1971) and Enclosure B (1976), the *Anabaena–Aphanizomenon* assemblage [6] was initially dominant, though generally it succeeded [3]. Summer-mixing episodes were frequently followed by periods in which the diatom–desmid groupings [7, 8] dominated, especially later in the year, when they sometimes merged, through further transitions, into phases of dominance by *Oscillatoria* [12] and vernal cryptomonad–diatom assemblages [1, 2, Y].

The remaining 23 transitions occurred during the summer and autumn, independently of changes in physical stability. These included a number of sub-sequences common to more than one lake or one year: for instance, from chrysophytes [5] to the *Sphaerocystis* assemblage [4] in Grasmere and in Enclosure A, 1977 (the sequence was reversed in Windermere, 1978) and from *Eudorina–Volvox* [3] to *Anabaena–Aphanizomenon* [6] in Crose Mere and Rostherne Mere (1973). In turn, many of these summer sequences eventually culminated in dominance by *Microcystis* or *Ceratium*, though sometimes briefly punctuated either by rapid phases of growth and attrition of opportunist species [X] and *Cryptomonas* [Y] or by short periods of dominance by diatoms [7, 8].

The principal periodic pathways that emerged from Reynolds' (1980*a*) analysis were the Volvocales [3] → Nostocales [6] → dinoflagellate/*Microcystis* [9/10] succession of eutrophic lakes, reverting through summer diatom–desmid assemblages [8] and the chrysophyte/*Sphaerocystis* [5/4] → dinoflagellate [10] sequence of more oligotrophic examples with similar diatom–desmid [7] reversions. Both series shifted in autumn towards diatoms [1, 2, 7, 8] and cryptomonads [Y] in shallower basins or to *Oscillatoria* [12] in the optically-deep examples.

Both the recognized assemblages and the proposed types of transitions that separate them provided an adequate means of rationalizing the 12 observed seasonal periodic sequences considered by Reynolds (1980*a*, Figure 86). It was moreover possible to summarize these possibilities within a hypothetical two-dimensional matrix (nutrient availability v. column stability), through which the separate pathways of autogenic succession, shift and reversion could be traced, in order to predict sequences of dominant phytoplankton. This matrix is redrawn in Figure 88.

Its construction was intended to provide a conceptual framework against which the seasonal periodicity of phytoplankton might be assessed. Future developments might be directed towards its expansion to include other types of lake and freshwater ecosystem, to adding other dimensions to accommodate other variables and, most importantly, to adding empirical evaluation of the axes that might then provide a truly predictive matrix of periodic events.

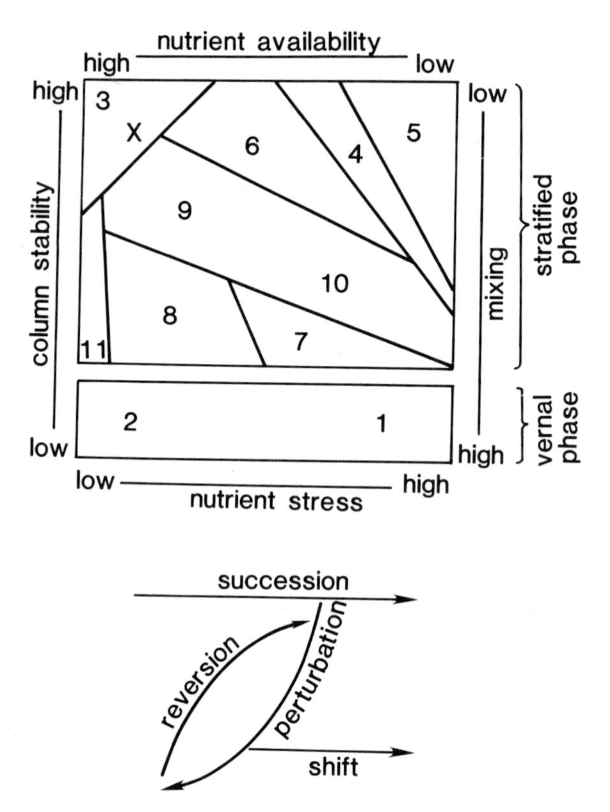

Figure 88. A hypothetical matrix, bounded by axes defining mixing/stability and (unspecified) nutrient concentration, accommodating several major 'algal' assemblages (1–11, X; as Figure 86). The lower part of the figure shows three kinds of periodic progressions from one dominant assemblage to another, and the directions that each would follow in the matrix. Thus, from a given starting point, *successional* changes are traced by a rightward horizontal progression (e.g. 3 → 6 → 4); mixing causes a downward (and, usually, leftward) progression, from which a '*shifted*' *succession* (if the mixing is sustained) or a '*reversion*' (if the column re-stratifies) may occur. Usage of the terms 'succession', 'shift' and 'reversion' is defined in the text. (Redrawn from Figure 7 of Reynolds, 1980*a*.)

For the present, it remains to stress that this theoretical approach to problems in phytoplankton periodicity complements and reinforces the conclusions reached earlier in this chapter, which were based on examination of the different morphological, physiological and behavioural adaptations of 'algae' to planktonic survival. In the end, the most successful organisms are likely to be those best fitted for life under contemporary conditions. Theory identifies the desirable adaptations; it is the evolutionary diversity of form and function among planktonic organisms that supplies them. It is therefore appropriate to conclude this section on periodicity of freshwater phytoplankton by stating that the general selective patterns are directly comparable with those recognized in marine environments by Margalef (e.g. 1958, 1978; see also Smayda's (1980) masterly review). The early stages are characterized by species with high *SA/V* ratios and high rates of potential increase; in later stages, larger species, with lower *SA/V* ratios and slower growth rates predominate, and the proportion of motile species increases. At the same time, productivity decreases and diversity increases. The parallels in freshwaters, especially in the instance of the diatom → chrysophyte → dinoflagellate sequence of more oligotrophic lakes, are readily apparent.

8.2 Longer-term floristic changes

In this concluding discussion, I attempt to relate changes in the composition and annual cycles of phytoplankton that become apparent over much longer periods than one year to the intrinsic biologies of the organisms involved. Such changes may be brought about by changing climates (especially with reference to the seasonal distribution of rainfall), catchment vegetation and land use and nutrient loadings, as well as by natural autochthonous limnological processes. These effects are often interrelated so that it is unnecessary to consider each separately. In different ways, their interactions are relevant to the problems of eutrophication (§5.7).

Before considering the responses, it is as well to examine the impact of these changes upon the growth environment, proceeding from the general to the more specialized aspects. Standing waters, it will be recalled, have been suggested to be special cases of hydrologically-flowing water. Because fluvial processes mould drainage channels to optimize flow (eroding here, depositing there), intermediate lakes and pools tend to become major sites of deposition of suspended sediment: where flows are slowed (as in channel basins), competence (i.e. carrying capacity) is reduced and suspended loads are deposited. In extremes (swift flowing, 'dirty' rivers entering large impoundments), deposition builds up in delta-fans that advance steadily into the infilling-lake. Wave-erosion may subsequently remove some of the deposited material to the deep-water sediments. Biological processes augment the river-borne (*allochthonous*) sediment, chiefly through recruitment of the detritus of pelagic (dead or dying remains of phytoplankton, zooplankton, nekton, and their faecal pellets) and littoral (derived from decomposition of fringing macrophytic, periphytic, and animal production) communities. The relative contributions of these biogenic (*autochthonous*) sources depend upon the size of the water body relative to its inflows, the nature of the catchment, and the total biological production within the impoundment. The biologically poorer the lake and the greater the relative discharge rates into it, the greater is the tendency to accumulate allochthonous sediment. Of greater relevance to the present discussion is that all lakes tend to become infilled by sediment, albeit at vastly differing rates, and ultimately, to disappear altogether. Throughout this process, *lakes become progressively smaller and shallower*.

For a constant annual nutrient-loading rate, it will be seen that the nutrient concentrations in the lake become diluted less in the diminishing lake volume: in line with the predictions of the Rast–Vollenweider models (see § 5.7) phytoplankton biomass becomes, on average, more concentrated. Moreover, reduction in mean depth ensures that suspended populations are carried near the surface proportionately more frequently. Thus, possible limitations imposed by nutrients and underwater light-climates stand to become ameliorated. Even if the overall mean phytoplankton biomass (expressed areally) does not change, it certainly becomes more concentrated. This process is sometimes called '*natural eutrophication*'. Its imitation by the results of other enrichments to the nutrient supply is superficial (see below), so the distinction from *anthropogenic* or *cultural eutrophication* is not entirely academic.

Superimposed upon this generally unidirectional reduction in lake volume are other reversible fluctuations. Long-term climatic fluctuations affect the hydraulic loading and throughflow characteristics of the lake. Diminishing flows erode and transport less allochthonous sediment to the lake and, generally, supply fewer leachates (including nutrients) to the lake basin. An extreme example of the impact of climatic change upon lake ecosystem has been investigated by P. W. Beales (reported in Reynolds, 1979*b*). Crose Mere, Shropshire, a 'closed' basin, now recharged by groundwater flow, seemingly became so hydrologically isolated during the relatively dry Boreal period (7000 to 9000 years B.P.) that, as those elements of its contemporary flora and fauna preserved in its sediments would indicate, it became markedly less alkaline and less productive (i.e. more oligotrophic) than at any prior or subsequent period of the post-glacial. Similarly, changes in land use affect the hydrology and the rates of erosion and nutrient-leaching from the catchment. Forests tend to 'damp' the sporadic ('flashy') run-off of precipitation from more open catchments and also protect their surface forms; nutrients are largely retained and recycled within the mature ecosystem rather than leached into the water reaching the lake. Clearance of forests and the imposition of cultivation on catchment soils, especially those to which inorganic fertilizers are applied, work in the opposite direction. This aspect of man's activities undoubtedly contributes to the nutrient-enrichment of recipient lakes (cf. Likens *et al.*, 1970).

Urbanization of catchments produces several further effects. Roofing, paving and improved drainage, designed to remove excess runoff, increase the fluctuations in water flow. Large human populations are fed at the expense of foodstuffs 'imported' from other catchments, but their wastes (sewage) are disposed of locally, so further exaggerating the supply of nutrients to lakes downstream. 'Improved' sanitation and sewage treatment ensure that nutrients (especially phosphorus) are mineralized and returned to the drainage in solution. Many of the 'unnatural' products consumed by man (e.g. detergents) yield additional loads of inorganic nutrients (in this case, phosphate) to recipient systems.

Together, these anthropogenic processes contribute more of the erstwhile limiting nutrients to water bodies, always magnifying (and, in view of the exponential increase in human populations, coupled with enormous social and technological advances, often greatly and rapidly so) the effects of natural eutrophication. Unlike the latter, however, cultural eutrophication is theoretically and practically reversible.

The immediate effects of additional free nutrients arriving in lakes are to stimulate *in situ* biological production (i.e. of phytoplankton) and to favour the maintenance of larger average standing crops. This much is well known and has been described by relatively sophisticated mathematical models (see §5.7). What is also apparent is (a) that increased 'average' crops of phytoplankton conceal exaggerated fluctuations in the standing stock (i.e. peaks become relatively more enhanced while intermediate minima do not), and (b) that, sooner or later, different 'algae' become dominant. These tendencies are, as yet unmodelled, their causes remain uncertain and their effects cannot be predicted with any real resolution.

The first type of response (a) is presumably related to the impact of the potentially increased crop on the growth medium. While nutrient concentrations, biomass and light-penetration are readily altered, the physical stability is not. Put another way, the greater is the alleviation of nutrient limitation, so the regulation moves towards physical processes: in much the same way as the differences in the seasonal fluctuations in phytoplankton biomass between oligotrophic and eutrophic lakes (see Figure 42) can be related to more frequent alternations in the light climate (§§3.4, 6.2), naturally abrupt changes in column mixing inevitably have a greater relative effect on the optical depth of surface layers rendered turbid by suspended phytoplankton and hence, on specific growth and loss rates and on competitive selection. More frequent variability in biomass is, probably, a direct consequence of rapid recovery of the nutrient-saturated growth rates on each occasion that hydraulic stability permits it.

The second type of response (b) is more complicated, though its basic mechanisms would appear to be fundamentally similar. Let us envisage a significant and sustained increase in the loadings of the characteristically limiting nutrient (say P) on a lake, whose phytoplankton already undergoes an annual periodic cycle of abundance and composition. The provisions of the biomass–P relation enables us to predict no more than that there will be 'more of the same': the species which already maintain stocks in the lake are best placed to exploit the enriched P supply. Efficient, biologically-mediated recycling processes may assist the stimulus to be carried through the entire annual sequence. If the augmentation of nutrient loading continues to rise slowly, or is sharply raised to a new level, several new responses may become manifest. First, the maximum uptake rates of the existing 'algae' will not remove the added nutrient sufficiently fast to prevent the excess from being exploited by faster-growing species that may arrive in the lake, or have been present in 'backgound' concentrations. Fertilization experiments in the Lund enclosures (reviewed in Lund & Reynolds, 1982) have repeatedly demonstrated that free nutrients do favour fast-growing species in this way, and not all of these species had been conspicuous in the enclosures in previous years, or in the natural lake (Blelham Tarn) outside (e.g. *Ankyra*, *Pedinomonas*, *Eudorina unicocca*, *Microcystis aeruginosa*). Once these species have experienced an opportunity to build up large populations (always favoured by virtue of their rapid growth) they or their propagules are poised to assume a leading position in subsequent years and actually depress by 'shading' the growth of the erstwhile competitors. It is one of the characteristics of compositional responses to more or less insidious or progressive eutrophication that they nevertheless occur quite abruptly.

Second, since many of the favoured fast-growing species are also small, they will presumably increase the potential food supply of the zooplankton. Besides supporting a potentially larger filter-feeding population that may well select dynamically against slower-growing but 'grazable' species of the oligotrophic sequence, a change in the dominant animal-species may well be portended. As enhanced community filtration is always likely ultimately to favour 'ungrazed' species, so their dominance becomes increasingly more likely (Nilssen, 1978).

Third, as eutrophication advances and biological production increases, so the greater is the sedimentation of biogenic material to the lake bottom. Hypolimnetic oxygen uptake is increased until it (and, in extremes, the sediment itself) becomes totally anoxic. While it is uncertain whether this selects against the successful perennation of benthic stocks and propagules of oligotrophic species, there is little doubt that it is tolerated by the 'resting' propagules of many eutrophic species (*Melosira*, *Stephanodiscus* spp., *Anabaena* akinetes, *Ceratium* cysts) and, indeed, may be conducive to the seasonal growth of *Microcystis* (Sirenko, 1972; Reynolds *et al.*, 1981).

Fourth, the maintenance of larger suspended stocks of phytoplankton reduces light penetration and may well increase average optical depth, biasing the outcome of interspecific competition in favour of 'shade' species. It is arguable that the tendency for even mildly-eutrophied, deep, alpine lakes to become dominated by *Oscillatoria* is a response directly attributable to their ability to maintain growth in mixed columns of greater optical depths than are tolerated by diatoms and almost any other potential competitors (see §8.1.2.1). The advantage is complicated, however, by the need to avoid near-surface radiation intensities when the water column is stratified. The formation of depth-maxima depends upon an adequate level of light penetrating to the depth of a stable metalimnion. In many smaller, shallower lakes (but which nevertheless still become thermally stratified), this condition cannot always be met. Frequent 'dislodgement' from the metalimnion is not well tolerated by *Oscillatoria* spp. (see, e.g., Lund & Reynolds, 1982) and leads to a reduction in the size of the inoculum persisting to exploit the onset of autumnal mixing. In this way, eutrophication has apparently led to a steady decline in the size of average *Oscillatoria agardhii* crops in Esthwaite Water (Lund, 1972*a*).

The increased severity of episodic vertical truncation of the euphotic zone may also contribute to changes in the characteristic phytoplankton periodicity: as discussed above (§8.1.2), such fluctuations are normal events in most lakes, to which the seasonal periodic cycles of dominance are closely related. Each episode tends to select in favour of species that are able to maintain themselves more continuously within the reduced depth of the euphotic zone. The more intense and the more frequent such episodes become, then the longer is the total period in which the appropriate species are able to grow. Mechanisms of this kind probably contribute indirectly to the long-term 'replacement' of *Sphaerocystis*, *Gemellicystis* and other colonial chlorophytes (which appear to be particularly sensitive to mixing through turbid columns: Reynolds *et al.*, 1983*b*) by self-regulating and motile species of dinoflagellates and cyanobacteria. Initially, the additional nutrient load stimulates the chlorophytes to produce larger standing crops but, as growth becomes limited by the photic conditions, the environment is opened to exploitation by representatives of the more 'eutrophic' species. In time, the latter become more firmly established, maintaining a fund of propagules in the lake, while population growth and perennation of the former becomes steadily weaker and, in extremes, the chlorophytes 'disappear' from the lake altogether.

Enrichment of erstwhile supplies of P may stress not only the photic resources of a lake, but also those of some other nutrients that were previously 'saturating' the (P-limited) requirements of 'algae' growth and which themselves are not necessarily increased in the same proportions, or even at all. These, too, contribute directly to long-term floristic changes: for instance, P-saturated 'algal' production leading to N-limitation must favour the ascendancy of nitrogen-fixing *Anabaena* spp.; with phosphorus enrichment that permits a more rapid growth and larger standing crops of diatoms, the availability of silicon is likely to become limiting (or to become limiting earlier in the year), with the result that species other than diatoms are able to compete more strongly for the available phosphorus (e.g. Schelske & Stoermer, 1971; Lund, 1972*b*); in soft-water lakes, particularly, a parallel effect invoking carbon may be discerned, where those 'algal' species that utilize sources of carbon other than dissolved CO_2 (e.g. bicarbonate) and are tolerant of pH levels ($\gg 8.5$) elevated by increased CO_2 demand (most cyanobacteria, some dinoflagellates and certain chlorophytes), are preferentially selected over 'intolerant' species of oligotrophic lakes (most chrysophytes, many desmids and, perhaps (Shapiro, 1973), some colonial representatives of the Chlorophyceae).

Some of these effects are evident in what are perhaps among the best-documented accounts of floristic responses to recent lake eutrophication: Lund's (1972*a*, *b*, 1978) weekly observations on the phytoplankton of the lakes of the Windermere drainage basin, spanning more than 30 years. In the south-basin of Windermere itself, the largest and deepest of the lakes monitored regularly, *Asterionella* crops have steadily declined (though total diatom production, now dominated by, *inter alia*, *Melosira italica* and *Fragilaria crotonensis*, has altered little) while *Oscillatoria* spp. have increased. The eutrophication of Blelham Tarn, the smallest of the lakes considered, is closely related to a series of known 'cultural' events (seasonally increased population, installation of mains water supply and sewage-treatment works). Average crops of *Aphanizomenon flos-aquae*, *Oscillatoria agardhii* var. *isothrix*, *Cryptomonas* spp., *Ceratium hirundinella*, *Asterionella formosa* and *Tabellaria flocculosa* var. *asterionelloides* have all increased by a factor of three or more while those of *Dinobryon divergens* and *Cyclotella praeterissima* have correspondingly decreased. During the same period, *Fragilaria crotonensis* (1962), *Anabaena solitaria* (1966), *Stephanodiscus* sp. (1973) and *Microcystis aeruginosa* (1975) each produced substantial populations in the lake for the first time and have continued to do so in subsequent years. In Esthwaite Water, the most eutrophic of the three lakes, the major response to further enrichment has been the production of much larger *Ceratium hirundinella* populations, which in some years attain light-limited proportions (see also Heaney & Talling, 1980*a*, *b*), and the reduction in *Oscillatoria*, referred to above.

In much the same way that relatively sharp increases in nutrient-loading are reversible (though often at very large financial costs, or in sacrifices of convenience), so the trends in increased abundance and altered species dominance can be put into reverse ('oligotrophication'). The initial experiments performed in the Blelham Enclosures (Lund, 1975; Lund & Reynolds, 1982) had the effect of isolating phytoplankton from the major source of limiting nutrients: not only was there a rapid (within one to two years) reduction in the algal biomass, but the scales of specific populations were of similar order to those of the Tarn 20–25 years previously. The species, however, were generally the same as those that grew contemporaneously in the outside lake (suggesting, once again, that floristic reversals, though favoured, are delayed because the species most likely to respond immediately are those already present in the system). Abrupt reductions in the average standing phytoplankton biomass in Lake Washington, following the diversion of all sewage outfalls, provide a more spectacular impression of the same principle (see e.g. Edmondson, 1970, 1972). Once the diversions began to take effect, the phytoplankton biomass, long dominated by *Oscillatoria agardhii*, was quickly reduced to below its 1933 level (the first year in which any quantitative observations were made), and there have been some indications of a change in relative species composition.

This spectacular 'recovery' of Lake Washington is an enduring example of what can be achieved when the eutrophication is still relatively mild (cf. Lorenzen, 1974); the results need not always be so impressive, especially if the nutrient loads (including internal recycles) are greatly in excess of the requirements of light-limited phytoplankton (Reynolds, 1978*c*, 1980*c*). Even if so much nutrient is eliminated that it is reduced to concentrations limiting a 'hypertrophic' plankton, there is a possibility that new or continuing problems will be experienced in the wake of a floral change. Of particular interest in this context are Clasen's (1979) experience in the Wahnbach Reservoir, Germany, where reduction in phosphorus loading has not been followed by a smooth transition in biomass to a new equilibrium. Instead, the more or less continuous dominance of *Oscillatoria rubescens* has been substituted by seasonally-fluctuating levels of (mainly) diatoms whose maximal concentrations continue to be unacceptably large.

Clasen's (1979) example provides the final point that I wish to emphasize. If the study of phytoplankton ecology is to have any relevance, beyond its intrinsic academic interest, it surely lies in its application to fundamental principles and design of management strategies for lakes and reservoirs, whether they are used for recreation, water supply or as commercial fisheries. Despite the considerable investment of resources and manpower in its investigation, it is humbling to realize that all our acquired knowledge scarcely allows us to make valid predictions about when and what species will be abundant in given waters. It has seemed to me that, since the early seventies, when the first quantitative expressions of the phosphorus–chlorophyll relationship began to appear, the greater body of researchers has either moved on to other problems or has become obsessed with elaborate attempts to quantify nutrient-loadings with incresing precision, presumably in the belief that once we know exactly what goes into a lake then we will be able to predict what the responses will be. To me and, I suspect, to the majority of water engineers, this approach seems misdirected. We are still faced with the dilemma that, regardless of the *average* level of biomass (give or take half-an-order of magnitude), it is the 'peaks' in biomass that cause the problems and they may be compounded by whether this or that species is dominant. And it is a dilemma which looks no nearer to solution than it was ten or twenty years previously.

There is a need to recognize that vastly different species have different morphological and physiological adaptations to survive and grow and that they will therefore respond in different ways to several dimensions of environmental variability. In short, we still need to know much more about the biologies of individual common *organisms*, not convenient correlatives of total biomass (carbon, chlorophyll). It is this philosophy that I have tried to project through the pages of this book. Should it serve as a catalyst for a new approach to the outstanding problems in the biology of aquatic ecosystems, I should be well satisfied.

Glossary of symbols

A carbon assimilation rate of individual animals.

a the ratio between the radius of a cell lacking any mucilaginous secretion to the radius of the same cell plus its mucilaginous covering (see §2.5.2.3).

a the total volume of water (as a fraction of the water body volume) filtered each day by the zooplankton. Derivation in §7.5.2.

b_i abundance or biomass of the *i*th species in a community (see §8.1.3).

$[chl]$ chlorophyll *a* concentration in water. Subscripts refer to (s) summer, (v) vernal or (y) year-round mean levels.

c_i mathematical expression of the composition of a community of *i* species.

D the demand, per unit time, of 'algae' on the available supply of resources (see §6.4).

d diameter (of a cell).

d_s species diversity index (defined in equation 21).

F filtration rate of an individual filter feeder, as a unit volume of water passed through the filtration apparatus per unit time.

F_d fraction of the total cells in a population that are dead.

F_N fraction of population, N (= 1), remaining at a given point during a process of continuous change (see e.g. §2.5.5.2).

$GALD$ greatest axial linear dimension (of a phytoplankton cell or colony).

g gravitational acceleration.

I underwater irradiance (given either relative to the irradiance obtaining immediately beneath the water surface, I_0' or as an absolute value; units: $\mu E\ m^{-1}\ s^{-1}$).

I_k the irradiance below which photosynthesis is light limited, such that $P \propto I$.

K_s the half-saturation concentration of a substrate that will sustain a dynamic cell process (e.g. nutrient uptake, growth) at half the theoretical maximum (saturated) rate.

K_z eddy diffusivity (defined in equation 2).

k' the exponential growth constant, the factor expressed as a logarithm to base e of increase in cells, dry weight, volume or carbon per unit time. (k'_2, k'_{10} are equivalent values as logarithm to the base 2, base 10, respectively.)

k'_{Si} the exponential growth constant for diatoms, calculated from known uptake of silicon from the environment.

k_d exponential death (or decomposition) rate.
k_g the exponential rate of removal of algae by filter feeders. It is related to the filtration rate by a conversion factor for selectivity.
k_L sum of rates of all loss processes detracting from growth and expressed in similar units.
k_n exponential rate of net increase (i.e. k' minus all loss rates, expressed in the same units).
k_s exponential rate of sinking loss.
k_w exponential rate of washout loss.
L horizontal length dimension of a turbulent eddy.
L_b body length (of a cladoceran).
L(P) specific loading rate of phosphorus per unit area of lake surface per year (see §5.7).
LDH Talling's (1957*a*) 'light-division-hours', an expression of the integral irradiance received by a body of water during the day (see equation 26).
M no. of mixings (see §2.3).
N population or concentration of (usually) algal cells, per unit volume of water. N_0 and N_t are corresponding populations separated by a period of time, t, during which a change in N is thought to have occurred.
N^2 the Brunt–Väsälä Frequency (equation 4).
$[N_\lambda]_w$ mean 'available' nitrogen concentration over the winter–spring period.
$[P_\lambda]$ mean 'available' phosphorus concentration over a given period – generally the winter to spring period ($[P_\lambda]_w$).
P specific photosynthetic rate (per unit chlorophyll, carbon, cell, etc).
P_{max} maximum specific photosynthesis rate.
PhAR photosynthetically active radiation.
Q_{10} the factor by which the rate of a given biochemical process increases per 10°C rise in temperature (or decreases per 10° fall).
q inflow volume (see §7.2).
q cell quota of a nutrient (see equation 36).
q_0 minimum cell quota, i.e. sustaining a zero growth rate (see equation 36).
R specific respiration rate (per unit chlorophyll, carbon, cell etc).
r (relating to regression equations) the coefficient of correlation between two variables.
r (elsewhere) radius (of a circle or sphere).
r_c critical radius of an oceanic 'patch' of plankton (§3.3.2).
r_s radius of a sphere with same volume as a non-spherical particle.
Re particle Reynolds number, expressing the ratio of inertial to viscous forces acting on a body moving through water (see equation 12).
R_f Reynolds number of flow (see equation 1).
Ri Richardson's Number, expressing the magnitude of work required to overcome a density gradient, relative to the energy available (see equation 3).
S concentration of a nutrient or substrate or of any other resource.
SA surface area (e.g. of a phytoplankton cell or colony).

S_H heat flux rate or rate of heat passage (per unit length, per unit time).
s no. of species present in an assemblage or community of organisms.
s^2 (statistical) variance.
t a period of time; t_0 = zero time, $t_{1,2,n}$ are fixed subsequent times.
t' clearance time of a static column by sedimenting particles (see §2.3).
t_e theoretical clearance time of a continuously mixed column by sedimenting particles; in §2.3, 2.5 it is evaluated as the time when $N_t = 0.05\ N_0$.
t_G cell generation time, the period that elapses between formation of a daughter cell and the point at which itself divides to produce two new cells.
U mean current velocity in the direction of flow.
u the horizontal velocity of a current.
u_o the horizontal velocity of surface flow.
u_s the horizontal velocity of advection (§3.3.3).
V volume (e.g. of a phytoplankton cell or colony). V_c is used specifically for cells to distinguish the volume contributed by mucilage (V_m). In Chapter 7 V is used to represent the volume of a body of water.
V_S nutrient uptake rate.
v' intrinsic terminal settling velocity of a particle.
v'_k settling velocity of killed 'algae'.
v_s terminal settling velocity of a spherical body.
W wind speed.
W_b weight of a planktonic animal.
W_c cell weight (dry and usually ash-free, unless otherwise specified in text).
w mean current velocity of convectional downwelling in a Langmuir rotation.
$\overset{*}{x}$ Lloyd's index of mean crowding (see §3.2.2).
z depth.
z_{eu} euphotic depth (approximated as the depth to which 1% of *PhAR* obtaining immediately beneath the water surface (I'_0) penetrates a water column).
z_m mixed depth (that portion of the water column turbulently or convectively mixed; generally characterized by a density gradient of < 0.02 kg m^4).
z_s depth of Secchi-disk extinction.
δ (mathematical) 'a small increase'; thus, $\delta\theta/\delta z$ represents the rate of change in temperature with depth, or 'the temperature gradient'.
Δ day length.
ε The vertical extinction coefficient, describing the attenuation of light with increasing depth. ε_a, ε_p and ε_w are the components of the (total) extinction coefficient (ε_t) due, respectively, to algae, other particles and the water itself. ε_s is the specific increment in vertical extinction coefficient per unit of chlorophyll *a*.
η viscosity (of a fluid).
θ temperature (customary, unless otherwise stated, or absolute).
Λ loading function term; please see §5.7.

ρ	density (of water).
ρ'	density (of a particle). The distinction is sometimes made between ρ'_c (the density of a cell) and ρ'_m (the density of surrounding mucilage).
Σ_i	'sum of' i variables.
Σk_L	sum of exponential rates of loss (see §7.1).
ΣNP	the instantaneous integral of community photosynthesis per unit surface area of water.
$\Sigma\Sigma NP$	the daily integral of community photosynthesis (i.e. photosynthesis per unit surface area per day).
$\Sigma\Sigma NR$	community respiration rate, per unit area of water surface per day.
σ	'succession rate' or index of rate of community compositional change.
σ_s	summed difference (SD) index.
τ_p	retention time (= mean residence time) of phosphorus in a lake.
τ_w	hydraulic retention time (= mean residence time of water in a lake or section of river).
ϕ	coefficient of 'selectivity' for the removal efficiency of algal particles from a filtration current.
ϕ_r	coefficient of form resistance of a non-spherical body (see equation 13).

References

Abeliovich, A. & Shilo, M. (1972). Photooxidative death in blue–green algae. *Journal of Bacteriology*, 111, 682–9.

Ahlgren, G. (1971). Limnological studies of Lake Norrviken, a eutrophicated Swedish lake. II. Phytoplankton and its production. *Schweizerishe Zeitschrift für Hydrologie*, 32, 354–96.

– (1978). Growth of *Oscillatoria agardhii* in chemostat culture. 2. Dependence of growth constants on temperature. *Mitteilungen der internationale Vereinigung für theoretische und angewandte Limnologie*, 21, 88–102.

Alcaraz, M., Paffenhofer, G.-A. & Strickler, J. R. (1980). Catching the algae: a first account of visual observations on filter-feeding calanoids. In *Evolution and Ecology of Zooplankton Communities*, ed. W. C. Kerfoot, pp. 241–8. Hanover, N.H.: University Press of New England.

Algéus, S. (1950). Further studies on the utilization of aspartic acid, succinamide and asparagine by green algae. *Physiologia Plantarum*, 3, 370–5.

Allen, M. B. (1956). Photosynthetic nitrogen fixation by blue–green alge. *Science Monographs, New York*, 83, 100–6.

Anderson, L. W. J. & Sweeney, B. M. (1978). Role of inorganic ions in controlling sedimentation rate of a marine centric diatom, *Ditylum brightwelli*. *Journal of Phycology*, 14, 204–14.

Andersson, G., Berggren, H., Cronberg, G. & Gelin, C. (1978). Effects of planktivorous and benthivorous fish on organisms and water chemistry in eutrophic lakes. *Hydrobiologia*, 59, 9–15.

Anon. (1968). *Water Quality Criteria.* Washington: Federal Water Pollution Control Administration.

Atkinson, K. M. (1980). Experiments in dispersal of phytoplankton by ducks. *British Phycological Journal*, 15, 49–58.

Bailey-Watts, A. E. (1974). The algal plankton of Loch Leven, Kinross, Scotland. *Proceedings of the Royal Society of Edinburgh*, B, 74, 135–56.

– (1976). Planktonic diatoms and some diatom–silica relations in a shallow eutrophic Scottish loch. *Freshwater Biology*, 6, 69–80.

– (1978). A nine-year study of the phytoplankton of the eutrophic and non-stratifying Loch Leven (Kinross, Scotland). *Journal of Ecology*, 66, 741–71.

Bailey-Watts, A. E. & Duncan, P. (1981). The phytoplankton. In *The ecology of Scotland's Largest Lochs*, ed. P. S. Maitland, pp. 91–118. Den Haag: Junk.

Bainbridge, R. (1957). The size and density of marine phytoplankton concentrations. *Biological Review of the Cambridge Philosophical Society*, 32, 91–115.

Baker, A. L. (1970). An inexpensive microsampler. *Limnology and Oceanography*, 15, 158–60.

Baker, A. L. & Baker, K. K. (1976). Estimation of planktonic wind drift by transmissometry. *Limnology and Oceanography*, 21, 447–52.

Banse, K. (1976). Rates of growth, respiration and photosynthesis of unicellular algae as related to cell size – a review. *Journal of Phycology*, 12, 135–40.

Belcher, J. H. (1968). Notes on the physiology of *Botryococcus braunii* Kützing. *Archiv für Mikrobiologie*, 61, 335–46.

Belcher, J. H. & Miller, J. D. A. (1960). Studies on the growth of Xanthophyceae in pure culture. IV. Nutritional types amongst Xanthophyceae. *Archiv für Mikrobiologie*, 36, 219–28.

Belcher, J. H. & Swale, E. M. F. (1979). *An Illustrated Guide to River Phytoplankton.* London: Her Majesty's Stationery Office.

Bellinger, E. (1974). A note on the use of algal sizes in estimates of population standing crops. *British Phycological Journal*, 9, 157–61.

Berger, C. (1975). Occurrence of *Oscillatoria agardhii* Gomont in some shallow eutrophic lakes. *Verhandlungen der internationale Vereinigung für theoretische und angewandte Limnologie*, 19, 2689–97.

Berman, T. (1976*a*). Release of dissolved organic matter by photosynthesizing algae in L. Kinneret, Israel. *Freshwater Biology*, 6, 13–18.

– (1976*b*). Light penetrance in Lake Kinneret. *Hydrobiologia*, 49, 41–8.

Berman, T., Hadas, O. & Kaplan, B. (1977). Uptake and respiration of organic compounds and heterotrophic growth in *Pediastrum duplex* (Meyen). *Freshwater Biology*, 7, 495–502.

Berman, T. & Holm-Hansen, O. (1974). Release of photoassimilated carbon as dissolved organic matter by marine phytoplankton. *Marine Biology*, 28, 305–10.

Bernhard, M. & Rampi, L. (1965). Horizontal microdistribution of marine phytoplankton in the Ligurian Sea. *Botanica gothoburgiensis*, 3, 13–24.

Besch, W. K., Ricard, M. & Cantin, R. (1972). Benthic diatoms as indicators of mining pollution in the Northwest Miramichi River system, New Brunswick, Canada. *Internationale Revue der gesamten Hydrobiologie*, 57, 39–74.

Bienfang, P. K. (1979). A new phytoplankton sinking rate method suitable for field use. *Deep-Sea Research*, 26A, 719–29.

Bienfang, P. K., Harrison, P. J. & Quarmby, L. M. (1982). Sinking rate response to depletion of nitrate, phosphate and silicate in four marine diatoms. *Marine Biology*, 67, 295–302.

Bienfang, P., Laws, E. & Johnson, W. (1977). Phytoplankton sinking rate determination: technical and theoretical aspects, an improved methodology. *Journal of Experimental Marine Biology and Ecology*, 30, 283–300.

Bindloss, M. E. (1974). Primary productivity of phytoplankton in Loch Leven, Kinross. *Proceedings of the Royal Society of Edinburgh*, B, 74, 157–81.

– (1976). The light-climate of Loch Leven, a shallow Scottish lake, in relation to primary production by phytoplankton. *Freshwater Biology*, 6, 510–18.

Birge, E. A. & Juday, C. (1922). The inland lakes of Wisconsin. The plankton. I. Its quantity and chemical composition. *Bulletin of the Wisconsin Geological and Natural History Survery*, No. 64, 222+ix pp.

Bishop, C. T., Anet, E. F. L. J. & Gorham, P. R. (1959). Isolation and identification of the fast-death factor in *Microcystis aeruginosa* NRC-1. *Canadian Journal of Biochemistry and Physiology*, 37, 453–71.

Bloesch, J. & Burns, N. M. (1980). A critical review of sedimentation trap technique. *Schweizerische Zeitschrift für Hydrologie*, 42, 15–55.

Booker, M. & Walsby, A. E. (1979). The relative form resistance of straight and helical blue–green algal filaments. *British phycological Journal*, 14, 141–50.

Bourrelly, P. (1966). *Les Algues d'Eau Douce. I. Algues Vertes.* Paris: Boubée.

– (1968). *Les Algues d'Eau Douce. II. Les Algues Jaunes et Brunes: Chrysophycées, Pheophycées. Xanthophycées et Diatomées.* Paris: Boubée.

– (1970). *Les Algues d'Eau Douce. III. Les Algues Bleues et Rouges, les Eugléniens, Peridiniens et Cryptomonadines.* Paris: Boubée.

Bowden, K. F. (1970). Turbulence. *Oceanographic and Marine Biological Annual Reviews*, 8, 11–32.

Bowers, J. A. (1979). Zooplankton grazing in similation models: the role of vertical migration. In *Perspectives on Lake Ecosystem Modelling*, ed. D. Scavia & A. Robertson, pp. 53–73. Ann Arbor, Mich.: Ann Arbor Science.

Brezonik, P. L. (1972). Nitrogen: sources and transformations in natural waters. In *Nutrients in Natural Waters*, ed. H. E. Allen & J. R. Kramer, pp. 1–47. New York: Wiley.

Brock, T. D. (1979). *Biology of Microorganisms* (3rd edition). Englewood Cliffs: Prentice-Hall.

Brook, A. J. (1971). The phytoplankton of Minnesota lakes – a preliminary survey. *Bulletin of the Water Resources Centre, University of Michigan Graduate School* No. 36. 12pp.

– (1981). *The Biology of Desmids.* Oxford: Blackwell.

Brook, A. J., Baker, A. L. & Klemer, A. R. (1971). The use of turbidimetry in studies of the population dynamics of phytoplankton populations, with special reference to *Oscillatoria agardhii* var. *isothrix. Mitteilungen der internationale Vereinigung für theoretische und angewandte Limnologie*, 19, 244–52.

Brooks, J. L. & Dodson, S. I. (1965). Predation, body size and composition of the plankton. *Science*, 150, 28–35.

Brown, E. J., Harris, R. F. & Koonie, J. F. (1978). Kinetics of phosphate uptake by aquatic micro-organisms: deviations from a simple Michaelis–Menten equation. *Limnology and Oceanography*, 23, 26–34.

Bruno, S. F. & McLaughlin, J. J. A. (1977). The nutrition of the freshwater dinoflagellate *Ceratium hirundinella. Journal of Protozoology*, 24, 548–53.

Buchanan, R. E. & Gibbons, N. E. (1974). *Bergey's Manual of Determinative Bacteriology*, 8th edition. Baltimore: Williams & Wilkins.

Burns, C. W. (1968*a*). The relationship between body size of filter-feeding cladocera and the maximum size of particle ingested. *Limnology and Oceanography*, 23, 675–8.

– (1968*b*). Direct observations of mechanisms regulating feeding behaviour of *Daphnia* in lakewater. *Internationale Revue der gesamten Hydrobiologie und Hydrographie*, 53, 83–100.

– (1969). Relation between filtering rate, temperature and body size in four species of *Daphnia. Limnology and Oceanography*, 154, 693–700.

Burns, N. M. & Rosa, F. (1980). *In situ* measurement of the settling velocity of organic carbon particles and 10 species of phytoplankton. *Limnology and Oceanography*, 25, 855–64.

Canter, H. M. (1972). A guide to the fungi occurring on planktonic blue–green algae. In *Proceedings of the Symposium on Taxonomy and Biology of Blue–green Algae*, ed. T. V. Desikachary, pp. 145–58. Madras: University of Madras.

– (1973). A new primitive protozoan devouring centric diatoms in the plankton. *Zoological Journal of the Linnean Society*, 52, 63–83.

– (1979) Fungal and protozoan parasites and their importance in the ecology of the phytoplankton. *Report, Freshwater Biological Association*, 47, 43–50.

Canter, H. M. & Jaworski, G. H. M. (1978). The isolation, maintenance and host range studies of a chytrid, *Rhizophydium planktonicum* Canter emend., parasitic on *Asterionella formosa* Hassal. *Annals of Botany*, 42, 967–79.

– (1979). The occurrence of a hypersensitive reaction in the planktonic diatom *Asterionella formosa* Hassal, parasitized by the chytrid *Rhizophydium planktonicum* Canter emend., in culture. *New Phytologist*, 82, 187–206.

– (1980). Some general observations on zoospores of the chytrid *Rhizophydium planktonicum* Canter emend. *New Phytologist*, 84, 515–31.

Canter, H. M. & Lund, J. W. G. (1953). Studies on plankton parasites. II. The parasitism of diatoms with special reference to lakes in the English Lake District. *Transactions of the British Mycological Society*, 27, 93–6.

– (1968). The importance of protozoa in controlling the abundance of planktonic algae in lakes. *Proceedings of the Linnean Society, London*, 179, 203–19.

Caperon, J. & Meyer, J. (1972). Nitrogen limited growth of marine phytoplankton. II. Uptake kinetics and their role in nutrient-limited growth of phytoplankton. *Deep-Sea Research*, 19, 619–32.

Carmichael, W. W. (1981). Freshwater blue–green algae (Cyanobacteria) toxins – a review. In *The Water Environment: Algal Toxins and Health*, ed. W. W. Carmichael, pp. 1–13. New York: Plenum.

Carmichael, W. W. and Gorham, P. R. (1978). Anatoxins from clones of *Anabaena flos-aquae* isolated from lakes of western Canada. *Mitteilungen der internationale Vereinigung für theoretische und angewandte Limnologie*, 21, 285–95.

Carpenter, E. J. & McCarthy, J. J. (1975). Nitrogen fixation and uptake of combined nitrogenous nutrients by *Oscillatoria* (*Trichodesmium*) *thiebautii* in the western Sargasso Sea. *Limnology and Oceanography*, 20, 389–401.

Carpenter, E. J. & Price, C. C. (1976). Marine *Oscillatoria* (*Trichodesmium*): explantion for aerobic nitrogen fixation without heterocysts. *Science*, 191, 1278–80.

Cassie, R. M. (1968). Sample design. In *Zooplankton Sampling. Part I: Review on Zooplankton Sampling Methods*, ed. D. J. Tranter, pp. 105–21. Paris: UNESCO.

Chapman, D. V., Dodge, J. D. & Heaney, S. I. (1980). Light and electron microscope observations on cysts and cyst formation in *Ceratium hirundinella*. *British Phycological Journal*, 15, 193.

Chapman, D. V., Livingstone, D. & Dodge, J. D. (1981). An electron-microscope study of the excystment and early development of the dinoflagellate, *Ceratium hirundinella*. *British Phycological Journal*, 16, 183–94.

Christensen, T. (1962). *Systematisk Botanik: Alger*. København: Munksgaard.

Chu, S. P. (1942). The influence of the mineral composition of the medium on the growth of planktonic algae. I. Methods and culture media. *Journal of Ecology*, 30, 284–325.

– (1943). The influence of the mineral composition of the medium on the growth of planktonic algae. II. The influence of the concentrations of inorganic nitrogen and phosphorus. *Journal of Ecology*, 31, 109–48.

Clasen, J. (1979). Das Ziel der Phosphoreliminierung am Zulauf der Wahnbachtalsperre im Hinblick auf die Olilgotrophierung dieses Gewässers. *Zeitschrift für Wasser – und Abwasser – Forschung*, 12, 65–77.

Clendenning, K. A., Brown, T. E. & Eyster, H. C. (1956). Comparative studies of photosynthesis in *Nostoc muscorum* and *Chlorella pyrenoidosa*. *Canadian Journal of Botany*, 34, 943–66.

Clesceri, N. L. & Lee, G. F. (1965). Hydrolysis of condensed polyphosphates. I: non-sterile environment. *Air and Water Pollution*, 9, 723–42.

Cloern, J. E. (1977). Effects of light intensity and temperature on *Cryptomonas ovata* (Cryptophyceae) growth and nutrient uptake. *Journal of Phycology*, 13, 389–95.

Colebrook, J. M. (1960). Plankton and water movements in Windermere. *Journal of Animal Ecology*, 29, 217–40.

Collins, M. (1978). Algal toxins. *Microbiological Reviews*, 42, 725–46.

Conway, H. L. (1977). Interactions of inorganic nitrogen in the uptake and assimilation by marine phytoplankton. *Marine Biology*, 39, 221–32.

Conway, K. & Trainor, F. R. (1972). *Scenedesmus* morphology and flotation. *Journal of Phycology*, 8, 138–43.

Cordoba-Molina, J. F., Hudgins, R. R. & Silverston, P. L. (1978). Settling in continuous sedimentation tanks. *Journal of the Environmental Division of the American Society of Civil Engineers*, 104 (**EE6**) 1263–75.

Coveney, M. F., Cronberg, G., Enell, M., Larsson, K. & Olofsson, L. (1977). Phytoplankton, zooplankton and bacteria-standing crop and production relationships in a eutrophic lake. *Oikos*, 29, 5–21.

Cronberg, G., Gelin, C. & Larsson, K. (1975). Lake Trummen restoration project. II. Bacteria, phytoplankton and phytoplankton productivity. *Verhandlungen der internationale Vereinigung für theoretische und angewandte Limnologie*, 19, 1088–96.

Daft, M. J., Begg, J. & Stewart, W. D. P. (1970). A virus of blue–green algae from freshwater habitats in Scotland. *New Phytologists*, 69, 1029–38.

Daft, M. J., McCord, S. B. & Stewart, W. D. P. (1975). Ecological studies on algal lysing bacteria in fresh waters. *Freshwater Biology*, 5, 577–96.

Daft, M. J. & Stewart, W. D. P. (1971). Bacterial pathogens of freshwater blue–green algae. *New Phytologist*, 70, 819–29.

Davison, W. (1980). Studies of chemical speciation in naturally anoxic basins. *Report of the Freshwater Biological Association*, 48, 53–9.

Denman, K. L. & Platt, T. (1975). Coherences in the horizontal distribution of phytoplankton and temperature in the upper ocean. *Memoires du Société Royale des Sciences de Liège, 6e serie*, 7, 19–30.

– (1976). The variance spectrum of phytoplankton in a turbulent ocean. *Journal of Marine Research*, 34, 593–601.

De Pinto, J. V., Bierman, V. J. & Verhoff, F. H. (1976). Seasonal phytoplankton succession as a function of species competition for phosphorus and nitrogen. In *Modelling Biochemical Processes in Aquatic Ecosystems*, ed. R. P. Canale, pp. 141–69. Ann Arbor, Mich.: Ann Arbor Science.

Dillon, P. J. & Rigler, F. H. (1974). The phosphorus–chlorophyll relationship in lakes. *Limnology and Oceanography*, 19, 767–73.

– (1975). A simple method for predicting the capacity of a lake for development based on lake trophic status. *Journal of the Fisheries Research Board of Canada*, 32, 1519–31.

Dinsdale, M. T. & Walsby, A. E. (1972). The interrelations of cell turgor pressure, gas-vacuolation and buoyancy in a blue–green alga. *Journal of Experimental Botany*, 23, 561–70.

Dobbins, W. E. (1944). Effect of turbulence on sedimentation. *Transactions of the American Society of Civil Engineers*, 109, 629–56.

Dodge, J. D. & Crawford, R. M. (1970). The morphology and fine structure of *Ceratium hirundinella* (Dinophyceae). *Journal of Phycology*, 6, 137–49.

Dokulil, M. (1979). Seasonal pattern of phytoplankton. In: *Neusiedlersee – the Limnology of a Shallow Lake in Central Europe*, ed. H. Löffler, pp. 203–31. Den Haag: Junk.

Dring, M. J. & Jewson, D. H. (1979). What does ^{14}C-uptake by phytoplankton really measure? A fresh approach using a theoretical method. *British Phycological Journal*, 14, 122–3.

Droop, M. R. (1973). Some thoughts on nutrient limitation in algae. *Journal of Phycology*, 9, 264–72.

– (1974). The nutrient status of algae cells in continuous culture. *Journal of the Marine Biological Association of the United Kingdom*, 54, 825–55.

Dubinsky, Z. & Berman, T. (1979). Seasonal changes in the spectral composition of downwelling irradiance in Lake Kinneret (Israel). *Limnology and Oceanography*, 24, 652–63.

Dugdale, R. C. (1967). Nutrient limitation in the sea: dynamics identification and significance. *Limnology and Oceanography*, 12, 685–95.

Dugdale, R. C., Dugdale, V. A., Neess, J. C. & Goering, J. J. (1959). Nitrogen fixation in lakes. *Science*, 130, 859–60.

Dumont, H. (1972). The biological cycle of molybdenum in relation to primary production. *Verhandlungen der internationale Vereinigung für theoretische und angewandte Limnologie*, 18, 84–92.

Duthie, H. C. (1965). Some observations on the ecology of desmids. *Journal of Ecology*, 53, 695–703.

Eberley, W. R. (1959). The metalimnetic oxygen maximum in Myers Lake. *Investigations of Indiana Lakes and Streams*, 5, 1–46.

– (1964). Primary production in the metalimnion of McLish Lake (Northern Indiana), an extreme plus-heterograde lake. *Verhandlungen der internationale Vereinigung für theoretische und angewandte Limnologie*, 15, 394–401.

Edmondson, W. T. (1965). Reproductive rate of planktonic rotifers as related to food and temperature in nature. *Ecological Monographs*, 35, 61–111.

– (1970). Phosphorus, nitrogen and algae in Lake Washington after diversion of sewage. *Science*, 169, 690–1.

– (1972). The present condition of Lake Washington. *Verhandlungen der internationalen Vereinigung für theoretische und angewandte Limnologie*, 19, 606–15.

– (1980). Secchi disk and chlorophyll. *Limnology and Oceanography*, 25, 378–9.

Einsele, W. & Grim, J. (1938). Über den Kieselsäuregehalt planktischer Diatomeen und dessen Bedeutung für einige Frage ihrer Ökologie. *Zeitschrift für Botanik*, 32, 545–90.

Elder, J. N. (1977). Iron uptake by freshwater algae and its diel variation. In *Biological Implications of Metals in the Environment*, pp. 346–57. Springfield, Va.: U.S. Energy Research and Development Administration.

Eppley, R. W., Holmes, R. W. & Strickland, J. D. H. (1967). Sinking rates of marine phytoplankton measured with a fluorometer. *Journal of Experimental Marine Biology and Ecology*, 1, 191–208.

Eppley, R. W., Rogers, J. N. & McCarthy, J. J. (1969). Half-saturation constants for uptake of nitrate and ammonium by marine phytoplankton. *Limnology and Oceanography*, 14, 912–20.

Evans, G. T. & Taylor, F. J. R. (1980). Phytoplankton accumulation in Langmuir cells. *Limnology and Oceanography*, 25, 840–5.

Faafeng, B. A. & Nilssen, J. P. (1981). A twenty-year study of eutrophication in a deep, soft water lake. *Verhandlungen der internationale Vereinigung für theoretische und angewandte Limnologie*, 21, 412–24.

Faller, A. J. (1971). Oceanic turbulence and the Langmuir circulation. *Annual Review of Ecology and Systematics*, 2, 201–34.

Farmer, D. M. & Takahashi, M. (1982). Effects of vertical mixing on photosynthetic responses. *Japanese Journal of Limnology*, 43, 173–81.

Fasham, M. J. & Pugh, P. R. (1976). Observations on the horizontal coherence of chlorophyll *a* and temperature. *Deep Sea Research*, 23, 527–38.

Fay, P., Stewart, W. D. P., Walsby, A. E. & Fogg, G. E. (1968). Is the heterocyst the site of nitrogen fixation in blue–green algae? *Nature*, 220, 810–2.

Fee, E. J. (1976). The vertical and seasonal distribution of chlorophyll in lakes of the Experimental Lakes Area, north western Ontario: implications for primary production estimates. *Limnology and Oceanography*, 21, 767–83.

Felföldy, L. J. M. (1960). The role of age and training in carbonate assimilation of unicellular algae. *Acta Biologica Academiae Sciencias Hungariae*, 11, 175–85.

– (1962). On the role of pH and inorganic carbon sources in photosynthesis in unicellular algae. *Acta Biologica Academiae Sciencias Hungariae*, 13, 207–14.

Fenchel, T. (1974). Intrinsic rate of natural increase: the relationship with body size. *Oecologia*, 14, 317–26.

Ferguson, A. J. D., Thompson, J. M. & Reynolds, C. S. (1982). Structure and dynamics of zooplankton communities maintained in closed systems, with special reference to the algal food supply. *Journal of Planktonic Research*, 4, 523–43.

Ferrante, J. G. & Parker, J. I. (1977). Transport of diatom frustules by copepod fecal pellets to the sediments of Lake Michigan. *Limnology and Oceanography*, 22, 92–8.

Findenegg, I. (1943). Untersuchungen über die Ökologie und die Produktionsverhältnisse des Planktons im Kärnter Seengebeite. *Internationale Revue des gesamten Hydrobiologie*, 43, 368–429.

– (1947). Über die Lichtanspruche planktischer Süsswasseralgen. *Sitzungsberichte der Akademie der Wissenschaften in Wien*, 155, 159–71.

– (1966*a*). Relationship between standing crop and primary productivity. *Memorie dell'Istituto italiano del Idrobiologia*, 18 (Suppl.), 271–89.

– (1966*b*). Factors controlling primary productivity especially with regard to water replenishment, stratification and mixing. *Memorie dell'Istituto italiano di Idrobiologia*, 18 (Suppl.), 105–19.

Fogg, G. E. (1949). Growth and heterocyst production in *Anabaena cylindrica* Lemm. II. In relation to carbon and nitrogen metabolism. *Annals of Botany (new series)*, 13, 241–59.

– (1966). The extracellular products of algae. *Oceanographic and Marine Biology Annual Review*, 4, 195–212.

– (1971). Extracellular products of algae in freshwater. *Ergebnisse der Limnologie*, 5, 1–25.

– (1975). *Algal Cultures and Phytoplankton Ecology*. 2nd edition. London: University of London.

Fogg, G. E. & Belcher, J. H. (1961). Physiological studies on a planktonic 'μ-alga'. *Verhandlungen der internationalen Vereinigung für theoretische und angewandte Limnologie*, 14, 893–6.

Fogg, G. E., Stewart, W. D. P., Fay, P. & Walsby, A. E. (1973). *The Blue–green Algae*. London: Academic Press.

Fott, B. (1959). *Algenkunde*. Jena: Gustav Fischer.

Foy, R. H. (1980). The influence of surface to volume ratio on the growth rates of planktonic blue–green algae. *British Phycological Journal*, 15, 279–89.

Foy, R. H., Gibson, C. E. & Smith, R. V. (1976). The influence of daylength, light intensity and temperature on the growth rates of planktonic blue–green algae. *British Phycological Journal*, 11, 151–63.

Friedman, M. M. (1980). Comparative morphology and functional significance of copepod receptors and oral structures. In *Evolution and Ecology of Zooplankton Communities*, ed. W. C. Kerfoot, pp. 185–97. Hanover, N.H.: University Press of New England.

Fritz, F. (1935). Uber die sinkgeschwindigkeit einiger Planktonorganismen. *Internationale Revue des gesamten Hydrobiologie und Hydrographie*, 32, 424–31.

Fuhs, G. W., Demmerle, S. D., Canelli, E. & Miu Chiu (1972). Characterization of phosphorus-limited plankton algae. *Special Symposia of the American Society of Limnology and Oceanography*, 1, 113–33.

Ganf, G. G. (1974*a*). Diurnal mixing and the vertical distribution of phytoplankton in a shallow equatorial lake (Lake George, Uganda). *Journal of Ecology*, 62, 611–29.

– (1974*b*). Incident solar radiation and underwater light penetration as factors controlling the chlorophyll *a* content of a shallow equatorial lake (L. George, Uganda). *Journal of Ecology*, 62, 593–609.

– (1975). Photosynthetic production and irradiance-photosynthesis relationships of the phytoplankton from a shallow equatorial lake (L. George, Uganda). *Oecologia*, 18, 165–83.

Ganf, G. G. & Viner, A. B. (1973). Ecological stability in a shallow equatorial lake (L. George, Uganda). *Proceedings of the Royal Society, London*, B 184, 321–46.

Gardiner, A. C. (1941). Silicon and phosphorus as factors limiting development of diatoms. *Journal of the Society for the Chemical Industry, London*, 60, 73–8.

Gavis, J. & Ferguson, J. F. (1975). Kinetics of carbon dioxide uptake by phytoplankton at high pH. *Limnology and Oceanography*, 20, 211–21.

Gelin, C. (1975). Nutrients, biomass and primary productivity of nannoplankton in eutrophic Lake Vombsjön, Sweden. *Oikos*, 26, 121–39.

George, D. G. (1976). A pumping system for collecting horizontal samples and recording continuously sampling depth, water temperature, turbidity and *in vivo* chlorophyll. *Freshwater Biology*, 6, 413–19.

– (1981*a*). Wind-induced water movements in the South Basin of Windermere. *Freshwater Biology*, 11, 37–60.

– (1981*b*). The spatial distribution of nutrients in the South Basin of Windermere. *Freshwater Biology*, 11, 405–24.

– (1981*c*). Zooplankton patchiness. *Report of the Freshwater Biological Association*, 49, 32–44.

– (1983). Interactions between zooplankton and phytoplankton distribution profiles in two large limnetic enclosures. *Journal of Plankton Research*, 5, 457–75.

George, D. G. & Edwards, R. W. (1973). *Daphnia* distributions within Langmuir circulations. *Limnology and Oceanography*, 18, 798–800.

– (1976). The effect of wind on the distribution of chlorophyll *a* and crustacean plankton in a shallow eutrophic reservoir. *Journal of Applied Ecology*, 13, 667–90.

George, D. G. & Heaney, S. I. (1978). Factors influencing the spatial distribution of phytoplankton in a small productive lake. *Journal of Ecology*, 66, 133–55.

Gerloff, G. C., Fitzgerald, G. P. & Skoog, F. (1952). The mineral nutrition of *Microcystis aeruginosa*. *American Journal of Botany*, 39, 26–32.

Gerloff, G. E. & Skoog, F. (1954). Cell contents of nitrogen and phosphorus as a measure of their availability for growth in *Microcystis aeruginosa*. *Ecology*, 35, 348–53.
– (1957). Nitrogen as a limiting factor for the growth of *Microcystis aeruginosa* in southern Wisconsin lakes. *Ecology*, 38, 556–61.
Gessner, F. (1955). *Hydrobotanik*, vol. 1: *Energiehausalt*. Berlin: VEB Deutscher Verlag der Wissenschaften.
Gibbons, N. E. & Murray, R. G. E. (1978). Proposals concerning the higher taxa of Bacteria. *International Journal of Systematic Bacteriology*, 28, 1–6.
Gibbs, M., Latzko, E., Harvey, M. J., Plaut, Z. & Shain, Y. (1976). Photosynthesis in the algae. *Annals of the New York Academy of Sciences*, 175, 541–54.
Gibson, C. E. (1971). Nutrient limitation. *Journal of the Water Pollution Control Federation*, 43, 2436–40.
– (1978). Field and laboratory observations on the temporal and spatial variation of carbohydrate content in planktonic blue–green algae in Lough Neagh, Northern Ireland. *Journal of Ecology*, 66, 97–115.
– (1981). Silica budgets and the ecology of planktonic diatoms in an unstratified lake (Lough Neagh, N. Ireland). *Internationale Revue des gesamten Hydrobiologie*, 66, 641–64.
Gibson, C. E., Wood, R. B., Dickson, E. L. & Jewson, D. M. (1971). The succession of phytoplankton in L. Neagh, 1968–1970. *Mitteilungen der internationale Vereinigung für theoretische und angewandte Limnologie*, 19, 146–60.
Gliwicz, Z. M. (1969). The food sources of lake zooplankton. *Ekologia polska, Seria B*, 15, 205–23.
– (1970). Calculation of food ration of zooplankton community as an example of using laboratory data for field conditions. *Polskie Archiwum Hydrobiologii*, 17, 169–75.
– (1975). Effect of zooplankton grazing on photosynthetic activity and composition of phytoplankton. *Verhandlungen der internationale Vereinigung für theoretische und angewandte Limnologie*, 19, 1490–7.
– (1977). Food size selection and seasonal succession of filter feeding zooplankton in an eutrophic lake. *Ekologia polska Seria A*, 25, 179–225.
– (1980). Filtering rates, food size selection and feeding rates in cladocerans – another aspect of interspecific competition in filter-feeding zooplankton. In *Evolution and Ecology of Zooplankton Communities*, ed. W. C. Kerfoot, pp. 282–91. Hanover, N.H.: University Press of New England.
Gliwicz, Z. M., Ghilarov, A. & Pijanowska, J. (1981). Food and predation as major factors limiting two natural populations of *Daphnia cucullata* Sars. *Hydrobiologia*, 80 205–18.
Gliwicz, Z. M. & Hillbricht-Ilkowska, A. (1972). Efficiency of utilization of nanoplankton primary production by communities of filter-feeding animals measured *in situ*. *Verhandlungen der internationale Vereinigung für theoretische und angewandte Limnologie*, 18, 197–203.
Gliwicz, Z. M. & Preis, A. (1977). Can planktivorous fish keep in check planktonic crustacean populations? A test of size-efficiency hypothesis. *Ekologia polska Seria A*, 25, 567–91.
Glooschenko, W. A., Moore, J. E. & Vollenweider, R. A. (1974). Spatial and temporal distribution of chlorophyll *a* and pheopigments in surface waters of Lake Erie. *Journal of the Fisheries Research Board of Canada*, 31, 265–74.

Goldman, C. R. (1960). Molybdenum as a factor limiting primary productivity in Castle Lake, California. *Science*, 132, 1016–17.

– (1964). Primary productivity and micro-nutrient limiting factors in some north American and New Zealand lakes. *Verhandlungen der internationale Vereinigung für theoretische und angewandte Limnologie*, 15, 365–74.

Goldstein, M. (1964). Speciation and mating behaviour in *Eudorina*. *Journal of Protozoology*, 11, 317–44.

Golterman, H. L., Bakels, C. C. & Jakobs-Möglin, J. (1969). Availability of mud phosphates for the growth of algae. *Verhandlungen der internationale Vereinigung für theoretische und angewandte Limnologie*, 17, 467–79.

Goryushin, V. A. & Chaplinskaya, S. M. (1968). Finding the viruses lysing blue–green algae (In Russian). In *Tsveteniye Vody*, ed. A. V. Topachevskii *et al*., pp. 171–4. Kiev: Naukova Dumka.

Gotham, I. J. & Rhee, G.-Y. (1981). Comparative kinetic studies of phosphate-limited growth and phosphate uptake in phytoplankton in continuous culture. *Journal of Phycology*, 17, 257–65.

Granhall, U. (1978). Environmental role of nitrogen-fixing blue–green algae and asymbiotic bacteria. *Ecological Bulletins* (*Stockholm*), No. 26, 391 pp.

Grant, N. G. & Walsby, A. E. (1977). The contribution of photosynthate to turgor pressure rise in the planktonic blue–green alga, *Anabaena flos-aquae*. *Journal of Experimental Botany*, 28, 409–15.

Griffiths, B. M. (1939). Early references to waterblooms in British lakes. *Proceedings of the Linnean Society of London*, 151, 12–19.

Grim, J. (1939). Beobachtungen am Phytoplankton des Bodensees (Obersee) sowie deren rechnerische Auswertung. *Internationale Revue des gesamten Hydrobiologie und Hydrographie*, 39, 193–315.

– (1950). Versuche zur Ermittung der Produktions koeffizienten einiger Planktophyten in einem flachen See. *Biologischen Zentralblatt*, 69, 147–73.

Gross, F. & Zeuthen, E. (1948). The buoyancy of planktonic diatoms: a problem of cell physiology. *Proceedings of the Royal Society of London*, B 135, 382–9.

Grünberg, H. (1968). Über das Zeta-Potential zweier synchron kultivierter *Chlorella*-Stämme. *Archiv für Hydrobiologie* (*Supplementband*), 33, 331–62.

Haffner, G. D., Harris, G. P. & Jarai, M. K. (1980). Physical variability and phytoplankton communities. III. Vertical structure in phytoplankton populations. *Archiv für Hydrobiologie*, 89, 363–81.

Haines, D. A. & Bryson, R. A. (1961). An empirical study of wind factor in Lake Mendota, *Limnology and Oceanography*, 6, 356–64.

Hairston, N. G., Tinkle, D. W. & Wilbur, H. M. (1970). Natural selection and the parameters for population growth. *Journal of Wildlife Management*, 34, 681–90.

Halbach, U. & Halbach-Keup, G. (1974). Quantitative Beziehungen zwischen Phytoplankton und der Populations dynamik des Rotators *Brachionus calyciflorus* Pallas. Befunde aus Laboratoriumsexperimenten und Freiland untersuchungen. *Archiv für Hydrobiologie*, 73, 273–309.

Hall, D. J., Threlkeld, S. T., Burns, C. W. & Crowley, P. H. (1976). The size efficiency hypothesis and the structure of zooplankton communities. *Annual Review of Ecology and Systematics*, 7, 177–208.

Halldal, P. (1970). *Photobiology of Microorganisms*. New York: Wiley.

Hammer, U. T., Walker, K. F. & Williams, W. D. (1973). Derivation of daily phytoplankton production estimates from short-term experiments in some shallow

eutrophic Australian saline lakes. *Australian Journal of Marine and Freshwater Research*, 24, 259–66.
Haney, J. F. (1971). An *in situ* method for the measurement of zooplankton grazing rates. *Limnology and Oceanography*, 16, 971–7.
– (1973). An *in situ* examination of the grazing activities of natural zooplankton communities. *Archiv für Hydrobiolgie*, 72, 88–132.
Hardin, G. (1960). The competitive exclusion principle. *Science*, 131, 1292–7.
Hardy, A. (1936). The continuous plankton recorder. *Discovery Reports*, 11, 457–510.
Hardy, R. W. F. (1977). *Dinitrogen Fixation.* New York: Wiley.
Harris, G. P. (1973). Diel and annual cycles of net plankton photosynthesis in Lake Ontario. *Journal of the Fisheries Research Board of Canada*, 30, 1779–87.
– (1978). Photosynthesis, productivity and growth: the physiological ecology of phytoplankton. *Ergebnisse der Limnologie*, 10, 1–163.
– (1980*a*). Temporal and spatial scales in phytoplankton ecology. Mechanisms, methods, models and management. *Canadian Journal of Fisheries and Aquatic Sciences*, 37, 877–900.
– (1980*b*). The measurement of photosynthesis in natural populations of phytoplankton. In *The Physiological Ecology of Phytoplankton*, ed. I. Morris, pp. 129–87. Oxford: Blackwell.
Harris, G. P., Heaney, S. I. & Talling, J. F. (1979). Physiological and environmental constraints in the ecology of the planktonic dinoflagellate *Ceratium hirundinella. Freshwater Biology*, 9, 413–28.
Harris, G. P. & Lott, J. N. A. (1973). Observations of Langmuir circulations in Lake Ontario. *Limnology and Oceanography*, 18, 584–9.
Harris, G. P. & Piccinin, B. B. (1977). Photosynthesis by natural phytoplankton populations. *Archiv für Hydrobiologie*, 80, 405–57.
– (1980). Physical variability and phytoplankton communities. IV. Temporal changes in the phytoplankton community of a physically variable lake. *Archiv für Hydrobiologie*, 89, 447–73.
Harris, G. P. & Smith, R. H. (1977). Observations of small scale spatial patterns of phytoplankton populations. *Limnology and Oceanography*, 22, 887–99.
Haworth, E. Y. (1969). The diatoms of a sediment core from Blea Tarn, Langdale. *Journal of Ecology*, 57, 429–39.
– (1980). Comparison of continuous phytoplankton records with the diatom stratigraphy in the recent sediments of Blelham Tarn. *Limnology and Oceanography*, 25, 1093–103.
Healey, F. P. (1973). Characteristics of phosphorus deficiency in *Anabaena. Journal of Phycology*, 9, 383–94.
Heaney, S. I. (1970). The toxicity of *Microcystis aeruginosa* Kütz. from some English reservoirs. *Water Treatment and Examination*, 20, 235–44.
– (1976). Temporal and spatial distribution of the dinoflagellate *Ceratium hirundinella* O. F. Müller within a small productive lake. *Freshwater Biology*, 6, 531–42.
– (1978). Some observations on the use of the *in vivo* fluorescence technique to determine chlorophyll *a* in natural populations and cultures of freshwater phytoplankton. *Freshwater Biology*, 8, 115–26.
Heaney, S. I., Chapman, D. V. & Morison, H. R. (1981). The importance of the cyst stage in the seasonal growth of the dinoflagellate, *Ceratium hirundinella*, in a small, productive lake. *British Phycological Journal*, 16, 136.
Heaney, S. I. & Eppley, R. W. (1981). Light, temperature and nitrogen as interacting

factors affecting diel vertical migrations of dinoflagellates in culture. *Journal of Plankton Research*, 3, 331–44.

Heaney, S. I. & Furnass, T. I. (1980). Laboratory models of diel vertical migration in the dinoflagellate *Ceratium hirundinella. Freshwater Biology*, 10, 163–70.

Heaney, S. I. & Talling, J. F. (1980*a*). Dynamic aspects of dinoflagellate distribution patterns in a small, productive lake. *Journal of Ecology*, 68, 75–94.

– (1980*b*). *Ceratium hirundinella* – ecology of a complex, mobile and successful plant. *Report of the Freshwater Biological Association*, 48, 27–40.

Hecky, R. E. & Kling, H. J. (1981). The phytoplankton and protozooplankton of the euphotic zone of Lake Tanganyika; species composition, biomass, chlorophyll content and spatio-temporal distribution. *Limnology and Oceanography*, 26, 548–64.

Hegewald, E. (1972). Untersuchungen zum Zetapotential von Planktonalgen. *Archiv für Hydrobiologie* (*Supplementband*), 42, 14–90.

Hensen, V. (1887). Über die Bestimmung des Planktons oder des in Meere treibenden Matierials an Pflanzen und Tieren. *Bericht des Deutschen wissenschaftlichen Kommission für Meereforschung*, 5, 1–109.

Herman, A. W. & Denman, K. L. (1977). Rapid underway profiling of chlorophyll with an *in situ* fluorometer mounted on a 'Batfish' vehicle. *Deep-sea Research*, 24, 385–97.

Hesslein, R. & Quay, P. (1973). Vertical eddy diffusion studies in the thermocline of a small stratified lake. *Journal of the Fisheries Research Board of Canada*, 30, 1495–500.

Hill, R. & Bendall, F. (1960). Function of two cytochrome components in chloroplasts: a working hypothesis. *Nature*, 186, 136–7.

Hillbricht-Ilkowska, A., Spodniewska, I. & Węgleńska, T. (1979). Changes in the phytoplankton–zooplankton relationship connected with the eutrophication of lakes. *Symposia Biologica Hungariae*, 19, 59–75.

Höll, K. (1928). Ökologie der Peridineen. Studien über den Einfluss chemischer und physikalischer Faktoren auf die Verbreitung der Dinoflagellaten im Süsswasser. *Pflanzenforsuchungen, Jena*, No. 11, 105 pp.

Hoogenhout, H. & Amesz, J. (1965). Growth rates of photosynthetic microorganisms in laboratory cultures. *Archiv für Mikrobiologie*, 50, 10–25.

Horne, A. J. & Goldman, C. R. (1972). Nitrogen fixation in Clear Lake, California. I. Seasonal variation and the role of heterocysts. *Limnology and Oceanography*, 17, 678–92.

Horne, A. J. & Wrigley, R. C. (1975). The use of remote sensing to detect how wind influences planktonic blue–green algal distribution. *Verhandlungen der internationale Vereinigung für theoretische und angewandte Limnologie*, 19, 784–91.

Horwood, J. W. (1978). Observations on spatial heterogeneity of surface chlorophyll in one and two dimensions. *Journal of the Marine Biological Association of the United Kingdom*, 58, 487–502.

Huber, G. & Nipkow, F. (1922). Experimentelle Untersuchungen über die Entwicklung von *Ceratium hirundinella. Zeitschrift für Botanik*, 14, 337–71.

Hughes, E. O., Gorham, P. R. & Zehnder, A. (1958). Toxicity of a unialgal culture of *Microcystis. Canadian Journal of Microbiology*, 4, 225–36.

Hughes, J. C. & Lund, J. W. G. (1962). The rate of growth of *Asterionella formosa* Hass. in relation to its ecology. *Archiv für Mikrobiologie*, 42, 117–29.

Huntsman, S. A. & Sunda, W. G. (1980). The role of trace metals in regulating phytoplankton growth. In *The Physiological Ecology of Phytoplankton*, ed. I. Morris, pp. 285–328. Oxford: Blackwell.

Hutchinson, G. E. (1944). Limnological studies in Connecticut. VII. A critical

examination of the supposed relationship between phytoplankton periodicity and chemical changes in lake waters. *Ecology*, 25, 3–26.
– (1961). The paradox of the plankton. *American Naturalist*, 95, 137–46.
– (1967). *A Treatise on Limnology, Vol. 2. Introduction to Lake Biology and the Limnoplankton*. New York: Wiley.
Irish, A. E. (1980). A modified 1-m Friedinger sampler – a description and some selected results. *Freshwater Biology*, 10, 135–9.
Ives, K. J. (1956). Electrokinetic phenomena of planktonic algae. *Proceedings of the Society for Water Treatment and Examination*, 5, 41–58.
Järnefelt, H. (1955). Uber die Sedimentation des Sestons. *Verhandlungen der internationale Vereinigung für theoretische und angewandte Limnologie*, 12, 144–58.
Jassby, A. D. & Goldman, C. R. (1974*a*). Loss rates from a phytoplankton community. *Limnology and Oceanography*, 19, 618–27.
– (1974*b*). A quantitative measure of succession rate and its application to the phytoplankton of lakes. *American Naturalist*, 108, 688–93.
Jaworski, G. H. M., Talling, J. F. & Heaney, S. I. (1981). The influence of carbon dioxide-depletion on growth and sinking rate of two planktonic diatoms in culture. *British Phycological Journal*, 16, 395–410.
Jensen, A. & Sakshaug, E. (1973). Studies on the phytoplankton ecology of the Trondheimsfjord. II. Chloroplast pigments in relation to the abundance and physiological state of the phytoplankton. *Journal of Experimental Marine Biology and Ecology*, 11, 137–55.
Jewson, D. H. (1976). The interaction of components controlling net phytoplankton photosynthesis in a well-mixed lake (Lough Neagh, Northern Ireland). *Freshwater Biology*, 6, 551–76.
– (1977). Light penetration in relation to phytoplankton content of the euphotic zone of Lough Neagh, N. Ireland. *Oikos*, 28, 74–83.
Jewson, D. H., Rippey, B. H. & Gilmore, W. K. (1981). Loss rates from sedimentation, parasitism and grazing during the growth, nutrient limitation and dormancy of a diatom crop. *Limnology and Oceanography*, 26, 1045–56.
Jewson, D. H. & Wood, R. B. (1975). Some effects on integral photosynthesis of artificial circulation of phytoplankton through light gradients. *Verhandlungen der internationale Vereinigung für theoretische und angewandte Limnologie*, 19, 1037–44.
Jónasson, P. & Adalsteinsson, H. (1979). Phytoplankton production in the shallow eutrophic Lake Mývatn, Iceland. *Oikos*, 32, 113–38.
Jones, H. R., Lack, T. J. & Jones, C. S. (1979). Population dynamics and production of *Daphnia hyalina* var. *lacustris* in Farmoor I, a shallow eutrophic reservoir. *Journal of Plankton Research*, 1, 45–65.
Jones, J. G. (1977). The study of aquatic microbial communities. In *Aquatic microbiology*, ed. F. A. Skinner & J. M. Shewan, pp. 1–30. London: Academic.
– (1979). A guide to methods for estimating microbial numbers and biomass in freshwater. *Scientific Publications of the Freshwater Biological Association*, No. 39, 112 pp.
Jones, J. G. & Simon, B. M. (1980). Decomposition processes in the profundal region of Blelham Tarn and the Lund Tubes. *Journal of Ecology*, 68, 493–512.
Jones, R. A. & Lee, G. F. (1980). Application of U.S. OECD eutrophication study results to deep lakes. *Progress in Water Technology*, 12(2), 81–101.
Jørgensen, S. E. (1964). Adaptation to different light intensities in the diatom *Cyclotella meneghiniana* Kütz. *Physiologia Plantarum*, 17, 136–45.

Joseph, J. & Sendner, H. (1958). Über die horizontale Diffusion im Meere. *Deutsches hydrographische Zeitschrift*, 11, 51–77.

Kahn, N. & Swift, E. (1978). Positive buoyancy through ionic control in the non-motile marine dinoflagellate *Pyrocystis noctiluca* Murray ex Schuett. *Limnology and Oceanography*, 23, 649–58.

Kajak, Z. M., Hillbricht-Ilkowska, A. & Pieczynska, A. (1972). The production processes in several Polish lakes. In *Productivity Problems of Freshwaters*, ed. Z. M. Kajak & A. Hillbricht-Ilkowska, pp. 129–47. Warszawa and Kraków: PWN.

Kalff, J. (1972). Net plankton and nanoplankton production and biomass in a north temperate zone lake. *Limnology and Oceanography*, 17, 712–20.

Kalff, J., Kling, H. J., Holmgren, S. H. & Welch, H. E. (1975). Phytoplankton, phytoplankton growth and biomass cycles in an unpolluted and a polluted polar lake. *Verhandlungen der internationale Vereinigung für theoretische und angewandte Limnologie*, 19, 487–95.

Kalff, J. & Knoechel. R. (1978). Phytoplankton and their dynamics in oligotrophic and trophic lakes. *Annual Review of Ecology and Systematics*, 9, 475–95.

Kalff, J. & Welch, H. E. (1974). Plankton production in Char Lake, a natural polar lake, and in Meretta Lake, a polluted lake, Cornwallis Island, Northwest Territories. *Journal of the Fisheries Research Board of Canada*, 31, 621–36.

Keating, K. I. (1977). Allelopathic influence on blue–green bloom sequence in a eutrophic lake. *Science*, 196, 885–6.

Kelly, M. G., Church, M. R. & Hornberger, G. M. (1974). A solution of the inorganic carbon mass balance equation and its relation to algal growth rates. *Water Resources Research*, 10, 493–7.

Kelly, M. G., Hornberger, G. M. & Cosby, B. J. (1974). Continuous automated measurement of rates of photosynthesis and respiration in an undisturbed river community. *Limnology and Oceanography*, 19, 305–12.

Kerfoot, W. C. (1980). *Evolution and Ecology of Zooplankton Communities*. Hanover, N.H.: University Press of New England.

Ketchum, B. H. & Redfield, A. C. (1949). Some physical and chemical characteristics of algae growth in mass cultures. *Journal of Cellular and Comparative Physiology*, 13, 373–81.

Kibby, H. V. (1971). Energetics and population dynamics of *Diaptomus gracilis*. *Ecological Monographs*, 41, 311–27.

Kiefer, D. A., Olson, R. J. & Wilson, W. H. (1979). Reflectance spectroscopy of marine phytoplankton. I. Optical properties related to age and growth rate. *Limnology and Oceanography*, 24, 664–72.

Kierstead, H. & Slobodkin, L. B. (1953). The size of water masses containing plankton blooms. *Journal of Marine Research*, 12, 141–7.

Kilham, P. & Kilham, S. S. (1980). The evolutionary ecology of phytoplankton. In *The Physiological Ecology of Phytoplankton*, ed. I. Morris, pp. 571–97. Oxford: Blackwell.

Kilham, P. & Titman, D. (1976). Some biological effects of atmospheric inputs to lakes: nutrient ratios and competitive interactions between phytoplankton. *Journal of Great Lakes Research*, 2 (supplement 1), 187–91.

Kilham, S. S. (1978). Nutrient kinetics of freshwater planktonic algae using batch and semicontinuous methods. *Mitteilungen der internationale Vereinigung für theoretische und angewandte Limnologie*, 21, 147–57.

Kirk, J. T. O. (1974). The contribution of phytoplankton to the attenuation of light within natural waters. In *Proceedings of the 3rd International Congress on Photosynthesis*, ed. M. Avron, pp. 245–53. Amsterdam: Elsevier.

– (1975*a*). A theoretical analysis of the contribution of phytoplankton to the attenuation of light within natural waters. I. General treatment of suspensions of pigmented cells. *New Phytologist*, 75, 11–20.
– (1975*b*). A theoretical analysis of the contribution of algal cells to the attenuation of light within natural waters. II. Spherical cells. *New Phytologist*, 75, 21–36.
– (1976). Yellow substance (Gelbstoff) and its contribution to the attenuation of photosynthetically active radiation in some inland and coastal south-eastern Australian waters. *Australian Journal of Marine and Freshwater Research*, 27, 61–72.
Klebahn, H. (1895). Gasvakuolen, ein Bestandtiel der Zellen der Wasserblütebildenden Phycochromaceen. *Flora, Jena*, 80, 241–82.
Klemer, A. R. (1976). The vertical distribution of *Oscillatoria agardhii* var. *isothrix*. *Archiv für Hydrobiologie*, 78, 343–62.
Kling, H. (1975). Phytoplankton successions and species distributions in prairie ponds of the Erickson–Elphinstone District, south-western Manitoba. *Technical Reports of the Fisheries and Marine Services of Canada*, No. 512, 31 pp.
Knoechel, R. & Kalff, J. (1975). Algal sedimentation: the cause of a diatom – blue–green succession. *Verhandlungen der internationale Vereinigung für theoretische und angewandte Limnologie*, 19, 745–54.
– (1976). Track autoradiography, a method for the determintion of phytoplankton species productivity. *Limnology and Oceanography*, 21, 590–6.
– (1978). An *in situ* study of the productivity and population dynamics of five freshwater plankton diatom species. *Limnology and Oceanography*, 23, 195–218.
Knudsen, B. M. (1953). The diatom genus *Tabellaria* II. Taxonomy and morphology of the plankton varieties. *Annals of Botany*, 17, 131–55.
Kolkwitz, R. (1911). Über das Kammerplankton des Süsswassers und der Meere. *Bericht der Deutschen botanischen Gesellschaft*, 29, 386–402.
– (1912). Plankton und Seston. *Bericht der Deutschen botanischen Gesellschaft*, 30, 334–46.
Kolmogoroff, A. N. (1941). The local structure of turbulence in incompressible viscous fluid for very large Reynolds Numbers. (English Summary of paper in Russian). *Doklady Akademii nauk SSSR*, 30, 301.
Konopka, A. E. & Brock, T. D. (1978). Effect of temperature on blue–green algae (Cyanobacteria) in Lake Mendota. *Applied and Environmental Microbiology*, 36, 572–6.
Konopka, A. E., Brock, T. D. & Walsby, A. E. (1978). Buoyancy regulation by *Aphanizomenon* in Lake Mendota. *Archiv für Hydrobiologie*, 83, 524–37.
Kowalczewski, A. & Lack, T. J. (1971). Primary production and respiration of the phytoplankton of the River Thames and Kennet at Reading. *Freshwater Biology*, 1, 197–212.
Kratz, W. A. & Myers, J. (1955). Nutrition and growth of several blue–green algae. *American Journal of Botany*, 42, 282–7.
Krüger, G. H. J. & Eloff, J. N. (1977). The influence of light intensity on the growth of different *Microcystis* isolates. *Journal of the Limnological Society of Southern Africa*, 3, 21–5.
Kuentzel, L. E. (1969). Bacteria, carbon dioxide and algal blooms. *Journal of the Water Pollution Control Federation*, 41, 1737–47.
Kuznetsov, S. I. (1968). Recent studies on the role of microorganisms in the cycling of substances in lakes. *Limnology and Oceanography*, 13, 211–23.
Lam, C. W. Y. & Silvester, W. B. (1979). Growth interactions among blue–green (*Anabaena oscillaroides, Microcystis aeruginosa*) and green (*Chlorella* sp.) algae. *Hydrobiologia*, 63, 135–43.

Lampert, W. (1977*a*). Studies on the carbon balance of *Daphnia pulex* De Geer as related to environmental conditions. II. The dependence of carbon assimilation on animal size, temperature, food concentration and diet species. *Archiv für Hydrobiologie* (*Supplementband*), 48, 310–35.

– (1977*b*). Ibid. III. Production and production efficiency. *Archiv für Hydrobiologie* (*Supplementband*), 48, 336–60.

– (1977*c*). Ibid. IV. Determination of the 'threshold' concentration as a factor controlling the abundance of zooplankton species. *Archiv für Hydrobiologie* (*Supplementband*), 48, 361–8.

Langmuir, I. (1938). Surface motion of water induced by wind. *Science*, 87, 119–23.

Lapage, S. P., Sneath, P. H. A., Lessel, E. F., Skerman, V. B. D., Seeliger, P. R. & Clark, W. A. (1975). *International Code of Nomenclature of Bacteria*. Washington: American Society for Microbiology.

Laws, E. A. (1975). The importance of respiration losses in controlling the size distribution of marine phytoplankton. *Ecology*, 56, 419–26.

Leach, J. H. (1972). Distribution of chlorophyll *a* and related variables in Ontario waters of Lake St Clair. *Proceedings of the 15th Conference on Great Lakes Research*, 80–6.

Leah, R. T., Moss, B. & Forrest, D. E. (1980). The role of predation in causing major changes in the limnology of a hyper-eutrophic lake. *Internationale Revue der gesamten Hydrobiologie und Hydrographie*, 65, 223–47.

Lean, D. R. S. (1973). Movements of phosphorus between its biologically important forms in lake water. *Journal of the Fisheries Research Board of Canada*, 30, 1525–36.

Le Cren, E. D. (1981). Report of the Director. *Report of the Freshwater Biological Association*, 49, 23–5.

Lee, G. F., Rast, W. & Jones, R. A. (1978). Eutrophication of water bodies: insights for an age-old problem. *Environmental Science and Technology*, 12, 900–8.

Lehman, J. T. (1976). Ecological and nutritional studies on *Dinobryon* Ehrenb.: seasonal periodicity and the phosphate toxicity problem. *Limnology and Oceanography*, 21, 646–58.

– (1978). Enhanced transport of inorganic carbon into algal cells and its implications for the biological fixation of carbon. *Journal of Phycology*, 14, 33–42.

– (1979). Physical and chemical factors affecting the seasonal abundance of *Asterionella formosa* Hass. in a small temperate lake. *Archiv für Hydrobiologie*, 87, 274–303.

– (1980). Nutrient recycling as an interface between algae and grazers in freshwater communities. In *Evolution and Ecology of Zooplankton Communities*, ed. W. C. Kerfoot, pp. 251–63. Hanover, N.H.: University Press of New England.

Lehman, J. T., Botkin, D. B. & Likens, G. E. (1975*a*). Lake eutrophication and the limiting CO_2 concept: a simulation study. *Verhandlungen der internationale Vereinigung für theoretische und angewandte Limnologie*, 19, 300–7.

– (1975*b*). The assumptions and rationales of a computer model of phytoplankton population dynamics. *Limnology and Oceanography*, 20, 343–64.

Lekan, J. F. & Wilson, R. E. (1978). Spatial variability of phytoplankton biomass in the surface waters of Long Island. *Estuarine and Coastal Marine Science*, 6, 239–51.

Lemmermann, E. (1910). *Algen. I. Schizophyceen, Flagellaten, Peridineen*. Leipzig: Brandenburg.

Lévêque, C., Carmouze, J. P., Dejoux, C., Durand, J. R., Gras, R., Iltis, A., Lemoalle, J., Loubens, G., Lauzanne, L. & Saint-Jean, L. (1972). Recherches sur les biomasses et la productivité du Lac Tchad. In *Productivity Problems of Freshwaters*, ed. Z. Kajak & A. Hillbricht-Ilkowska, pp. 165–81. Warszawa & Kraków. PWN.

Levich, V. G. (1962). *Physicochemical Hydromechanics*. Englewood Cliffs, N.J.: Prentice-Hall.

Lewis, W. M. (1976). Surface/volume ratio: implication for phytoplankton morphology. *Science*, 192, 885–7.

– (1978*a*). Analysis of succession in a tropical phytoplankton community and a new measure of succession rate. *American Naturalist*, 112, 401–14.

– (1978*b*). Dynamics and succession of the phytoplankton in a tropical lake: Lake Lanao, Philippines. *Journal of Ecology*, 66, 849–80.

Liere, L. van (1979). *On* Oscillatoria agardhii *Gomont. Experimental Ecology and Physiology of a Nuisance Bloomforming Cyanobacterium*. Zeist: De Nieuwe Schouw.

Liere, L. van, Zevenboom, W. & Mur, L. R. (1975). Nitrogen as a limiting factor for the growth of the blue–green alga *Oscillatoria agardhii*. *Proceedings of the Conference on Nitrogen as a Water Pollutant*. København: IAWPR.

Likens, G. E., Bormann, F. H., Johnson, N. M., Fisher, D. W. & Pierce, R. S. (1970). Effects of forest cutting and herbicide treatment on nutrient budgets in the Hubbard Brook watershed-ecosystem. *Ecological Monographs*, 40, 23–47.

Lin, C. K. (1972). Phytoplankton succession in a eutrophic lake with special reference to blue–green algal blooms. *Hydrobiologia*, 39, 321–34.

Lingeman, R. & Vermij, S. (1980). Estimation of primary productivity in aquatic systems using free oxygen measurements. *Water Resources Bulletin*, 16, 745–8.

Livingstone, D. & Jaworski, G. H. M. (1980). The viability of akinetes of blue–green algae recovered from the sediments of Rostherne Mere. *British Phycological Journal*, 15, 357–64.

Lloyd, D. (1974). Dark respiration. In *Algal Physiology and Biochemistry*, ed. W. D. P. Stewart, pp. 505–29. Oxford: Blackwell.

Lloyd, M. (1967). 'Mean crowding'. *Journal of Animal Ecology*, 36, 1–30.

Lohmann, H. (1911). Über das Nannoplankton und die Zentrifugierung Kleinster Wasserproben zur Gewinnung desselben in lebendem Zustande. *Internationale Revue der gesamten Hydrobiologie und Hydrographie*, 4, 1–38.

Lorenzen, C. J. (1966). A method for the continuous measurement of *in vivo* chlorophyll concentration. *Deep Sea Research*, 13, 223–7.

– (1971). Continuity in the distribution of surface chlorophyll. *Journal du Conseil internationale pour l'Exploration de la Mer*, 34, 18–23.

Lorenzen, M. W. (1974). Predicting the effect of nutrient diversion on lake recovery. In *Modelling the Eutrophication Process*, ed. E. J. Middlebrooks, D. H. Falkenborg & T. E. Maloney, pp. 205–21. Ann Arbor: Ann Arbor Science.

Luftig, R. & Haselkorn, R. (1967). Morphology of a virus of blue–green algae and properties of its deoxyribonucleic acid. *Journal of Virology*, 1, 344–61.

Lund, J. W. G. (1949). Studies on *Asterionella*. I. The origin and nature of the cells producing seasonal maxima. *Journal of Ecology*, 37, 389–419.

– (1950). Studies on *Asterionella formosa* Hass. II. Nutrient depletion and the spring maximum. *Journal of Ecology*, 38, 1–35.

– (1954). The seasonal cycle of the planktonic diatom *Melosira italica* (Ehr.) Kütz. subsp. *subarctica* O. Mull. *Journal of Ecology*, 42, 151–79.

– (1957). Fungal diseases of plankton algae. In *Biological Aspects of the Transmission of Disease*, ed. C. Horton-Smith, pp. 19–24. Edinburgh: Oliver & Boyd.

– (1959). Buoyancy in relation to the ecology of the freshwater phytoplankton. *British Phycological Bulletin*, 1(7), 1–17.

– (1961). The algae of the Malham Tarn district. *Field Studies*, 1(3), 85–115.

– (1964). Primary production and periodicity of phytoplankton. *Verhandlungen der internationale Vereinigung für theoretische und angewandte Limnologie*, 15, 37–56.
– (1965). The ecology of the freshwater phytoplankton. *Biological Reviews of the Cambridge Philosophical Society*, 40, 231–93.
– (1966*a*). The importance of turbulence in the periodicity of certain freshwater species of the genus *Melosira* (In Russian). *Botanicheskĭ Zhurnal SSSR*, 51, 176–87.
– (1966*b*). Synecological problems – summation. In *Marine Biology II*, ed. C. H. Oppenheimer, pp. 227–49. New York: New York Academy of Sciences.
– (1970). Primary production. *Water Treatment and Examination*, 19, 332–58.
– (1971). An artificial alteration of the seasonal cycle of the plankton diatom *Melosira italica* subsp. *subarctica* in an English lake. *Journal of Ecology*, 59, 521–33.
– (1972*a*). Changes in the biomass of blue–green and other algae in an English lake from 1945–1969. In *Proceedings of the Symposium on Taxonomy and Biology of Blue–green Algae*, ed. T. V. Desikachary, pp. 305–27. Madras: University of Madras Press.
– (1972*b*). Eutrophication. *Proceedings of the Royal Society of London*, B 180, 371–82.
– (1975). The use of large experimental tubes in lakes. In *The Effects of Storage on Water Quality*, ed. R. E. Youngman, pp. 291–311. Medmenham: Water Research Centre.
– (1978). Changes in the phytoplankton of an English lake, 1945–1977. *Hydrobiological Journal*, 14(1), 6–21.
– (1981). Investigations on phytoplankton, with special reference to water usage. *Occasional Publications of the Freshwater Biological Association*, No. 13, 64 pp.
Lund, J. W. G., Jaworski, G. H. M. & Butterwick, C. (1975). Algal bioassay of water from Blelham Tarn, English Lake District, and the growth of planktonic diatoms. *Archiv für Hydrobiologie* (*Supplementband*), 49, 49–69.
Lund, J. W. G., Mackereth, F. J. H. & Mortimer, C. H. (1963). Changes in depth and time of certain chemical and physical conditions and of the standing crop of *Asterionella formosa* Hass. in the north basin of Windermere in 1947. *Philosophical Transactions of the Royal Society of London*, B 246, 255–90.
Lund, J. W. G. & Reynolds, C. S. (1982). The development and operation of large limnetic enclosures in Blelham Tarn, English Lake District, and their contribution to phytoplankton ecology. *Progress in Phycological Research*, 1, 1–65.
Macan, T. T. (1970). *Biological Studies on the English Lakes*. London: Longman.
MacArthur, R. H. & Wilson, E. O. (1967). *The Theory of Island Biogeography*. Princeton: Princeton University Press.
Mackereth, F. J. H. (1953). Phosphorus utilization by *Asterionella formosa* Hass. *Journal of Experimental Botany*, 4, 296–313.
Mague, T. H. (1977). Ecological aspects of dinitrogen fixation by blue–green algae. In *Dinitrogen Fixation*, Vol. II., ed. R. W. F. Hardy, pp. 85–140. New York: Wiley.
Maloney, T. E., Miller, W. R. & Blind, N. L. (1973). Use of algal assays in studying eutrophication problems. In *Advances in Water Pollution Research*, ed. S. H. Jenkins, pp. 205–15. Oxford: Pergamon.
Manny, B. A. (1972). Seasonal changes in organic nitrogen content of net- and nanoplankton in two hardwater lakes. *Archiv für Hydrobiologie*, 71, 103–23.
Margalef, R. (1957). Nuevos aspectos del problema de la suspension en los organismos planctonicos. *Investigación Pesquera*, 7, 105–16.
– (1958). Temporal succession and spatial heterogeneity in phytoplankton. In *Perspectives in Marine Biology*, ed. A. A. Buzzati-Traverso, pp. 323–49. Berkeley: University of California Press.
– (1964). Correspondence between the classic types of lakes and the structural and

dynamic properties of their populations. *Verhandlungen der internationale Vereinigung für theoretische und angewandte Limnologie*, 15, 169–75.
– (1968). *Perspectives in Ecological Theory*. Chicago: University of Chicago Press.
– (1977). *Ecologia*. Barcelona: Omega.
– (1978). Life-forms of phytoplankton as survival alternatives in an unstable environment. *Oceanologia Acta*, 1, 493–509.
Marra, L. (1978*a*). Effect of short term variations in light intensity on photosynthesis of a marine phytoplankter – laboratory simulation study. *Marine Biology*, 46, 191–202.
– (1978*b*). Phytoplankton photosynthetic response to vertical movement in a mixed layer. *Marine Biology*, 46, 203–8.
Masters, M. J. (1971). The ecology of *Chytridium deltanum* and other fungus parasites on *Oocystis* spp. *Canadian Journal of Botany*, 49, 75–87.
May, L. (1980). On the ecology of *Notholca squamula* Müller in Loch Leven, Kinross, Scotland. *Hydrobiologia*, 73, 177–80.
McAlice, B. J. (1970). Observations on the small-scale distribution of estuarine phytoplankton. *Marine Biology*, 7, 100–11.
McCarthy, J. J. (1980). Nitrogen. In *The physiological Ecology of Phytoplankton*, ed. I. Morris, pp. 191–233. Oxford: Blackwell.
McCarthy, J. J., Taylor, W. R. & Taft, J. L. (1975). The dynamics of nitrogen and phosphorus cycling in the open waters of Chesapeake Bay. In *Marine Chemistry in the Coastal Environment*, ed. T. M. Church, pp. 664–81. Washington: American Chemical Society.
McCombie, A. M. (1953). Factors influencing the growth of phytoplankton. *Journal of the Fisheries Research Board of Canada*, 10, 253–82.
McLachlan, J. & Gorham, P. R. (1961). Growth of *Microcystis aeruginosa* Kütz in a precipitate-free medium buffered with tris. *Canadian Journal of Microbiology*, 7, 869–82.
McMahon, J. W. E. & Rigler, F. H. (1965). Feeding rate of *Daphnia magna* (Strauss) in different foods labelled with radioactive phosphorus. *Limnology and Oceanography*, 10, 105–13.
McNaught, D. C. (1979). Considerations of scale in modeling large aquatic ecosystems. In *Perspectives on Lake Ecosystem Modeling*, ed. D. Scavia & A. Robertson, pp. 3–24. Ann Arbor: Ann Arbor Science.
McNown, J. S. & Malaika, J. (1950). Effect of particle shape on settling velocity at low Reynolds numbers. *Transactions of the American Geophysical Union*, 31, 74–82.
McQueen, D. J. (1970). Grazing rates and food selection in *Diaptomus oregonensis* from Marion Lake, B.C. *Journal of the Fisheries Research Board of Canada*, 27, 13–20.
Meffert, M.-E. (1971). Cultivation and growth of two planktonic *Oscillatoria* species. *Mitteilungen der internationalen Vereinigung für theoretische und angewandte Limnologie*, 19, 189–205.
Melack, J. M. (1979). Temporal variability of phytoplankton in tropical lakes. *Oecologia*, 44, 1–7.
Moed, J. R. (1973). Effect of combined action of light and silicon depletion on *Asterionella formosa* Hass. *Verhandlungen der internationale Vereinigung für theoretische und angewandte Limnologie*, 18, 1367–74.
Moll, R. A. & Rohlf, F. J. (1981). Analysis of temporal and spatial phytoplankton variability in a Long Island salt marsh. *Journal of Experimental Marine Biology and Ecology*, 51, 133–44.
Morgan, K. C. & Kalff, J. (1979). Effect of light and temperature interactions on the growth of *Cryptomonas erosa* (Cryptophyceae). *Journal of Phycology*, 15, 127–34.

Morris, I. (1974). Nitrogen assimilation and protein synthesis. In *Algal Physiology and Biochemistry*, ed. W. D. P. Stewart, pp. 583–609. Oxford: Blackwell.

Mortimer, C. H. (1974). Lake hydrodynamics. *Mitteilungen der internationale Vereinigung für theoretische und angewandte Limnologie*, 20, 124–97.

Moss, B. (1967). Vertical heterogeneity in the water column of Abbot's Pond. II. The influence of physical and chemical conditions on the spatial and temporal distribution of the phytoplankton and of a community of epipelic algae. *Journal of Ecology*, 57, 397–414.

– (1973*a*). The influence of environmental factors on the distribution of freshwater algae: an experimental study. II. The role of pH and the carbon dioxide-bicarbonate system. *Journal of Ecology*, 61, 157–77.

– (1973*b*). The influence of environmental factors on the distribution of freshwater algae: an experimental study. III. Effects of temperature, vitamin requirements and inorganic nitrogen compounds on growth. *Journal of Ecology*, 61, 179–92.

Mullin, J. B. & Riley, J. P. (1955). The colorimetric determination of silicate with special reference to sea and natural waters. *Analytica chimica Acta*, 12, 162–76.

Mullin, M. M., Sloan, P. R. & Eppley, R. W. (1966). Relationship between carbon content, cell volume and area in phytoplankton. *Limnology and Oceanography*, 11, 307–11.

Munawar, M. & Munawar, I. F. (1975). Some observations on the growth of diatoms in Lake Ontario, with emphasis on *Melosira binderana* Kütz, during thermal bar conditions. *Archiv für Hydrobiologie* 75, 490–9.

Mur, L., Gons, H. J., Liere, L. van (1978). Competition of the green alga *Scenedesmus* and the blue–green alga *Oscillatoria*. *Mitteilungen der internationale Vereinigung für theoretische und angewandte Limnologie*, 21, 473–9.

Murphy, T. P., Lean, D. R. S. & Nalewajko, C. (1976). Blue–green algae: their excretion of iron-selective chelators enables them to dominate other algae. *Science*, 192, 900–2.

Nadin-Hurley, C. M. & Duncan, A. (1976). A comparison of Daphnid gut particles with the sestonic particles in two Thames Valley reservoirs throughout 1970 and 1971. *Freshwater Biology*, 6, 109–23.

Nakamoto, N. (1975). A freshwater red tide on a water reservoir. *Japanese Journal of Limnology*, 36, 55–64.

Nalewajko, C. (1966). Dry weight, ash and volume data for some freshwater planktonic algae. *Journal of the Fisheries Research Board of Canada*, 23, 1285–8.

Nalewajko, C. & Lean, D. R. S. (1978). Phosphorus kinetics–algal growth relationships in batch cultures. *Mitteilungen der internationale Vereinigung für theoretische und angewandte Limnologie*, 21, 184–92.

– (1980). Phosphorus. In *The Physiological Ecology of Phytoplankton*, ed. I. Morris, pp. 325–58. Oxford: Blackwell.

Nasev, D., Nasev, S. & Guiard, V. (1978). Statistische Auswertung von Planktonuntersuchungen. Teil II. Räumliche Verteilung des Planktons. Konfidenzintervalle für die Individuendichte (Ind./l) des Phytoplanktons. *Wissenschaftliche Zeitschrift der Wilhelm-Pieck-Universität Rostock*, 27, 357–61.

Naumann, E. (1919). Några synpunkter angående limnoplanktons ökologi med särskild hänsyn till fytoplankton. *Svensk Botanisk Tidskrift*, 13, 129–63.

Nauwerck, A. (1963). Die Beziehungen zwischen zooplankton und phytoplankton in See Erken. *Symbolae Botanicae Upsaliensis*, 17(5), 1–163.

Neilands, J. B. (1973). Microbial iron transport compounds (siderochromes). In *Inorganic Biochemistry, Vol. I* (ed. B. L. Eichhorn), pp. 167–202. Amsterdam: Elsevier.

Nilssen, J. P. (1978). Eutrophication, minute algae and inefficient grazers. *Memorie dell'Istituto italiano di Idrobiologia*, 36, 121–38.
O'Brien, W. J. (1974). The dynamics of nutrient limitation of phytoplankton algae: a model reconsidered. *Ecology*, 55, 135–41.
Odum, E. P. (1971). *Fundamentals of Ecology*. Philadelphia: Saunders.
Oglesby, R. T. (1977). Phytoplankton summer standing crops and annual productivity as functions of phosphorus loading and various physical factors. *Journal of the Fisheries Research Board of Canada*, 34, 2255–70.
Oglesby, R. T. & Schaffner, W. R. (1975). The response of lakes to phosphorus. In *Nitrogen and Phosphorus. Food Production, Waste and the Environment*, ed. K. S. Porter, pp. 23–57. Ann Arbor, Mich.: Ann Arbor Science.
– (1978). Phosphorus loadings to lakes and some of their responses. Part 2. Regression models of summer phytoplankton standing crops, winter total P and transparency of New York lakes with known phosphorus loadings. *Limnology and Oceanography*, 23, 135–45.
Okubo, A. (1971). Oceanic diffusion diagrams. *Deep-Sea Research*, 18, 789–802.
– (1978). Horizontal dispersion and critical scales for phytoplankton patches. In *Spatial Pattern in Plankton Communities*, NATO Conference Series IV, No. 3, ed. J. H. Steele, pp. 21–42. New York: Plenum Press.
Oliver, R. K., Kinnear, A. J. & Ganf, G. G. (1981). Measurements of cell density of three freshwater phytoplankters by density gradient centrifugation. *Limnology and Oceanography*, 26, 285–94.
Olrik, K. (1981). Succession of phytoplankton in response to environmental factors in Lake Arresø, North Zealand, Denmark. *Schweizerische Zeitschrift für Hydrologie*, 43, 6–15.
Oren, A. & Padan, E. (1978). Induction of anaerobic, photoautotrophic growth in the cyanobacterium, *Oscillatoria limnetica. Journal of Bacteriology*, 133, 558–63.
Osborne, P. L. & Moss, B. (1977). Palaeolimnology and trends in the phosphorus and iron budgets of an old, man-made lake, Barton Broad, Norfolk. *Freshwater Biology*, 7, 213–33.
Overbeck, J. (1968). Prinzipielles zum Vorkommen der Bakterien im See. *Mitteilungen der internationale Vereinigung für theoretische und angewandte Limnologie*, 14, 134–44.
Owens, M. (1974). In running waters. In *A Manual on Methods for Measuring Primary Production in Aquatic Environments*, ed. R. A. Vollenweider (IBP Handbook No. 12, Second Edition), pp. 111–19. Oxford. Blackwell.
Owens, O. v. H. & Esaias, W. E. (1976). Physiological responses of phytoplankton to major environmental factors. *Annual Review of Plant Physiology*, 27, 461–83.
Ozmidov, R. V. (1958). On the calculation of horizontal diffusion of the pollutant patches in the sea (In Russian). *Doklady Akademii Nauk, SSSR*, 120, 761–3.
Paasche, E. (1960). On the relationship between primary production and standing stock of phytoplankton. *Journal du Conseil internationale pour l'Exploration de la Mer*, 26, 33–48.
– (1980). Silicon. In *The Physiological Ecology of Phytoplankton*, ed. I. Morris, pp. 259–84. Oxford: Blackwell.
Padan, E. & Cohen, Y. (1982). Anoxygenic photosynthesis. In *The Biology of the Cyanobacteria*, ed. by N. G. Carr & B. A. Whitton, pp. 215–35. Oxford: Blackwell.
Parsons, T. R. & Takahashi, M. (1973). *Biological Oceanographic Processes*. New York: Pergamon.

Patten, B. C. (1968). Mathematical models of plankton production. *Internationale Revue der gesamten Hydrobiologie*, 53, 357–408.

Pavoni, M. (1963). Die Bedeutung des Nannoplanktons im Vergleich zum Netzplankton. *Schweizerische Zeitschrift für Hydrologie*, 25, 219–341.

Pearsall, W. H. (1922). A suggestion as to factors influencing the distribution of free-floating vegetation. *Journal of Ecology*, 9, 241–53.

– (1930). Phytoplankton in the English Lakes. 1. The proportions in the water of some dissolved substances of biological importance. *Journal of Ecology*, 18, 306–20.

– (1932). Phytoplankton in the English Lakes. 2. The composition of the phytoplankton in relation to dissolved substances. *Journal of Ecology*, 20, 241–62.

Pennington, W. T. (1978). The impact of man on some English lakes: rates of change. *Polskie Archiwum Hydrobiologii*, 25, 429–37.

Petersen, R. (1975). The paradox of the plankton: an equilibrium hypothesis. *American Naturalist*, 109, 35–49.

Peterson, B. J., Hobbie, J. E. & Haney, J. F. (1978). *Daphnia* grazing on natural bacteria. *Limnology and Oceanography*, 23, 1039–44.

Pianka, E. R. (1970). On *r*- and *K*-selection. *American Naturalist*, 104, 592–7.

– (1972). *r*- and *K*-selection or *b*- and *d*-selection. *American Naturalist*, 106, 581–8.

Pielou, E. C. (1966). Species-diversity and pattern diversity in the study of ecological succession. *Journal of Theoretical Biology*, 10, 370–83.

Pingree, R. D., Holligan, P. M., Mardell, G. T. & Head, R. N. (1976). The influence of physical stability on spring, summer and autumn phytoplankton blooms in the Celtic Sea. *Journal of the Marine Biological Association of the United Kingdom*, 56, 845–873.

Platt, T. (1972). Local phytoplankton abundance and turbulence. *Deep-Sea Research*, 19, 183–7.

– (1975). The physical environment and spatial structure of phytoplankton populations. *Memoires du Société Royale des Sciences de Liège*, *6ᵉ série*, 7, 9–17.

Platt, T. & Denman, K. L. (1975). Spectral analysis in ecology. *Annual Review of Ecology and Systematics*, 6, 189–210.

– (1980). Patchiness in phytoplankton distribution. In *The Physiological Ecology of Phytoplankton*, ed. I. Morris, pp. 413–31. Oxford: Blackwell.

Platt, T., Dickie, L. M. & Trites, R. W. (1970). Spatial heterogeneity of phytoplankton in a near-shore environment. *Journal of the Fisheries Research Board of Canada*, 27, 1453–73.

Platt, T. & Jassby, A. D. (1976). The relationship between photosynthesis and light for natural assemblages of marine phytoplankton. *Journal of Phycology*, 12, 421–30.

Pollingher, U. & Berman, T. (1975). Temporal and spatial patterns of dinoflagellate blooms in Lake Kinneret, Israel (1969–1974). *Verhandlungen der internationale Vereinigung für theoretische und angewandte Limnologie*, 19, 1370–82.

Porter, J. & Jost, M. (1973). Light-shielding by gas vacuoles in *Microcystis aeruginosa*. *Journal of General Microbiology*, 75, xxii.

– (1976). Physiological effects of the presence and absence of gas vacuoles in the blue–green alga, *Microcystis aeruginosa* Kuetz. emend. Elenkin. *Archives of Microbiology*, 110, 225–31.

Porter, K. G. (1973). Selective grazing and differential digestion of algae by zooplankton. *Nature*, 244, 179–80.

– (1977). The plant–animal interface in freshwater ecosystems. *American Scientist*, 65, 159–70.

Pourriot, R. (1977). Food and feeding habits of Rotifera. *Ergebnisse der Limnologie*, 8, 243–60.

Powell, T. & Jassby, A. D. (1974). The estimation of vertical eddy diffusivities below the thermocline in lakes. *Water Resources Research*, 10, 191–8.

Powell, T. M., Richerson, P. J., Dillon, T. M., Agee, B. A., Dozier, B. J., Godden, D. A. & Myrup, L. O. (1975). Spatial scales of current speed and phytoplankton biomass fluctuations in Lake Tahoe. *Science*, 189, 1088–90.

Pratt, D. M. (1965). The winter–spring diatom flowering in Narrangansett Bay. *Limnology and Oceanography*, 15, 113–20.

Provasoli, L., McLaughlin, J. J. A. & Pintner, I. A. (1954). Relative and limiting constituents for the growth of algal flagellates. *Transactions of the New York Academy of Science* (*Series* 2), 16, 412–17.

Prowse, G. A. & Talling, J. F. (1958). The seasonal growth and succession of plankton algae in the White Nile. *Limnology and Oceanography*, 3, 222–38.

Quay, P. D., Broecker, W. S., Hesslein, R. H. & Schindler, D. W. (1980). Vertical diffusion rates determined by tritium tracer experiments in the thermocline and hypolimnion of two lakes. *Limnology and Oceanography*, 25, 201–18.

Ramsbottom, A. E. (1976). Depth charts of the Cumbrian lakes. *Scientific Publications of the Freshwater Biological Association*, No. 33. 40 pp.

Raven, J. A. (1974). Carbon dioxide fixation. In *Algal Physiology and Biochemistry*, ed. W. D. P. Stewart, pp. 391–423. Oxford: Blackwell.

Ravera, O. & Vollenweider, R. A. (1968). *Oscillatoria rubescens* D.C. as an indicator of pollution. *Schweizerische Zeitschrift für Hydrologie*, 30, 374–80.

Redfield, A. C. (1958). The biological control of chemical factors in the environment. *American Scientist*, 46, 205–21.

Reynolds, C. S. (1971). The ecology of the planktonic blue–green algae in the north Shropshire meres. *Field Studies*, 3, 409–32.

– (1972). Growth, gas vacuolation and buoyancy in a natural population of a planktonic blue–green alga. *Freshwater Biology*, 2, 87–106.

– (1973*a*). The seasonal periodicity of planktonic diatoms in a shallow eutrophic lake. *Freshwater Biology*, 3, 89–110.

– (1973*b*). Growth and buoyancy of *Microcystis aeruginosa* Kütz. emend. Elenkin in a shallow eutrophic lake. *Proceedings of the Royal Society of London*, B 184, 29–50.

– (1973*c*). Phytoplankton periodicity of some north Shropshire meres. *British Phycological Journal*, 8, 301–20.

– (1973*d*). The phytoplankton of Crose Mere, Shropshire. *British Phycological Journal*, 8, 153–62.

– (1975*a*). Interrelations of photosynthetic behaviour and buoyancy regulation in a natural population of a blue–green alga. *Freshwater Biology*, 5, 323–38.

– (1975*b*). Temperature and nutrient concentration in the characterization of the water supply to a small kataglacial lake basin. *Freshwater Biology*, 5, 339–56.

– (1976*a*). Succession and vertical distribution of phytoplankton in response to thermal stratification in a lowland lake, with special reference to nutrient availability. *Journal of Ecology*, 64, 529–51.

– (1976*b*). Sinking movements of phytoplankton indicated by a simple trapping method. I. A *Fragilaria* population. *British Phycological Journal*, 11, 279–91.

– (1976*c*). Sinking movements of phytoplankton indicated by a simple trapping method. II. Vertical activity ranges in a stratified lake. *British Phycological Journal*, 11, 293–303.

– (1978*a*). Stratification in natural populations of bloom-forming blue–green algae. *Verhandlungen der internationale Vereinigung für theoretische und angewandte Limnologie*, 20, 2285–92.
– (1978*b*). The plankton of the north-west Midland meres. *Occasional Papers of the Caradoc and Severn Valley Field Club*, No. 2. 36+xxiii pp.
– (1978*c*). Phosphorus and the eutrophication of lakes – a personal view. In *Phosphorus in the Environment: Its Chemistry and Biochemistry*, ed. R. Porter & D. FitzSimons (CIBA Foundation Symposium No. 57, new series), pp. 201–28. Amsterdam: Excerpta Medica.
– (1978*d*). Notes on the phytoplankton periodicity of Rostherne Mere, Cheshire, 1967–1977. *British Phycological Journal*, 13, 329–35.
– (1979*a*). Seston sedimentation: experiments with *Lycopodium* spores in a closed system. *Freshwater Biology*, 9, 55–76.
– (1979*b*). The limnology of the eutrophic meres of the Shropshire–Cheshire Plain: a review. *Field Studies*, 5, 93–173.
– (1979*c*). Eutrophication explained. *Salmon & Trout Magazine* No. 215, 43–7.
– (1980*a*). Phytoplankton assemblages and their periodicity in stratifying lake systems. *Holarctic Ecology*, 3, 141–59.
– (1980*b*). *Estudos sobre o fitoplancton en lagos*. São Carlos: Universidade de São Carlos.
– (1980*c*). Processes controlling the quantities of biogenic materials in lakes and reservoirs subject to cultural eutrophication. *Pollution Reports of the Department of the Environment*, No. 8, 45–62.
– (1982). Phytoplankton periodicity: its motivation, mechanisms and manipulation. *Report of the Freshwater Biological Association*, 50, 60–75.
Reynolds, C. S. & Allen, S. E. (1968). Changes in the phytoplankton of Oak Mere following the introduction of base-rich water. *British Phycological Bulletin*, 3, 451–62.
Reynolds, C. S. & Butterwick, C. (1979). Algal bioassay of unfertilized and artificially fertilized lake water, maintained in Lund Tubes. *Archiv für Hydrobiologie* (*Supplementband*), 56, 166–83.
Reynolds, C. S. & Jaworski, G. H. M. (1978). Enumeration of natural *Microcystis* populations. *British Phycological Journal*, 13, 269–77.
Reynolds, C. S., Jaworski, G. H. M., Cmiech, H. A. & Leedale, G. F. (1981). On the annual cycle of the blue–green alga *Microcystis aeruginosa* Kütz. emend. Elenkin. *Philosophical Transactions of the Royal Society of London*, B 293, 419–77.
Reynolds, C. S., Morison, H. R. & Butterwick, C. (1982*a*). The sedimentary flux of phytoplankton in the south basin of Windermere. *Limnology and Oceanography*, 27, 1162–75.
Reynolds, C. S. & Rodgers, M. W. (1983). Cell- and colony-division in *Eudorina* (Chlorophyta: Volvocales) and some ecological implications. *British Phycological Journal*, 18, 111–19.
Reynolds, C. S. & Rogers, D. A. (1976). Seasonal variations in the vertical distribution and buoyancy of *Microcystis aeruginosa* Kütz. emend. Elenkin in Rostherne Mere, England. *Hydrobiologia*, 48, 17–23.
Reynolds, C. S., Thompson, J. M., Ferguson, A. J. D. & Wiseman, S. W. (1982*b*). Loss processes in the population dynamics of phytoplankton maintained in closed systems. *Journal of Plankton Research*, 4, 561–600.
Reynolds, C. S., Tundisi, J. G. & Hino, K. (1983*a*). Observations on a metalimnetic *Lyngbya* population in a stably stratified tropical lake (Lagoa Carioca, Eastern Brasil). *Archiv für Hydrobiologie*, 97, 7–17.

Reynolds, C. S. & Walsby, A. E. (1975). Water-blooms. *Biological Reviews of the Cambridge Philosophical Society*, 50, 437–81.

Reynolds, C. S. & Wiseman, S. W. (1982). Sinking losses of phytoplankton in closed limnetic systems. *Journal of Plankton Research*, 4, 489–522.

Reynolds, C. S., Wiseman, S. W., Godfrey, B. M. & Butterwick, C. (1983*b*). Some effects of artificial mixing on the dynamics of phytoplankton populations in large limnetic enclosures. *Journal of Plankton Research*, 5, 203–34.

Reynolds, J. H., Middlebrooks, E. J., Porcella, D. B. & Grenney, W. J. (1975). Effects of temperature on growth constants of *Selenastrum capricornutum*. *Journal of the Water Pollution Control Federation*, 47, 2420–36.

Rhee, G.-Y. (1973). A continuous culture study of phosphate uptake, growth rate and polyphosphate in *Scenedesmus* sp. *Journal of Phycology*, 9, 495–506.

– (1978). Effects of N:P atomic ratios and nitrate limitation on algal growth, cell composition and nitrate uptake. *Limnology and Oceanography*, 23, 10–25.

Rhee, G.-Y. & Gotham, I. J. (1980). Optimum N:P ratios and co-existence of planktonic algae. *Journal of Phycology*, 16, 486–9.

Rhee, G.-Y. & Gotham, I. J. & Chisholm, S. W. (1981). Use of cyclostat cultures to study phytoplankton ecology. In *Continuous Cultures of Cells, Vol. II*, ed. P. H. Calcott, pp. 159–86. Boca Raton, Fla: CRC.

Richerson, P. & Armstrong, R. & Goldman, C. R. (1970). Contemporaneous disequilibrium, a new hypothesis to explain the paradox of the plankton. *Proceedings of the National Academy of Sciences*, 67, 1710–14.

Richerson, P. J., Powell, T. M., Leigh-Abbott, M. R. & Coil, J. A. (1978). Spatial heterogeneity in closed basins. In *Spatial Pattern in Plankton Communities* (NATO Conference Series IV, No. 3), ed. J. H. Steele, pp. 239–76. New York: Plenum Press.

Richman, S., Bohon, S. A. & Robbins, S. E. (1980). Grazing interactions among freshwater calanoid copepods. In *Evolution and Ecology and Zooplankton Communities*, ed. W. C. Kerfoot, pp. 219–33, Hanover, N.H.: University Press of New England.

Ridley, J. E. (1970). The biology and management of eutrophic reservoirs. *Water Treatment and Examination*, 19, 374–99.

Rigler, F. H. (1966). Radiobiological analysis of inorganic phosphorus in lakewater. *Verhandlungen der internationale Vereinigung für theoretische und angewandte Limnologie*, 16, 465–70.

– (1968). Further observations inconsistent with the hypothesis that the molybdenum blue method measures orthophosphate in lakewater. *Limnology and Oceanography*, 13, 7–13.

Riley, G. A., Stommel, H. & Bumpus, D. F. (1949). Quantitative ecology of the plankton of the western North Atlantic. *Bulletin of the Bingham Oceanographic Collection*, 12, 1–169.

Rippka, R., Neilson, A., Kunisawa, R. & Cohen-Bazire, C. (1971). Nitrogen fixation by unicellular blue–green algae. *Archiv für Mikrobiologie*, 76, 341–8.

Rodhe, W. (1948). Environmental requirements of freshwater plankton algae: experimental studies in the ecology of phytoplankton. *Symbolae Botanicae Upsaliensis*, 10, 5–149.

– (1955). Can plankton production proceed during winter darkness in sub-arctic lakes? *Verhandlungen der internationale Vereinigung für theoretische und angewandte Limnologie*, 12, 117–22.

Rodhe, W., Vollenweider, R. A. & Nauwerck, A. (1958). The primary production and standing crop of phytoplankton. In *Perspectives in Marine Biology*, ed. A. A. Buzzati-Traverso, pp. 299–322. Berkeley: University of California Press.

Roelofs, T. D. & Oglesby, R. T. (1970). Ecological observations on the planktonic cyanophyte *Gloeotrichia echinulata*. *Limnology and Oceanography*, 15, 244–9.

Rohlich, G. A. (1969). *Eutrophication: Causes, Consequences, Correctives*. Washington: U.S. National Academy of Sciences.

Rother, J. A. & Fay, P. (1977). Sporulation and the development of planktonic blue–green algae in two Salopian meres. *Proceedings of the Royal Society of London*, B 196, 317–32.

– (1979*a*). Some physiological-biochemical characteristics of planktonic blue–green algae during bloom-formation in three Salopian meres. *Freshwater Biology*, 9, 369–79.

– (1979*b*). Blue–green algal growth and sporulation in response to simulated surface-bloom conditions. *British Phycological Journal*, 14, 59–68.

Round, F. E. (1965). *The Biology of the Algae*. London: Arnold.

– (1971). The growth and succession of algal populations in freshwaters. *Mitteilungen der internationale Vereinigung für theoretische und angewandte Limnologie*, 19, 70–99.

Ruttner, F. (1930). Das Plankton des Lunzer Untersees: seine Verteilung in Raum und Zeit wärend der Jahre 1908–1913. *Internationale Revue der gesamten Hydrobiologie und Hydrographie*, 23, 1–138 and 161–287.

– (1938). Limnologische Studien an einigen Seen der Östalpen. *Archiv für Hydrobiologie*, 32, 167–319.

– (1953). *Fundamentals of Limnology* (Translation of *Grundriss der Limnologie* by D. G. Frey & F. E. D. Fry). Toronto: University of Toronto.

Ryther, J. H. & Guillard, R. R. L. (1962). Marine planktonic diatoms. IV. Some effects of temperature on respiration of five species. *Canadian Journal of Microbiology*, 8, 447–53.

Ryther, J. H., Menzel, D. W., Hulburt, E. M., Lorenzen, C. J. & Corwin, N. (1971). The production and utilization of organic matter in the Peru coastal current. *Investigación Pesquera* 35, 43–59.

Safferman, R. S. & Morris, M. E. (1963). Algal virus: isolation. *Science*, 140, 679–80.

Sakamoto, M. (1966). Primary production by phytoplankton community in some Japanese lakes and its dependence on lake depth. *Archiv für Hydrobiologie*, 62, 1–28.

Sakshaug, E. (1980). Problems in the methodology of studying phytoplankton. In *The Physiological Ecology of Phytoplankton*, ed. by I. Morris, pp. 57–91. Oxford: Blackwell.

Sandusky, J. C. & Horne, A. J. (1978). A pattern analysis of Clear Lake phytoplankton. *Limnology and Oceanography*, 23, 636–48.

Sargent, J. R. (1976). The structure, metabolism and function of lipids in marine organisms. In *Biochemical and Biophysical Perspectives in Marine Biology*, ed. by D. C. Malins & J. R. Sargent, pp. 150–212. London: Academic.

Schaffner, W. R. & Oglesby, R. T. (1978). Phosphorus loadings to lakes and some of their responses. Part 1. A new calculation of phosphorus loading and its application to 13 New York lakes. *Limnology and Oceanography*, 23, 120–34.

Schelske, C. L. & Stoermer, E. F. (1971). Eutrophication, silica depletion and predicted changes in the algal quality in Lake Michigan. *Science*, 172, 423–4.

Schindler, D. W. (1971). Carbon, nitrogen and phosphorus and the eutrophication of freshwater lakes. *Journal of Phycology*, 7, 321–9.

– (1977). Evolution of phosphorus limitation in lakes. *Science*, 196, 260–2.

– (1978). Factors regulating phytoplankton production and standing crop in the world's freshwaters. *Limnology and Oceanography*, 23, 478–86.

Schindler, D. W., Fee, E. J. & Ruszczynski, T. (1978). Phosphorus input and its consequences for phytoplankton standing crop and production in the Experimental Lakes Area and similar lakes. *Journal of the Fisheries Research Board of Canada*, 35, 190–6.

Schindler, D. W., Kalff, J., Welch, H. E., Brunskill, G. J., Kling, H. & Kritsch, N. (1974). Eutrophication in the high arctic – Meretta Lake, Cornwallis Island (75°N lat.). *Journal of the Fisheries Research Board of Canada*, 31, 647–62.

Schmidt, W. (1925). *Des Massenaustausch in freier Luft und verwandte Erscheinungen.* (Probleme der Kosmischen Physik, Vol. 7). Hamburg: Grand.

Scott, J. T., Myer, G. E., Stewart, R. & Walther, E. G. (1969). On the mechanism of Langmuir circulations and their role in epilimnion mixing. *Limnology and Oceanography*, 14, 493–503.

Shapiro, J. (1967). Iron available to algae. In *Chemical Environment in the Aquatic Habitat* (ed. H. L. Golterman & R. S. Clymo), pp. 219–28. Amsterdam: Noord-Hollandsche.

– (1973). Blue–green algae: why they become dominant. *Science*, 179, 382–4.

Sharp, J. H. (1977). Excretion of organic matter by marine phytoplankton. Do healthy cells do it? *Limnology and Oceanography*, 22, 381–99.

Shilo, M. (1970). Lysis of blue–green algae by *Myxobacter*. *Journal of Bacteriology*, 104, 453–61.

Simpson, F. B. & Neilands, J. B. (1976). Siderochromes in cyanophyceae: isolation and characterization of schizokinen from *Anabaena* sp. *Journal of Phycology*, 12, 44–8.

Sirenko, L. A. (1972). *Physiological Basis of Multiplication of Blue–green Algae in Reservoirs* (In Russian). Kiev: Naukova Dumka.

Sirenko, L. A., Stetsenko, N. M., Arendarchuk, V. V. & Kuz′menko, M. I. (1968). Role of oxygen conditions in the vital activity of certain blue–gree algae. *Microbiology*, 37, 199–202.

Skabiachevskiĭ, A. P. (1960). *Planktonic Diatom Algae of Freshwater of the USSR: Systematics, Ecology and Distribution* (In Russian). Moskva: Izdatel'stvo Moskovsogo Universiteta.

Skellam, J. G. (1951). Random dispersal in theoretical populations. *Biometrika*, 78, 196–218.

Skulberg, O. M. (1964). Algal problems related to eutrophication of European water supplies and a bioassay method to assess fertilizing influences of pollution in inland waters. In *Algae and Man*, ed. D. F. Jackson, pp. 262–99. New York: Plenum.

Small, L. F. (1963). Effect of wind on the distribution of chlorophyll *a* in Clear Lake, Iowa. *Limnology and Oceanography*, 8, 426–32.

Smayda, T. J. (1970). The suspension and sinking of phytoplankton in the sea. *Annual Review of Oceanography and Marine Biology*, 8, 353–414.

– (1974). Some experiments on the sinking characteristics of two freshwater diatoms. *Limnology and Oceanography*, 19, 628–35.

– (1980). Phytoplankton species succession. In *The physiological Ecology of Phytoplankton*, ed. I. Morris, pp. 493–570. Oxford: Blackwell.

Smayda, T. J. & Boleyn, B. J. (1965). Experimental observations on the flotation of marine diatoms. I. *Thalassiosira* cf. *nana*, *Thalassiosira rotula* and *Nitzschia seriata*. *Limnology and Oceanography*, 10, 499–509.

– (1966). Experimental observations on the flotation of marine diatoms. II. *Skeletonema costatum* amd *Rhizosolenia setigera*. *Limnology and Oceanography*, 11, 18–34.

Smith, G. M. (1917). The vertical distribution of *Volvox* in the plankton of Lake Monona. *American Journal of Botany*, 5, 178–85.

Smith, I. R. (1975). Turbulence in lakes and rivers. *Scientific Publications of the Freshwater Biological Association*, No. 29. 79 pp.
– (1982). A simple theory of algal deposition. *Freshwater Biology*, 12, 445–9.
Sommer, U. (1981). The role of *r*- and *K*-selection in the succession of phytoplankton in Lake Constance. *Acta Oecologica*, 2, 327–42.
Sournia, A. (1978). *Phytoplankton Manual*. Paris: UNESCO.
Sparrow, F. K. (1960). *Aquatic Phycomycetes*, 2nd edn, Ann Arbor: University of Michigan.
Stanier, R. Y. Sistrom, W. R., Hansen, T. A., Whitton, B. A., Castenholz, R. W., Pfennig, N., Gorlenko, V. N., Kondrat'eva, E. N., Eimhjeleen, K. E., Whittenburg, R., Gherna, R. L. & Truper, H. G. (1978). Proposal to place the nomenclature of the cyanobacteria (blue–green algae) under the rules of the International Code of Nomenclature of Bacteria. *International Journal of Systematic Bacteriology*, 28, 335–6.
Starkweather, P. L., Gilbert, J. J. & Frost, T. M. (1979). Bacteria feeding by *Brachionus calyciflorus*. Clearance and ingestion rates, behaviour and population dynamics. *Oecologia*, 44, 26–30.
Steel, J. A. (1972). The application of fundamental limnological research in water supply and management. *Symposia of the Zoological Society of London*, 29, 41–67.
– (1973). Reservoir algal productivity. In *The Use of Mathematical Models in Water Pollution Control*, ed. A. James, pp. 107–35. Newcastle: University of Newcastle-upon-Tyne.
Steele, J. H. (1976). Patchiness. In *The Ecology of the Seas*, ed. D. H. Cushing & J. J. Walsh, pp. 98–115. Philadelphia: Saunders.
Steele, J. H. & Yentsch, C. S. (1960). The vertical distribution of chlorophyll. *Journal of the Marine Biological Association of the United Kingdom*, 39, 217–26.
Steeman-Nielsen, E., Hansen, V. K. & Jørgensen, E. G. (1962). The adaptation to different light intensities in *Chlorella vulgaris* and the time dependence on transfer to a new light intensity. *Physiologia Plantarum*, 15, 505–17.
Steeman-Nielsen, E. & Jensen, P. K. (1958). Concentration of CO_2 and rate of photosynthesis in *Chlorella pynenoidosa*. *Physiologia Plantarum*, 21, 647–54.
Steeman-Nielsen, E. & Jørgensen, E. G. (1968). The adaptation of planktonic algae. III. With special consideration of the importance in nature. *Physiologia Plantarum*, 21, 647–54.
Stewart, K. M. (1972). Isotherms under ice. *Verhandlungen der internationale Vereinigung für theoretische und angewandte Limnologie*, 18, 303–11.
– (1976). Oxygen deficits, clarity and eutrophication in some Madison lakes. *Internationale Revue des gesamten Hydrobiologie*, 61, 563–79.
Stewart, W. D. P. (1973). Nitrogen fixation. In *The Biology of the Blue–green Algae*, ed. N. Carr & B. A. Whitton, pp. 260–78. Oxford: Blackwell.
– (1974). *Algal Physiology and Biochemistry*. Oxford: Blackwell.
Stewart, W. D. P., Fitzgerald, G. P. & Burris, R. H. (1967). *In situ* studies on N_2 fixation using the acetylene reduction technique. *Proceedings of the National Academy of Sciences of the USA*, 58, 2071–8.
Stewart, W. D. P. & Lex, M. (1970). Nitrogenase activity in the blue–green alga *Plectonema boryanum* strain 594. *Archiv für Mikrobiologie*, 73, 250–60.
Stewart, W. D. P. & Pearson, H. W. (1970). Effects of aerobic and anaerobic conditions on growth and metabolism of blue–green algae. *Proceedings of the Royal Society of London*, B 175, 293–311.

Strickland, J. D. H. (1960). Measuring the production of marine phytoplankton. *Bulletin of the Fisheries Research Board of Canada*, No. 122. 172+viii pp.
– (1965). Production of organic matter in the primary stages of the marine food chain. In *Chemical Oceanography Vol. I.*, ed. J. P. Riley & G. Skirrow, pp. 477–610. London: Academic Press.
Strickland, J. D. H., Holm-Hansen, O., Eppley, R. W. & Linn, R. J. (1969). The use of a deep tank in plankton ecology. I. Studies of the growth and composition of phytoplankton at low nutrient levels. *Limnology and Oceanography*, 14, 23–34.
Stumm, W. & Morgan, J. J. (1970). *Aquatic Chemistry*. New York: Wiley.
Sutcliffe, W. H., Baylor, E. R. & Menzel, D. W. (1963). Sea surface chemistry and Langmuir circulation. *Deep-Sea Research*, 10, 233–43.
Swale, E. M. F. (1963). Notes on *Stephanodiscus hantzschii* Grun. in culture. *Archiv für Mikrobiologie*, 45, 210–16.
– (1968). The phytoplankton of Oak Mere, Cheshire, 1963–1966. *British Phycological Bulletin*, 3, 441–9.
– (1969). Phytoplankton in two English rivers. *Journal of Ecology*, 57, 1–23.
Sze, P. (1980). Seasonal succession of phytoplankton in Onondaga Lake, New York (U.S.A.). *Phycologia*, 19, 54–9.
Sze, P. & Kingsbury, J. M. (1972). Distribution of phytoplankton in a polluted saline lake, Onondaga Lake, New York. *Journal of Phycology*, 8, 25–37.
Szemes, G. (1967). Das Phytoplankton der Donau. In *Limnologie der Donau*, ed. by R. Liepolt, pp. V, 158–79. Stuttgart: Schweizerbart'sche.
Talling, J. F. (1951). The element of chance in pond populations. *Naturalist, Hull*, October–December 1951, 157–70.
– (1957*a*). The phytoplankton population as a compound photosynthetic system. *New Phytologist*, 56, 133–49.
– (1957*b*). Photosynthetic characteristics of some freshwater plankton diatoms in relation to underwater radiation. *New Phytologist*, 56, 29–50.
– (1960). Self-shading effects in natural populations of a planktonic diatom. *Wetter und Leben*, 12, 235–42.
– (1965). The photosynthetic activity of phytoplankton in East African lakes. *Internationale Revue der gesamten Hydrobiologie*, 50, 1–32.
– (1971). The underwater light climate as a controllinhg factor in the production ecology of freshwater phytoplankton. *Mitteilungen der internationale Vereinigung für theoretische und angewandte Limnologie*, 19, 214–43.
– (1974). In standing waters. In *A manual on Methods for Measuring Primary Production in Aquatic Environments*, ed. R. A. Vollenweider (IBP Handbook No. 12, 2nd edn), pp. 119–23. Oxford: Blackwell.
– (1976*a*). The depletion of carbon dioxide from lake water by phytoplankton. *Journal of Ecology*, 64, 79–121.
– (1976*b*). Phytoplankton: composition, development and productivity. In *The Nile: Biology of an Ancient River*, ed. J. Rzóska, pp. 385–406. Den Haag: Junk.
– (1980). Phytoplankton. In *Euphrates and Tigris*, ed. J. Rzóska, pp. 81–6. Den Haag: Junk.
– (1981). The development of attenuance depth-profiling to follow the distribution of phytoplankton and other particulate material in a productive English lake. *Archiv für Hydrobiologie*, 93, 1–20.
Talling, J. F. & Talling, I. B. (1965). The chemical composition of African lake waters. *Internationale Revue der gesamten Hydrobiologie*, 50, 421–63.

Talling, J. F., Wood, R. B., Prosser, M. V. & Baxter, R. M. (1973). The upper limit of photosynthetic productivity by phytoplankton: evidence from Ethiopian soda lakes. *Freshwater Biology*, 3, 53–76.

Tamiya, H., Iwamura, T., Shibata, K., Hase, E. & Nihei, T. (1953). Correlation between photosynthesis and light independent metabolism in the growth of *Chlorella. Biochimica et Biophysica Acta*, 12, 23–40.

Taylor, F. J. R. (1980). Basic biological features of phytoplankton cells. In *Physiological Ecology of Phytoplankton*, ed. I. Morris, pp. 3–55. Oxford: Blackwell.

Tel-Or, E. & Stewart, W. D. P. (1977). Photosynthetic components and activities of nitrogen-fixing isolated heterocysts of *Anabaena cylindrica. Proceedings of the Royal Society of London*, B 198, 61–86.

Tessenow, U. (1966). Untersuchungen über den Kieselsäuregehalt der Binnengewässer. *Archiv für Hydrobiologie* (*Supplementband*), 32, 1–136.

Therriault, J.-C., Lawrence, D. J. & Platt, T. (1978). Spatial variability of phytoplankton turnover in relation to physical processes in a coastal environment. *Limnology and Oceanography*, 23, 900–11.

Therriault, J.-C. & Platt, T. (1978). Spatial heterogeneity of phytoplankton biomass and related factors in the near-surface waters of an exposed coastal embayment. *Limnology and Oceanography*, 25, 888–99.

– (1981). Environmental control of phytoplankton patchiness. *Canadian Journal of Fisheries and aquatic Sciences*, 38, 638–41.

Thiennemann, A. (1918). Untersuchungen über die Beziehungen zwischen dem Sauerstoffgehalt des Wassers und der Zusammensetzung der Fauna in norddeutschen Seen. *Archiv für Hydrobiologie*, 12, 1–65.

Thomas, E. A. (1949). Sprungschichtneigung im Zurichsee durch Strum. *Schweizerische Zeitschrift für Hydrologie*, 11, 527–45.

– (1950). Auffällige biologische Folgen von Sprungschichtneigungen im Zürichsee. *Schweizerische Zeitschrift für Hydrologie*, 12, 1–24.

Thomas, W. H. & Dodson, A. N. (1975). On silicic acid limitation of diatoms in the near-surface waters of the eastern tropical Pacific Ocean. *Deep-Sea Research*, 22, 671–7

Thompson, J. M., Ferguson, A. J. D. & Reynolds, C. S. (1982). Natural filtration rates of zooplankton in a closed system: the derivation of a community grazing index. *Journal of Plankton Research*, 4, 545–60.

Tilman, D. (1977). Resource competition between planktonic algae: an experimental and theoretical approach. *Ecology*, 58, 338–48.

Tilman, D. & Kilham, S. S. (1976). Phosphate and silicate uptake and growth kinetics of the diatoms *Asterionella formosa* and *Cyclotella meneghiniana* in batch and semicontinuous culture. *Journal of Phycology*, 12, 375–83.

Titman, D. (1975). A fluorimetric technique for measuring sinking rates of freshwater phytoplankton. *Limnology and Oceanography*, 20, 869–75.

– (1976). Ecological competition between algae: experimental confirmation of resource-based competition theory. *Science*, 192, 463–5.

Titman, D. & Kilham, P. (1976). Sinking in freshwater phytoplankton: some ecological implications of cell nutrient status and physical mixing processes. *Limnology and Oceanograpny*, 21, 409–17.

Toerien, D. F., Hyman, K. L. & Bruwer, M. J. (1975). A preliminary trophic status classification of some South African impoundments. *Water, South Africa*, 1, 15–23.

Tolstoy, A. (1977). Chlorophyll *a* as a measure of phytoplankton biomass. *Acta Universitatis Upsaliensis* No. 416. 28 pp.

Trevisan, R. (1978). Noto sull 'uso dei volumi algali per la stima della biomassa. *Rivista di Idrobiologia*, 17, 345–58.

Tundisi, J. G., Tundisi, T. M., Rocha, O., Gentil, J. G. & Nakamoto, N. (1977). Primary production, standing stock of phytoplankton and ecological factors in a shallow tropical reservoir (Represa do Broa, São Carlos, Brasil). *Semana Medio Ambiente y Represas, Montevideo*, 1, 138–72.

Tyagi, V. V. S. (1975). The heterocysts of blue–green algae (Myxophyceae). *Biological Reviews of the Cambridge Philosophical Society*, 50, 247–84.

Uhlmann, D. (1971). Influence of dilution, sinking and grazing rate on phytoplankton populations of hyperfertilized ponds and micro-ecosystems. *Mitteilungen der internationale Vereinigung für theoretische und angewandte Limnologie*, 19, 100–24.

Ütermohl, H. (1931). Neue Wege in der quantitativen Erfassung des Planktons. *Verhandlungen der internationale Vereinigung für theoretische und angewandte Limnologie*, 5, 567–96.

Vance, B. D. (1965). Composition and succession of cyanophycean water blooms. *Journal of Phycology*, 1, 81–6.

Vanderhoef, L. N., Huang, C.-Y., Musil, R. & Williams, J. (1974). Nitrogen fixation (acetylene reduction) by phytoplankton in Green Bay, Lake Michigan, in relation to nutrient concentrations. *Limnology and Oceanography*, 19, 119–25.

Vanderhoef, L. N., Leibson, P. J., Musil, R. J., Huang, C.-Y., Fiehweg, R. E., Williams, J. W., Wackwitz, D. L. & Mason, K. T. (1975). Diurnal variation in algal acetylene reduction (nitrogen fixation) *in situ*. *Plant Physiology, Lancaster*, 55, 273–6.

Van Vlymen, C. D. (1979). The natural history of Slapton Ley Nature Reserve. XII. The water balance of Slapton Ley. *Field Studies*, 5, 59–84.

Venrick, E. L. (1978). Systematic sampling in a planktonic ecosystem. *Fishery Bulletin*, 76, 617–27.

Verduin, J. (1951). A comparison of phytoplankton data obtained by a mobile sampling method with those obtained from a single station. *American Journal of Botany*, 38, 5–11.

Vetter, H. (1937). Limnologische Untersuchungen über das Phytoplankton und seine Beziehüngen zur Ernahrung des Zooplanktons im Schleinsee bei Lagenargen am Bodensee. *Internationale Revue gesamten Hydrobiologie und Hydrographie*, 34, 499–561.

Vincent, W. F. (1980*a*). The physiological ecology of a *Scenedesmus* population in the hypolimnion of a hypertrophic pond. I. Photoautotrophy. *British Phycological Journal*, 15, 27–34.

– (1980*b*). The physiological ecology of a *Scenedesmus* population in the hypolimnion of a hypertrophic pond. II. Heterotrophy. *British Phycological Journal*, 15, 35–41.

Vincent, W. F. & Silvester, W. B. (1979). Growth of blue–green algae in the Manuku (N.Z.) oxidation ponds. II. Experimental studies on algal interaction. *Water Research*, 13, 717–23.

Vollenweider, R. A. (1960). Beiträge zur Kenntnis optischer Eigenschaften der Gewässer und Primärproduktion. *Memorie dell'Istituto italiano di Idrobiologia*, 11, 241–64.

– (1965). Calculation models of photosynthesis-depth curves and some implications regarding day rate estimates in primary production. *Memorie dell'Istituto italiano di Idrobiologia*, 18, Suppl., 425–57.

– (1968). *Scientific Fundamentals of the Eutrophication of Lakes and Flowing Waters, with Particular Reference to Nitrogen and Phosphorus as Factors in Eutrophication*. Paris: OECD.

– (1974). *A Manual on Methods for Measuring Primary Production in Aquatic Environments* (IBP Handbook 12). Oxford: Blackwell.

– (1976). Advances in defining critical loading levels for phosphorus in lake eutrophication. *Memorie dell'Istituto italiano di Idrobiologia*, 33, 53–83.

Vollenweider, R. A. & Kerekes, J. (1980). The loading concept as basis for controlling eutrophication philosophy and preliminary results of the OECD programme on eutrophication. *Progress in Water Technology*, 12(2), 5–38.

Volodin, B. B. (1970). The part played by sulfur in the vital activities of the blue–green algae of the genus *Microcystis*. *Hydrobiological Journal*, 6(6), 29–37.

Walsby, A. E. (1971). The pressure relationships of gas vacuoles. *Proceedings of the Royal Society of London*, B 178, 301–26.

– (1972). Structure and function of gas vacuoles. *Bacteriological Reviews*, 36, 1–32.

– (1978*a*). The gas vesicles of aquatic prokaryotes. In *Relations between Structure and Function in the Prokaryotic Cell*, ed. R. Y. Stanier *et al.*, pp. 327–57. Cambridge: University Press.

– (1978*b*). The properties and buoyancy-providing role of gas vacuoles in *Trichodesmium* Ehrenberg. *British Phycological Journal*, 13, 103–16.

Walsby, A. E. & Klemer, A. R. (1974). The role of gas vacuoles in the microstratification of a population of *Oscillatoria agardhii* var. *isothrix* in Deming Lake, Minnesota. *Archiv für Hydrobiologie*, 74, 375–92.

Walsby, A. E. & Reynolds, C. S. (1980). Sinking and floating. In *The Physiological Ecology of Phytoplankton*, ed. I. Morris, pp. 371–412. Oxford: Blackwell.

Walsby, A. E. & Xypolyta, A. (1977). The form resistance of chitan fibres attached to the cells of *Thalassiosira fluviatilis* Hustedt. *British Phycological Journal*, 12, 215–23.

Watson, S. W. & Kalff, J. (1981). Relationships between nanoplankton and lake trophic status. *Canadian Journal of Fisheries and aquatic Sciences*, 38, 960–7.

Watt, W. D. (1966). The release of dissolved organic material from the cells of phytoplankton populations. *Proceedings of the Royal Society of London*, B 164, 521–51.

Werner, D. (1977). Silicate metabolism. In *The Biology of Diatoms*, ed. D. Werner, pp. 110–49. Oxford: Blackwell.

Westlake, D. F. (1974). Measurements on non-isolated natural communities. In *A Manual on Methods for Measuring Primary Production in Aquatic Environments*, ed. R. A. Vollenweider (IBP Handbook No. 12, 2nd edn), pp. 110–11. Oxford: Blackwell.

Wetzel, R. G. (1964). A comparative study of the primary productivity of higher aquatic plants, periphyton and the phytoplankton in a large shallow lake. *Internationale Revue der gesamten Hydrobiologie*, 49, 1–61.

Whittingham, C. P. (1974). *The Mechanism of Photosynthesis*. London: Arnold.

Whitton, B. A. & Peat, A. (1969). On *Oscillatoria redekei* Van Goor. *Archiv für Mikrobiologie*, 68, 362–76.

Wildman, R., Loescher, J. H. & Winger, C. L. (1975). Development and germination of akinetes of *Aphanizomenon flos-aquae*. *Journal of Phycology*, 11, 96–104.

Willén, E. (1976). A simplified method of phytoplankton counting. *British Phycological Journal*, 11, 265–78.

Willoughby, L. G. (1976). Actinomyctes and actinophage in fresh water. *Report of the Freshwater Biological Association*, 44, 53–8.

Wiseman, S. W. & Reynolds, C. S. (1981). Sinking rate and electrophoretic mobility of the freshwater diatom *Asterionella formosa*: an experimental investigation. *British Phycological Journal*, 16, 357–61.

Yamagishi, H. & Aoyama, K. (1972). Ecological studies on dissolved oxygen and bloom of *Microcystis* in Lake Suwa. 1. Horizontal distribution of dissolved oxygen in relation to drifting of *Microcystis* by wind. *Bulletin of the Japanese Society of Scientific Fisheries*, 38, 9–16.

Yentsch, C. S. (1980). Light attenuation and phytoplankton photosynthesis. In *The Physiological Ecology of Phytoplankton*, ed. I. Morris, pp. 95–127. Oxford: Blackwell.

Yentsch, C. S. & Ryther, J. H. (1959). Relative significance of the net phytoplankton and nanoplankton in the water of Vineyard Sound. *Journal du Conseil international pour l'Exploration de la Mer*, 24, 231–8.

Zevenboom, W. & Mur, L. R. (1981). Simultaneous short-term uptake of nitrate and ammonium by *Oscillatoria agardhii* grown in nitrate- or light-limited continuous culture. *Journal of General Microbiology*, 126, 355–63.

Zhuravleva, V. I. & Matskevich, Ye. S. (1974). Electrokinetic properties of *Microcystis aeruginosa. Hydrobiological Journal*, 10(1), 55–7.

Zimmermann, U. (1969). Ökologische und physiologische Untersuchungen an der planktonischen Blaualge *Oscillatoria rubescens* D.C. unter besonderen Berücksichtigung von Licht und Temperatur. *Schweizerische Zeitschrift für Hydrologie*, 31, 1–58.

Index to lakes and rivers

with latitudes and longitudes

Ant, River. Norfolk, UK. 52° 44′ N, 1° 29′ E. 148
Arresø. Frederiksborg, Sjælland, Denmark. 55° 59′ N, 12° 7′ E. 213

Barton Broad. Norfolk, UK. 52° 43′ N, 1° 29′ E. 148
Baykal, Ozero. R.S.F.S.R., USSR. 53° N, 108° E. 7
Blelham Tarn. Cumbria, UK. 54° 24′ N, 2° 58′ W. 116, 171, 187, 203, 232, 288, 303, 304, 324, 326
Blelham Enclosures A, B, C: in Blelham Tarn, Cumbria, UK. (*q.v.*). 31, 88, 89, 90, 97, 98, 101, 102, 104, 114, 115, 116, 117, 121, 122, 171, 172, 187, 203–210, 212, 218, 239, 240, 242, 243, 247, 248, 260, 261, 264, 271, 272, 273, 274, 275, 287, 306, 312, 313, 314, 324, 327
Bodensee. Germany/Switzerland. 47° 39′ N, 9° 11′ E. 113, 116, 118, 203, 204, 223, 239
Broa, Represa de. São Paulo, Brasil. 20° 11′ S, 47° 48′ W. 294

Carioca, Lagoa. Minas Gerais, Brasil. 19° 10′ S, 42° 1′ W. 103
Castle Lake. California, USA. 41° 14′ N, 122° 23′ W. 120, 227, 248, 311
Char Lake. Cornwallis I., Northwest Territories, Canada. 74° 43′ N, 94° 59′ W. 150, 213, 283
Crose Mere. Shropshire, UK. 52° 53′ N, 2° 50′ W. 31, 54, 94, 99, 103, 114, 115, 117, 118, 121, 127, 142, 143, 144, 148, 149, 151, 152, 153, 154, 155, 156, 172, 182, 219, 231, 239, 251, 284, 285, 287, 289, 303, 304, 305, 314, 322

Drontermeer. Gelderland, Netherlands. 52° 16′ N, 5° 32′ E. 290

Eglwys Nynydd, West Glamorgan, UK. 51° 33′ N, 3° 44′ W. 95, 109, 110
Ennerdale Water. Cumbria, UK. 54° 32′ N, 3° 23′ W. 113, 118
Erie Lake. Canada/USA. 42° N, 81° E. 84
Erken, Sjön. Stockholm, Sweden. 59° 25′ N, 18° 15′ E. 87, 114, 117, 118
Esthwaite Water. Cumbria, UK. 54° 21′ N, 3° 0′ W. 109, 110, 218, 325, 327

Folly Pool. Shropshire, UK. 52° 44′ N, 2° 52′ W. 115, 289, 294

Gara, River. Devonshire, UK. 50° 19′ N, 3° 40′ W. 323
George, Lake. Uganda. 0° 0′ N, 30° 10′ E. 104, 105, 116, 148, 150, 155, 156, 213, 303
Gjersjøen. Akershus, Norway. 59° 48′ N, 10° 51′ E. 290
Grasmere. Cumbria, UK. 54° 27′ N, 3° 0′ W. 117, 118, 231, 313

Hamilton Harbour, L. Ontario. Ontario Canada. 43° 15′ N, 79° 51′ W. 114, 116, 117, 145, 213
Heart Lake. Ontario, Canada. 43° 46′ N, 79° 48′ W. 259
Hertel, Lac. Quebec, Canada. (Coordinates untraced: near Montreal.) 240, 241, 246

Kettle Mere. Shropshire, UK. 52° 54′ N, 2° 52′ W. 219
Kinneret (Lake). Israel. 32° 50′ N, 35° 28′ E. 84
Klopeinersee. Kärnten, Austria. 46° 38′ N, 14° 35′ E. 116, 290

Lanao, Lake. Mindanao, Philippines. 7° 52′ N, 124° 13′ E. 116, 213, 308, 311
Leven, Loch. Tayside, UK. 56° 20′ N, 3° 20′ W. 145, 171, 294
Linsley Pond. Connecticut, USA. 41° 19′ N, 72° 46′ W. 117
Lohjanjärvi. Uusimaa, Finland. 60° 15′ N, 23° 57′ E. 239
Lunzer Untersee. Nieder Osterreich, Austria. 47° 57′ N, 15° 4′ E. 116, 117, 118

Maggiore, Lago. Novara/Varese, Italy. 46° 0′ N, 8° 40′ E. 290
Malham Tarn. Yorkshire, UK. 54° 6′ N, 2° 4′ W. 173
Memphrémagog, Lac. Quebec, Canada/Vermont, USA. 45° 5′ N, 72° 16 W. 84
Meretta, Lake. Cornwallis I., Northwest Territories, Canada. 74° 42′ N, 94° 57′ W. 213, 283
Michigan, Lake. Michigan/Wisconsin, USA. 44° N, 87° W. 179, 287, 294
Mikołajske, Jezioro. Olsztyn, Poland. 53° 10′ N, 21° 33′ E. 259
Millstätersee. Kärnten, Austria. 46° 48′ N, 13° 33′ E. 113, 114, 116, 294
Mývatn. Iceland. 65° 35′ N, 17° 0′ W. 213

Neagh, Lough. Northern Ireland, UK. 54° 35′ N, 6° 30′ W. 148, 155, 156, 171, 172, 200, 212, 289, 304
Neusiedlersee. Burgenland, Austria/Gyor-Sopron, Hungary. 47° 47′ N, 16° 44′ E. 294
Norrviken, Sjön. Stockholm, Sweden. 59° 29′ N, 17° 58′ E. 290

Oak Mere. Cheshire, UK. 53° 13′ N, 2° 39′ W. 214
Onondaga Lake. New York, USA. 43° 5′ N, 76° 11′ W. 117, 213
Ontario, Lake. Canada/USA. 43° 50′ N, 78° W. 84, 131

Petén, Laguma de. Guatemala. 16° 55′ N, 89° 50′ W. 7
Priest Pot. Cumbria, UK. 54° 23′ N, 3° 0′ W. 289

Red Rock Tarn. Victoria, Australia. 38° 20′ S, 143° 30′ E. 150
Rostherne Mere. Cheshire, UK. 53° 21′ N, 2° 23′ W. 117, 148, 154, 155, 176, 177, 219, 284, 303, 304, 313
Rothay, River. Cumbria, UK. 54° 44′ N, 2° 59′ W. 231

St Clair, Lake. Ontario, Canada/Michigan, USA. 42° 25′ N, 82° 40′ W. 84
St George, Lake. Ontario, Canada. 43° 57′ N, 79° 28′ W. 238, 239
St James' Park Lake. London, UK. 51° 30′ N, 0° 9′ W. 290
Sanctuary Lake. Pennsylvania, USA. 41° 40′ N, 80° 26′ W. 167
Severn, River. UK. 52° 30′ N, 2° 30′ W. 171
Schleinsee. Baden Wurttemberg, West Germany. 47° 36′ N, 9° 31′ E. 117, 118
Slapton Lower Ley. Devonshire, UK. 50° 17′ N, 3° 38′ W. 232, 233
Solar Lake. Sinai, Egypt. 29° 26′ N, 35° 50′ E. 103

Tahoe, Lake. California/Nevada, USA. 39° N, 120° W. 84
Tchad, Lac. Chad. 12° 30′ N, 14° 30′ E. 150

Veluwemeer. Gelderland, Netherlands. 52° 25′ N, 5° 45′ E. 213, 290
Victoria, Lake. Kenya/Tanzania/Uganda. 1° S, 33° E. 177, 294
Vierwaldstattersee. Switzerland. 46° 58′ N, 8° 30′ E. 290

Wahnbach Talsperre. Nordhein-Westfalen, West Germany. 50° 50′ N, 7° 8′ E. 327
Washington, Lake. Washington, USA. 47° 35′ N, 122° 14′ W. 190, 290, 327
Wast Water. Cumbria, UK. 54° 26′ N, 3° 17′ E. 113, 118
White Mere. Shropshire, UK. 52° 53′ N, 2° 53′ W. 219
Windermere. Cumbria, UK. 54° 20′ N, 2° 57′ W. 92, 93, 114, 117, 118, 121, 122, 148, 154, 155, 170, 171, 172, 248, 284, 294, 303, 304, 313, 314, 326
Wolderwijd. Gelderland, Netherlands. 52° 29′ N, 5° 51′ E. 213, 290
Worthersee. Kärnten, Austria. 46° 37′ N, 14° 10′ E. 290

Zürichsee. Zurich, Switzerland. 47° 12′ N, 8° 42′ E. 283, 290

Index to genera/species
with authorities for 'algal' species

*taxon not 'algal' † taxon not freshwater

Acineta Ciliophora, Suctoria 6
Actinastrum Chlorophyta, Chlorococcales 68
**Actinoplanes* Actinomyceta 6
Anabaena Cyanobacteria, Nostocales: 9, 27, 94, 103, 117, 119, 129, 167, 180, 191, 198, 213, 214, 218, 219, 221, 249, 254, 267, 271, 289, 295, 296, 297, 299, 300, 302, 306, 307, 309, 313, 318, 319; *A. circinalis* Rabenh. *ex* Born. *et* Flah. 22, 30, 38, 62, 63, 103, 145, 205; *A. cylindrica* Lemm. 194; *A. flos-aquae* Bréb. *ex* Born. *et* Flah. 16, 28, 68, 160, 194, 199, 200, 204, 205, 206, 207, 208, 210, 211, 215, 216; *A. solitaria* Klebs 218, 326
Anabaenopsis Cyanobacteria, Nostocales 9
Anacystis Cyanobacteria, Chroococcales: *A. nidulans* (= *Synechococcus*) 194, 195
Ankistrodesmus Chlorophyta, Chlorococcales: 12, 117, 119, 160, 213. 214, 289; *A. braunii* (Näg.) Collins 195; *A. falcatus* (Corda) Ralfs 21, 31
Ankyra Chlorophyceae, Chlorococcales: 12, 96, 221, 255, 262, 265, 269, 270, 272, 295, 324; *A. judayi* (G.M.Sm.) Fott. 13, 21, 31, 38, 97, 119, 204, 205, 206, 207, 209, 210, 211, 215, 264, 268, 274, 275
Aphanizomenon Cyanobacteria, Nostocales: 9, 27, 28, 94, 103, 117, 167, 180, 198, 218, 219, 249, 254, 289, 295, 299, 313, 318, 319; *A. flos-aquae* Ralfs *ex*. Born. *et* Flah. 16, 22, 31, 38, 119, 194, 199, 200, 201, 204
Aphanocapsa Cyanobacteria, Chroococcales 9, 117, 119
Aphanothece Cyanobacteria, Chroococcales 9, 117, 119
**Aphelidium* Proteomyxidia 253
**Argulus* Crustacea, Branchiura 7
**Artemia* Crustacea, Anostraca 7
Arthrodesmus Chlorophyta, Zygnematales 12
**Ascomorpha* Rotatoria 6, 255
**Asplanchna* Rotatoria 6, 255
Asterionella Chrysophyta, Bacillariophyceae: 11, 25, 65, 72, 73, 78, 79, 81, 88, 89, 90, 92, 93, 117, 130, 133, 134, 135, 152, 153, 170, 171, 191, 198, 214, 220, 221, 249, 252, 255, 265, 267, 270, 285, 289, 313, 318, 326; *A. formosa* Hass. 15, 22, 30, 33, 34, 35, 36, 38, 57, 58, 68, 69, 71, 75, 76, 77, 78, 116, 119, 136, 145, 160, 164, 178, 179, 194, 199, 203, 204, 205, 206, 207, 208, 210, 211, 215, 232, 237, 238, 239, 240, 241, 242, 243, 244, 247, 251, 264, 268, 274, 275, 279, 282, 285, 287, 288, 299, 305, 326
**Asterocaelum* Rhizopoda 6, 253, 254
Attheya Chrysophyta, Bacillariophyceae 11
Azolla Pteridophyta 4

**Blastocladiella* Fungi, Chytridales 6
**Bosmina* Crustacea, Cladocera: 7, 255; *B. corregoni* 263; *B. longirostris* 258, 260
Botryococcus Chlorophyta, Chlorococcales: 12, 84; *B. braunii* Kütz. 58
**Brachionus* Rotatoria: 6, 254; *B. calyciflorus* 258
**Bythotrephes* Crustacea, Cladocera 7

Carteria Chlorophyta, Volvocles 12
Ceratium Pyrrhophyta, Dinophyceae: 9, 25, 73, 109, 116, 117, 133, 154, 156, 223, 224, 273, 289, 295, 296, 297, 299, 302, 306, 307, 309, 312, 318, 319; *C. hirundinella* O. F. Müll. 14, 21, 30, 38, 67, 99, 100, 110, 119, 136, 145, 194, 203, 204, 205, 206, 207, 209, 210, 215, 216, 217, 221, 300, 313, 326, 327
**Ceriodaphnia* Crustacea, Cladocera 7, 255
**Chaoborus* Arthropoda, Insecta, Diptera 4, 7
Chilomonas Cryptophyta, Cryptomonadales 9
Chlamydomonas Chlorophyta, Volvocales: 12, 195, 196, 310; *C. reinhardii* Dangeard 166

Chlorella Chlorophyta, Chlorococcales: 12, 21, 119, 136, 196, 214, 249, 255, 262, 270, 289, 295, 310; *C. pyrenoidosa* Chick 31, 195; *C. vulgaris* Beyerinck 57, 69, 195
Chlorococcum Chlorophyta, Chlorococcales 57, 69
**Chromatium* Bacteria 6
Chromulina Chrysophyta, Chrysophyceae 10, 21, 204, 264, 268
Chroomonas Cryptophyta, Cryptomonadales 9
Chrysochromulina Chrysophyta, Haptophyceae: 10, 21, 119, 295; *C. parvula* Lackey 21
Chrysococcus Chrysophyceae 10, 21, 283, 285
**Chydorus* Crustacea, Cladocera: *C. sphaericus* 258, 260, 261
Closterium Chlorophyta, Zygnematales: 12, 214, 238, 285, 306, 313; *C. aciculare* T. West 21, 31, 38, 119, 204, 306; *C. acutum* Bréb. 13, 119; *C. tortum* B. M. Griffiths 119
Coelastrum Chlorophyta, Chlorococcales 12, 67, 117, 119, 289, 301, 313
Coelosphaerium Cyanobacteria, Chroococcales 9, 94, 119, 213
Coenococcus Chlorophyta, Chlorococcales 12, 117, 119
**Coleps* Ciliophora 6
**Conochilus* Rotatoria 6, 254, 255
Coscinodiscus Chrysophyta, Bacillariophyceae: †*C. concinnus* W. Smith 58; †*C. wailesii* Gran & Angst 51, 53
Cosmarium Chlorophyta, Zygnematales: 12, 117, 256, 306; *C. abbreviatum* Racib. 119; *C. contractum* Kirchn. 119; *C. depressum* (Näg.) Lund 21, 119
**Craspedacusta* Coelenterata, Hydrozoa 6
Crucigenea Chlorophyta, Chlorococcales 12, 117, 119, 295
Cryptomonas Cryptophyta, Cryptomonadales: 9, 191, 205, 206, 207, 209, 210, 211, 215, 216, 217, 221, 255, 262, 265, 270, 274, 275, 276, 285, 289, 295, 302, 319, 326; *C. erosa* Ehrenb. 194, 199, 201; *C. ovata* Ehrenb. 14, 20, 31, 38, 194, 198, 199, 203, 204, 264, 268, 282
Cyclotella Chrysophyta, Bacillariophyceae: 11, 25, 170, 171, 283, 285, 286, 305, 309, 310; *C. comensis* Grun. 116, 119, 279, 288; *C. comta* (Ehrenb.) Kütz. 116, 119, 239; *C. glomerata* Bachmann 213; *C. melosiroides* Kirchner, *in* Schröter *et* Kirchner 239; *C. meneghiniana* Kütz. 21, 57, 69, 119, 164, 178, 179, 194, 287; †*C. nana* Hust. 53; *C. praeterissima* Lund 21, 57, 69, 75, 76; *C. pseudostelligera* Hust. 288
**Cypria* Crustacea, Ostracoda: 7; **C. javensis* 7; *C. petenensis* 7

Dactylococcopsis Cyanobacteria, Chroococcales 103
**Daphnia* Crustacea, Cladocera: 7, 73, 255, 257, 258, 262, 263, 264, 265, 266, 268, 269, 270, 271, 272; *D. cuculata* 263; *D. galeata* 258; *D. hyalina* var. *lacustris* 73, 111, 258, 259, 260, 261, 262, 263; *D. magna* 258; *D. middendorffiana* 258, 263; *D. pulex* 258, 267; *D. schoedleri* 258
**Desulfovibrio* Bacteria 6
**Diaptomus* Crustacea, Calanoidea: 255, 256, 257; *D. gracilis* (*see Eudiaptomus gracilis*); *D. oregonensis* 258, 263
Diatoma Chrysophyta, Bacillariophyceae 11, 117, 119, 285, 286, 287
Dictyosphaerium Chlorophyta, Chlorococcales: 12, 117; *D. pulchellum* Wood 17, 22
Didymogenes Chlorophyta, Chlorococcales 117
Dinobryon Chrysophyta, Chrysophyceae: 10, 22, 117, 119, 158, 180, 204, 213, 220, 285, 289, 295, 296, 302, 313, 318; *D. divergens* Imhof 159, 326
Ditylum Chrysophyta, Bacillariophyceae: †*D. brightwelli* (West) Grun. *in* Van Heurek 58, 59
**Dreissensia* Mollusca, Lamellibranchiata 7

**Eichhornia* Phanerophyta, Angiospermae 4
Elakatothrix Chlorophyta, Chlorococcales 12, 289
**Epistylis* Ciliophora 6
**Ergasilus* Crustacea, Cyclopoidea 7
Ethmodiscus Chrysophyta, Bacillariophyceae: †*E. rex* (Wallich *in* Rattray) Hendey *in* Wiseman *et* Hendey 51
Euastrum Chlorophyta, Zygnematales 12
**Eudiaptomus* Crustacea, Calanoidea: 7, 73, 255; *E. gracilis* 258, 260, 261, 263
Eudorina Chlorophyta, Chlorophyceae: 12, 26, 67, 79, 117, 119, 220, 221, 249, 262, 266, 270, 271, 289, 295, 296, 299, 300, 302, 313, 318; *E. elegans* Ehrenb. 17, 31, 38, 97; *E. unicocca* G. Sm. 23, 31, 62, 97, 98, 195, 203, 204, 205, 206, 207, 208, 210, 211, 212, 215, 216, 217, 264, 275, 324
Euglena Euglenophyta: 11, 14; *E. gracilis* Klebs 166, 195
**Eurytemora* Crustacea, Calanoidea 7
Exuviella Pyrrhophyta, Adinophyceae 9

Filinia Rotatoria 6, 254, 255
Fragilaria Chrysophyta, Bacillariophyceae: 11, 25, 81, 88, 117, 119, 133, 175, 220, 252, 270, 272, 285, 285, 313; *F. capucina* Desmaz. 30; *F. construens* (Ehrenb.) Grun. 213; *F. crotonensis* Kitton 15, 22, 30, 33, 35, 38, 57, 62, 65, 68, 69, 70, 73, 77, 119, 136, 194, 203, 204, 205, 206, 207, 208, 210, 215, 216, 238, 239, 242, 244, 247, 264, 266, 275, 276, 286, 288, 299, 305, 306, 326
Frontonia Ciliophora 6

Gastropus Rotatoria 255
Gemellicystis Chlorophyta, Tetrasporales: 12, 60, 117, 295, 325; *G. neglecta* Teiling 17, 62, 119
Geminella Chlorophyta, Ulotrichales 12
Glenodinium Pyrrhophyta, Dinophyceae 9
Gloeocapsa Cyanobacteria, Chroococcales 9, 167
Gloeocystis Chlorophyta, Tetrasporales: 12, 62; *G. gigas* Lagerheim 26
Gloeothece Cyanobacteria, Chroococcales 9
Gloeotrichia Cyanobacteria, Nostocales: 9, 61, 94, 119, 167, 218, 295, 299; *G. echinulata* J.E.Sm. *ex* R. Richt 16, 218
Gomphosphaeria Cyanobacteria, Chroococcales: 9, 94, 116, 117, 119, 168:
Goniochloris Chrysophyta, Xanthophyceae 11
Gonium Chlorophyta, Volvocales 12
Gonyaulax Pyrrhophyta, Dinophyceae 9
Gonyostomum Raphidophyta, Raphidomonadales 10
Gymnodinium Pyrrhophyta, Dinophyceae 9, 213, 283

Holopedium Crustacea, Cladocera 7
Hydrocharis Phanerophyta, Angiospermae 4

Kellicottia Rotatoria: 6, 254; *K. longispina* 255
Kephyrion Chrysophyta, Chrysophyceae 10, 21
Keratella Rotatoria: 6, 254, 255; *K. cochlearis* 255, 260, 262, 263; *K. quadrata* 255, 260, 262, 263
Kirchneriella Chlorophyta, Chlorococcales 12

Lagerheimia Chlorophyta, Chlorococcales 12, 214, 238
Lemna Phanerophyta, Angiospermae 4, 309
Lepocinclis Euglenophyta 11
Leptodora Crustacea, Cladocera 7
Limnocalanus Crustacea, Calanoidea 7, 255
Limnocnida Coelenterata, Hydrozoa 6
Lyngbya Cyanobacteria, Nostocales: 9, 309; *L. limnetica* Lemm. 103

Macrohectopus Crustacea, Amphipoda: 7; *M. branickii* 7
Mallomonas Chrysophyta, Chrysophyceae: 10, 25, 117, 119, 220, 285, 295, 318; *M. caudata* Iwanoff 14, 21, 119
Melosira Chrysophyta, Bacillariophyceae: 11, 25, 68, 88, 145, 170, 171, 172, 220, 285, 286, 309, 325; *M. binderana* Kütz. 30, 180, 283; *M. granulata* (Ehrenb.) Ralfs 23, 30, 33, 34, 38, 68, 117, 119, 199, 204, 213, 305, 313; *M. islandica* Otto Müller 213; *M. italica* (Ehr.) Kütz. subsp. *subarctica* Mull. 16, 23, 33, 57, 65, 66, 68, 69, 116, 119, 136, 213, 247, 279, 286, 288, 299, 305, 313, 326
Mesocyclops Crustacea, Cyclopoidea 7
Metallogenium Bacteria 6
Methylomonas Bacteria 6
Metopus Ciliophora 6
Micrasterias Chlorophyta, Zygnematales 72, 73
Microcystis Cyanobacteria, Chroococcales: 9, 19, 26, 59, 64, 68, 94, 95, 103, 109, 111, 117, 156, 168, 174, 176, 180, 213, 214, 220, 224, 249, 262, 266, 267, 270, 271, 272, 273, 297, 300, 302, 306, 307, 308, 314, 318, 325; *M. aeruginosa* Kütz. emend. Elenkin 17, 23, 31, 38, 56, 57, 62, 94, 104, 105, 110, 119, 136, 145, 160, 194, 196, 203, 204, 205, 206, 207, 209, 210, 215, 216, 221, 264, 275, 276, 300, 324, 326
Moina Crustacea, Cladocera 7
Monodus Chrysophyta, Xanthophyceae: 11, 21, 25; *M. subterraneus* Boye Petersen 194
Monoraphidium Chlorophyta, Chlorococcales 12, 117, 119, 295
Mysis Crustacea, Mysidacea 4, 7

Nassula Ciliophora 253, 254
Nephrodiella Chrysophyta, Xanthophyceae 11
Nitrosomonas Bacteria 6
Nitzschia Chrysophyta, Bacillariophyceae 11, 117, 310
Notholca Rotatoria: 6, 254; *N. squamula* 255
Notodroma Crustacea, Ostracoda 4

Ochrobium Bacteria 6
Ochromonas Chrysophyta, Chrysophyceae 11, 213

Oocystis Chlorophyta, Chlorococcales: 12, 117, 119, 131, 133, 201, 213, 289; *O. borgei* Snow 119, 213
Ophiocytium Chrysophyta, Xanthophyceae 11
**Ophryoglena* Ciliophora 6, 253, 254
Oscillatoria Cyanobacteria, Nostocales: 9, 25, 64, 100, 119, 168, 201, 262, 282, 294, 300, 306, 309, 319; *O. agardhii* Gom. 39, 100, 119, 145, 194, 199, 200, 201, 289, 291, 313, 325, 326; *O. agardhii* Gom. var. *isothrix* Skuja 16, 22, 31, 38, 101, 102, 204, 205, 206, 207, 209, 210, 211, 215, 326; *O. limnetica* Lemm. 100; *O. prolifica* Gom. 100, 289; *O. redekei* Van Goor 16, 39, 145; 194, 199, 200, 201, 289; *O. rubescens* D.C. *ex* Gom. 100, 194, 199, 289, 291, 326; †*O. thiebautii* (*see Trichodesmium thiebautii*) 62, 168

Pandorina Chlorophyta, Volvocales: 12, 117, 119, 289, 295, 299, 318; *P. morum* Bory 119, 204
Paulschulzia Chlorophyta, Tetrasporales 12
Pediastrum Chlorophyta, Chlorococcales: 12, 68, 117, 119, 213, 214, 285, 301, 308, 313; *P. boryanum* (Turp.) Meneghin 22, 119; *P. duplex* Meyen 15, 119; *P. tetras* (Ehrenb.) Ralfs 119
Pedinomonas Chlorophyta, Pedinomonadales 11, 324
**Pelomyxa* Rhizopoda 6, 253, 254
**Pelonema* Bacteria 6
**Peloploca* Bacteria 6
**Peranema* Mastigophora 6
Peridinium Pyrrhophyta, Dinophyceae: 9, 116, 119, 145, 213, 224, 273, 295, 309; *P. cinctum* (Müll.) Ehrenb. 204; *P. willei* Huitfeld-Kaas. 14, 119
Phacotus Chlorophyta, Volvocales 12, 174
Phacus Euglenophyta: 11; *P. longicauda* (Ehrenb.) Duj. 14
**Planktomyces* Bacteria 6
Plectonema Cyanobacteria, Chroococcales 167
**Podochytrium* Fungi, Chytridales 6
**Polyarthra* Rotatoria 6, 255
Prymnesium Chrysophyta, Haptophyceae 10
Pseudanabaena Cyanobacteria, Nostocales 9
Pseudopedinella Chrysophyta, Chrysophyceae 10
Pyramimonas Chlorophyta, Pyraminonadales 11
Pyrocystis Pyrrhophyta, Adinophyceae: 9; †*P. noctiluca* Murray ex Schutt. 59

Quadrigula Chlorophyta, Chlorococcales 12

Radiococcus Chlorophyta, Chlorococcales 12
Raphidonema Chlorophyta, Ulotrichales 12
**Rhizophydium* Fungi, Chytridiales: 6; *R. planktonicum* Canter emend. 252
Rhizosolenia Chrysophyta, Bacillariophyceae: 11, 119, 286; *R. eriensis* H. L. Smith 288; †*R. setigera* Brightwell 67
Rhodomonas Cryptophyta, Cryptophyceae: 9, 25, 221, 262, 270, 295, 312; *R. minuta* Skuja var. *nannoplanktica Skuja* (= *R. pusilla* (Bachm.) Jav.) 14, 20, 98, 204, 268
**Rozella* Fungi, Chytridiales 252

**Scapholeberis* Crustacea, Cladocera 4
Scenedesmus Chlorophyta, Chloroccales: 12, 67, 117, 119, 160, 196, 213, 238, 267, 285, 289, 294, 301, 310; *S. obliquus* (Turp.) Kütz. 160, 195; *S. quadricauda* (Turp.) Breb. 22, 31, 160, 195
Selenastrum Chlorophyta, Chlorococcales: 12, 238; *S. capricornutum* Printz: 195
Sorastrum Chlorophyta, Chlorococcales 308
Sphaerocystis Chlorophyta, Chlorococcales: 12, 60, 96, 117, 180, 271, 273, 295, 296, 302, 306, 307, 313, 318, 319, 325; *S. schroeteri* Chodat. 23, 62, 97, 119, 204, 205, 206, 207, 208, 210, 211, 212, 215, 216, 266, 300
Spirulina Cyanobacteria, Nostocales 9, 309
Spondylosium Chlorophyta, Zygnematales 12
Staurastrum Chlorophyta, Zygnematales: 12, 21, 25, 31, 67, 74, 117, 119, 213, 267, 285, 306, 309, 313; *S. brevispinum* Breb. (now referred to *Staurodesmus brevispina*?) 62; *S. cingulum* (W. et G. S. West) G. M. Smith 306; *S. paradoxum* Meyen 21; *S. pingue* Teiling 13, 21, 31, 38, 119, 204, 205, 206, 207, 209, 210, 215, 300, 306
Staurodesmus Chlorophyta, Zygnematales 12, 117, 119, 206
Stephanodiscus Chrysophyta, Bacillariophyceae: 11, 25, 116, 171, 172, 220, 257, 283, 289, 310, 318, 325, 326; *S. astraea* (Ehrenb.) Grun. 13, 21, 30, 31, 33, 34, 35, 37, 38, 51, 55, 57, 69, 77, 119, 145, 154, 282, 285, 286, 287, 288, 305; *S. hantzschii* Grun. 21, 31, 32, 33, 34, 35, 38, 117, 119, 160, 194, 213, 287
Stenocalyx Chrysophyta, Chrysophyceae 10

Stichococcus Chlorophyta, Ulotrichales 12
Stichogloea Chrysophyta, Chrysophyceae 10
Stylochromonas Chrysophyta, Chrysophyceae 10
Surirella Chrysophyta, Bacillariophyceae 11
**Synchaeta* Rotatoria 6
Synechococcus Cyanobacteria, Chroococcales 9, 13, 19, 20, 117, 145, 194, 195, 198, 199, 267
Synedra Chrysophyta, Bacillariophyceae: 11, 65, 119, 286, 310; *S. acus* Kütz. 57, 69; *S. radians* Kütz. 247, 286; *S. ulna* (Nitzsch) Ehrenb. 21
Synura Chrysophyta, Chrysophyceae 10, 175, 220, 255, 285, 289

Tabellaria Chrysophyta, Bacillariophyceae: 11, 68, 117, 285, 313; *T. fenestrata* (Lyngb.) Kütz. 116, 239, 240, 241, 246, 247, 286; *T. flocculosa* (Roth) Kütz. var. *asterionelloides* (Roth) Knuds. 15, 22, 31, 33, 34, 57, 69, 72, 76, 119, 194, 204, 206, 305, 326
Tetraëdron Chlorophyta, Chlorococcales 12, 117
Tetrastrum Chlorophyta, Chlorococcales 12, 119, 289, 295
Thalassiosira Chrysophyta, Bacillariophyceae: †*T. weisflogii* (Grunow) Fryxell et Hasle Syn. †*T. fluviatilis* Hust. 57, 67
**Thiobacillus* Bacteria 6
**Thiopedia* Bacteria 6
**Tintinnidium* Ciliophora 6
Trachelomonas Euglenophyta: 11, 98; *T. hispida* (Perty) Stein emend. Defl. 14
Tribonema Chrysophyta, Xanthophyceae: 11, 262; *T. aequale* Pascher 194
**Trichocerca* Rotatoria 6, 255
Trichodesmium Cyanobacteria, Nostocales (*see also Oscillatoria*): 9; †*T. thiebautii* Gomont 62, 168

Uroglena Chrysophyta, Chrysophyceae: 10, 17, 26, 117, 119, 205, 206, 207, 208, 210, 211, 213, 214, 215, 216, 217, 220, 221, 295, 302, 313, 318; *U. americana* Calkins 159; *U. lindii* Bourrelly 23, 204

**Vampyrella* Rhizopoda 254
Volvox Chlorophyta, Volvocales: 12, 26, 27, 117, 119, 220, 295, 299, 300, 313, 318; *V. aureus* Ehrenb. 17, 31, 32, 38
**Vorticella* Ciliophora 6

**Wolffia* Phanerophyta, Agiospermae 4
Woloszynskia Pyrrhophyta, Dinophyceae 9

Xanthidium Chlorophyta, Zygnematales 12, 117

**Zygorhizidium* Fungi, Chytridales: 6; *Z. affluens* Canter 251, 253

General index

Abioseston (= tripton), 1
Acetylene reduction, 167
Accessory photosynthetic pigments, 29, 124, 145
Actinomycetes, 5, 6
Adaptations of phytoplankton: for suspension, 2, 27, 40–81; to pelagic existence, 8, 18, 27, 40, 278; to low average light intensities, 130, 284
Adenosine triphosphate *see* ATP
Adinophyceae 9; *see also* Dinoflagellates
Advection, 108
Advective patchiness, 108–12
Advective velocities, 108–9
Aestivation in *Oscillatoria*, 292
Akinetes (of Cyanobacteria), 9, 16, 218–19, 297, 325; germination of, 219
'Algae', reason for the use of inverted commas in this work, 8
Allelopathy, 248–9
Allochthonous material in lakes, 321
Allogenic factors, in periodicity, 223, 316–17
Aluminium, 162
Ammonium: in natural waters, 164, 165–6, 182; preference to nitrate of phytoplankton, 166
Amoeboids, 6, 253
Anthropogenic eutrophication, 185, 322
Aphotic zone, 147
Arrhenius temperature plots, 135, 197, 199
Ash content of 'algae', 30, 31; as a fraction of dry weight of, 32
Ash-free dry-weights of 'algae', 29–33, 38, 160
Assemblages of phytoplankton species, 117–20, 180, 288, 313, 318; periodic sequences of, 318–21
Assimilatory products of 'algae', 9–12, 29; *see also* Condensates
ATP (adenosine triphosphate), 124, 125
Auftrieb, 1
Autochthonous material in lakes, 184, 321
Autogenic factors in (successional) change, 315–17
Autotrophy, 2, 13, 18, 37, 40; environmental requirement for, 18, 40, 123, 157; characteristics of, 37; *see also* Photoautotrophy
Auxospores, 16

Bacillariophyceae, 10, 30–1, 38–9, 194; *see also* Diatoms
Bacteria, 5, 6, 238; parasitic on algae, 245, 250; as a source of food for zooplankton, 255, 263, 268, 270
Behaviour, adaptive, 29
Behaviour of particles in flow, 48–50
Bicarbonate; in natural waters, 174, 175; as a source of photosynthetic carbon, 135–6, 300
Bioassay techniques, 162, 176
Biogeochemical cycling, 5
Biomass of phytoplankton: chlorophyll content as a correlative of, 37, 193; increase in, 40, 193; upper sustainable limits of, 152–6
Blooms *see* Water-blooms
Blue-green Algae, 5; *see also* Cyanobacteria
Boron, 157, 175
Boundary layers, 43–4, 46, 74, 235
Brackish waters, 10, 175
Brunt-Väsälä frequency, 47
Buoyancy regulation: in Cyanobacteria, 27, 63–4, 100–5, 233, 292, 295, 301, 307; in marine phytoplankton, 58–9

β-carotene, 29, 124
Calanoids, 7, 255; feeding in, 255; filtration apparatus of, 256; filtration rates of, 257–9; raptorial behaviour in, 256; food dependence in, 271
Calcification, 10, 28, 174
Calcium, 157; in relation to the CO_2-system, 173–4
Calvin, cycle, 124, 125, 133
Carbohydrates, density of, 56, 79
Carbon: algal content of, 36, 157, 160; minimum requirements of, 36; in relation to cell volume, 37; as an index

Carbon (*cont.*)
of biomass, 113, 150; sources of, 36, 136; fixation of, 40, 123, 150–2; losses from cells, 227, 247–8; as a limiting factor in growth, 136; as a factor in phytoplankton periodicity, 299–300
Carbonates, 28
Carbon-dioxide, system in natural waters, 136, 173–4; as a nutrient, 36; photosynthetic fixation of, 123–6, 129; photosynthetic depletion of, 136; as a limiting factor, 133, 135–7; as a regulating factor in species composition, 136, 173–4, 300
Carboxylation, 124, 136
Carinthian lakes, 113, 116, 118
Carotenoids, 10, 29, 38, 58
Catchments; as a source of nutrients, 162, 164–6, 168, 185, 322; of silt, 184, 322; changing land use in, 185, 323
Cell-generation time: quantatitive definition of, 195; as a unit of time, 86, 183
Cell-division, 193; phasing and synchrony in, 201–2
Cell nutrient quota, 161, 180
Cell sap, 28, 59
Cell walls, 28
Centric diatoms, 11, 29, 285; 310; size and shape in, 21–2, 25; sinking behaviour in, 53–6; distribution of, in relation to silicon, 171; as food for protozoans, 254; *see also*, Diatoms
Chain-formation, 53, 68
Chelating agents, 176
Chloride, 175
Chlorine, 37, 157
Chlorococcales, 12, 96; seasonal abundance of, 117, 285, 289, 295, 298, 301; nutrient preferences of, 180; cell division in, 193; growth rates in, 195; heterotrophy in, 289; in ponds, 309
Chloromonads, 10
Chlorophyceae, 12, 248; *see also* Chlorophyta, Green Algae
Chlorophyll; distribution among algal groups, 9–12; 'algal' content of, 30–1, 37–9; variation in, 130, 284, 291; as an index of biomass, 37, 113, 127; as a fraction of dry weight, 37–8; relative to cell volume, 38–9; relative to carbon content, 150; absorption spectrum of, 124; relation with light penetration in lakes, 141–5; relationship with magnesium, 175; lake concentrations of, in relation to nitrogen and phosphorus concentrations, 186, 190–1
Chlorophyll *a*, 29, 37–9, 124, 150
Chlorophyll *b*, 124
Chlorophyta (Green algae), 11, 12, 97; chlorophyll content of, 38–9, 124, 145; dry-weights of, 31, 32; buoyancy in, 96, 97; colony formation, 67; photosynthesis in, 131; elemental composition of, 160; phosphorus requirement of, 36; seasonal abundance of, 117–20, 199, 289, 295; nutrient preferences of, 180; growth rates in, 195; sedimentation and decomposition in, 248; parasites of, 253
Choanoflagellates, 10
Chromatic adaptation, 291
Chromatophores 5, 28; see also Plastids
Chrysolaminarin, 10, 29
Chrysomonads, 10, 168, 255
Chrysophyceae, 10, 180, 309; growth rates in, 194; cyst production in, 220; sensitivity to pH, 300
Chrysophyta, 10, 28, 29, 96; seasonal abundance of, 117–20, 283, 285, 295, 298; silicon requirement of, 157, 168; nutrient limitation of, 158; elemental composition of, 160; growth rates in, 194; in ponds, 309
Chrysose, 10, 29
Chytrids, 6, 250; parasitic attacks on algae, 250; epidemics of, 251–2; effects of, on host populations, 251
Ciliates, planktonic, 6, 253, 254
Cladocerans, planktonic; feeding in, 255; filtration apparatus of, 256; filtration rates of, 257–9; food selection in, 256, 259–66; dependence upon food supply, 266–73
Classification of 'algae', 8–12
Climate, Light-, 139–56, 183; effect on phytoplankton growth, 202; effect on phytoplankton periodicity, 214–17
Climate change, effect on lakes of, 184
Cobalt, 157, 177
Coccolithophorids, 10
Coccoliths, 174
Coelenterates, planktonic, 6
Coenobial phytoplankton, 15, 19, 196
Coexistence of species 120–2, 180
Colonial phytoplankton, 17, 19, 26, 67–72, 196–7
Colony formation, significance of, 67–72
Community change, 310–21; rate of, 310–15; correlation with diversity, 314; direction of, 315–21
Community filtration-rate, 259, 261, 270
Community grazing-index, 259
Compensation depth, *see also* Euphotic depth 14, 152
Competition, interspecific, 120; for limiting resources, 178–80, 296
Competitive exclusion principle, 120

Condensates, of assimilation, 28, 29, 56, 160
Contractile vacuoles, 28
Convection currents, 41, 47, 282, 303
Copepoda, planktonic, 7, 255; *see also* Crustacea, Zooplankton
Copper, 157, 175, 177
Corioli's Force, 15, 41, 46, 282, 303
Craspedophyceae, 10
Crustacea, planktonic, 4, 7, 254, 255; *see also* Zooplankton
Cryptomonads *see* Cryptophyta
Cryptophyta: taxonomic position of, 9; dry weights of, 28; chlorophyll content of, 38–9; photosynthetic pigments of, 124; vertical distribution of, 96; growth rates of, 194, 283; perennation in, 220–1; as a food source for zooplankton, 255, 268, 270; seasonal abundance in, 289, 295, 313
Cultural eutrophication, 185, 322
Culture media, 36, 196
Cyanobacteria: phylogeny and nomenclature of, 5, 7; taxonomic position of, 9; cytological organization in, 5, 28–9, 124; morphology of, 20–4, 27–8; dry weights of, 30, 32; chlorophyll content of, 38, 39; pigmentation of, 29, 124, 125; gas vacuoles of, 28, 61–4; buoyancy regulation in, 63, 64, 96, 100–5, 233, 292, 295, 301, 307; vertical distribution of, 84, 93, 100–5; bloom formation in, 94–5, 307; seasonal abundance of, 117, 180, 199, 211–12, 285, 289, 290, 295, 300; sensitivity to hydraulic mixing in, 301; photosynthesis in, 132; photooxidative death in, 132, 139; elemental composition of, 160; nitrogen fixation in, 167–8; preference of, for calcareous waters, 174; iron chemistry in, 176–8; effect of nutrient resource ratios on, 180; abundance of, in eutrophic lakes, 186; growth rates of, 194, 197, 200–1; as food for zooplankton, 254; sedimentation and decomposition of, 248
Cyanophyta, 5, 9; *see also* Cyanobacteria
Cyclopoids, 7, 25; *see also* Zooplankton
Cysts, of *Ceratium*, 14; formation of, 217, 325, 307; excystment from, 218, 297
Cysts, of Chrysophyceae, 220
Cytological organization in phytoplankton, 5, 8, 28–9; differences between groups, 9
Cytoplasm, 28, 54

Daphniids, 262, 265; *see also* Cladocerans
Day length: effect on photosynthetic integrals, 140–1, 149–52; effect on growth rates, 200–1; effect on periodicity, 284, 294, 303
DCMU-inhibition of photosynthesis, 236, 246
Death of phytoplankton cells, 132, 139, 181, 202, 225; and decomposition, 227, 245, 248; causes of, 245–6; rates of loss due to, 246–53
Decomposition of phytoplankton, 96, 227, 248
Densities: of fresh waters, 41, 45, 59; of sea water, 58; of algal cells, 27, 48, 57–64, 77–9; of cell components, 56–9; of mucilage, 59–61; relevance of, to sinking rates, 52, 54, 56–64, 77–9, 87–105; regulation of cell densities, 54, 58–64, 77–9
Density gradients, in natural waters, 43, 44–6, 47, 89–92, 104, 105, 234, 283; *see also* Temperature gradients
Desmophyceae, 9
Desmids; taxonomic position of, 12; seasonal abundance in, 117, 158, 306, 319
Diatoms: taxonomic positions of, 10, 11; dry weights of, 30–1; morphology of, 28; chlorophyll content of, 38–9; densities of, 57; silicon contents of, 33, 169; silicon requirements of, 157, 158, 168; silica-limitation of, 168–73; sinking behaviour in, 53–82, 88–93, 294; spatial distributions of, 88–93; photosynthetic behaviour of, 130, 139, 201; growth rates of, 194, 197, 201, 283; resource-competition among, 178–9; seasonal distribution of, 116–20, 198, 211–14; 283, 285–8, 305, 309; elimination of, from plankton, 294, 296; resting stages in, 220; parasitism of, 250–3
Diffusion gradients, 157
Diffusivity *see* Eddy diffusivity
Dinoflagellates, 2, 9, 20, 84, 96; dry weights of, 30; chlorophyll in, 38–9; growth rates of, 194; seasonal distribution of, 116–20, 211–13, 298, 319; eurytrophy in, 217
Dinophyceae, 9; *see also* Dinoflagellates
Dispersal mechanisms of phytoplankton, 220
Diversity, 120–2, 314, 317; *see also* Species diversity indices
Downwellings, of advective currents, 109
Droop model, of nutrient-limited growth, 161
Dry-weights of phytoplankton, 29–33; in relation to cell volume, 32; in relation to chlorophyll content, 38; in relation to silicon content, 33; as an index of

Dry-weights of phytoplankton (*cont.*)
natural biomass, 113; *see also* Ash-free dry-weights

Eddy diffusivity, 43, 46; quantification of, 43; effect on phytoplankton distribution of, 88, 90–1, 234, 282
Eddy spectrum, 42
Eddy viscosity, 42
Elemental composition of 'algae', 36, 37, 157–60
Electrophoretic mobility of cells, 81
Entrainment, of phytoplankton in flow, 18, 50, 74–7; of hypolimnetic water, 46, 303
Environmental variability, 83–7, 183, 278; perception of, 86, 105, 112, 183; effect on species diversity, 120; effect on specific growth, 202, 212
Environmental selectivity of phytoplankton, 281–310, 314
Enzyme systems, 36, 125, 157
Epilimnia, 45; formation of, 91; light climates of, 154; chemical properties of, 181; seasonal changes in, 182; comparison between those of oligotrophic and eutrophic lakes, 184
Euglenoids, 11, 98, 195
Eukaryotes, 5, 8, 9–12, 28–9, 124, 193
Euphotic content, 146, 152
Euphotic depth, 146; relation with extinction coefficient, 146; relation with Secchi-disk extinction, 147; relative to mixed depth, 147–9, 151, 152–6, 183, 288, 295, 300, 306
Euphotic zone, 40, 147
Eutrophic lakes: characteristics of, 184; phytoplankton cycles in, 114–16; 285, 299, 319–20; species in, 117; biomass in, 323; nutrients in, 162, 166, 182–90; light in, 325; hypolimnetic anoxia in, 162, 175, 182, 325; zooplankton of, 271–3
Eutrophication, 136, 183–91, 321–8; responses of phytoplankton abundance to, 186–91; responses to phytoplankton species composition to, 323–7; recovery from, 327–8; role of nitrogen in, 186, 190–1; role of phosphorus in, 186–90
Excretion, in phytoplankton, 85, 125
Exponential rate coefficients: of growth, 193–5, 225, 227, 240–5, 285, 294, 296; of net increase, in natural waters, 193, 202–4, 223–4, 225, 227, 240–5, 296; of loss, 202, 225, 227; of washout, 228–33, 283; of sedimentation, 228, 234–5, 240–5, 294–5; of death and decomposition, 228, 240–1, 246–7, 251–3; of grazing loss, 228, 259–66; 297; relative magnitudes of, 274–6; seasonal variations in, 285, 287, 296

Extracellular production and release, 132–3, 177

Faecal pellets, 78, 162, 181
Farm ponds, phytoplankton of, 115, 309–10
Fats and oils, accumulation of, 58; *see also* Lipids
Filamentous 'algae', 9, 11, 12, 16, 19
Filter-feeding, in zooplankton, 255; anatomical adaptations for, 256; filtration rates of, 257–9; limiting concentrations for effective, 257; size-selection of, 259–66; as a factor in seasonal periodicity, 296
Filtration rates: of individuals, 257–9; of communities, 258–9; *see also* Filter-feeding
Flagella, 26, 27
Flagellates, 9–12, 14, 17; suspension of, 77–8; vertical migrations of, 97–100; limnological conditions favouring, 283
Floristic change, 277, 321–8
Flotation in phytoplankton, 1, 2, 18, 51, 61–4, 94–5, 238–9
Fluid dynamics *see* Water movements
Fluorescence, 125, 183
Form resistance: significance of, 18, 64–7, 72–7, 260; quantification of, 65–72
Fulvic acids, 176
Fungi, 6; parasitic on phytoplankton, 245–50; epidemics of, 251–2

Gastrotrichs, 7
Gas vacuoles, of cyanobacteria, 16, 28, 61–4, 254; structure of, 62; effect on cell-densities of, 62–4; giving buoyancy to protozoan consumers, 284
Gas vesicles, 62, 94, 96; critical collapse pressures of, 63
Gause's hypothesis, 120
Gelbstoff, 143, 176
Geochemical cycling, 5
Gliwicz, Z. M., work of, 266
Glycolate: production of, 125; extracellular release of, 133
Glycogen, 29
Grazing, of zooplankton on phytoplankton, 27, 181, 202, 225; effects on populations, 253–73, 288, 294; role of, in eutrophication, 324; resistance of algae to, 72–3, 261–5
Greatest axial linear dimensions (*GALD*) of phytoplankton, 20–3, 25–6; *see also* Size
Green algae *see* Chlorophyta
Growth in phytoplankton, 40, 85; requirements for, 181; characteristics of,

Growth in phytoplankton (*cont.*) 192–5; quantification of, 193–5; distinction from 'increase' of, 202–3; relation to photosynthetic productivity, 196–7
Growth rates of phytoplankton: in light-saturated cultures, 193–9; nutrient-limitation of, 159–161, 170; light-limitation of, 199–202; temperature limitation of, 197–202; in natural lakes, 170, 178, 183, 226, 278; measurements of 226; in relation to photosynthetic productivity, 227; as related to size, 196–7; as related to SA/V, 197–8; effect of photoperiod on, 200–2; responses of, to environmental variability, 203, 205–9, 223, 301–3; in relation to survival strategies, 221–3
Growth strategies, 221–4, 278–9

Haptonema, 10
Haptophyceae, 10, 74
Heleoplankton, 3
Hensen, V., work of, 1–2
Heterocysts, 16; structure and function of, 167
'Heterogenous diversity', definition of, 86
Heterotrophy, 157; in Chlorococcales, 289
Hexose production, 125
Holomixis, 303
'Homogeneous diversity', definition of, 86
Horizontal distribution of phytoplankton, 84, 105–12
Humic acids, 176
Hydraulic retention time, 185, 189
Hydrogen, in phytoplankton cells, 157
Hypolimnia, 46; formation and isolation of, 91; inhabitants of, 95; nutrients in, 165, 182; chemical properties of, 181–3; seasonal anoxia in, 162, 175, 182, 184, 186, 325; comparison between those of eutrophic and oligotrophic lakes, 184

Ice cover, 281–3
Increase in phytoplankton: as distinct from growth, 193; in natural lakes, *see also* Growth, Population dynamics
'Inocula' of species, 192, 217, 231, 285; strategies for maintenance of, 220–1; importance of size of, 297
Insects, planktonic, 4, 7
Ionic density regulation, in phytoplankton, 58–9; *see also* Densities
Ion pumps, 157
Iron, 37, 157, 167, 175; as limiting nutrient, 176–7; and siderochromes, 177; precipitates of, 304
Irradiance *see* Light

Kelvin-Helmholtz Instability, 46, 303
K-selection *see* *r*- and *K*-selection
'KISS-model', of critical patch size, 108–9; *see also* Patchiness

Laminar flow, 42, 51, 89
Langmuir rotations, 41–2, 91, 106, 109
Lentic habitats, 3
Leucosin, 10
Liebig's Law of the Minimum, 159, 169
Light: surface incidence of, 130, 139, 140, 153, 181; duration of, 140, 149–56; 284; 293, 294; attenuation (extinction) of, underwater, 83, 126–8, 139, 141–7, 149–52, 182; diel variations in 140; seasonal variations in, 203; spectral composition of, underwater, 127, 129; absorbance of, by water, 143, 144; absorbance of, by algae, 124–5; 143–5; underwater fields of, 85, 123, 133, 141, 149; penetration, in relation to mixed depth, 139, 147–9, 152–6, 211; and silicon-limited growth of diatoms, 172; availability of in mixed columns, 183, 186; -deprivation as a cause of death, 245
Light climate, underwater, 139–56, 183; effect on growth, 202; as a regulator of phytoplankton periodicity, 214–17
'Light divisions', 141, 153–6, 294
Light inhibition, of photosynthesis, 128, 132–4; *see also* Photoinhibition
Light-limitation: of photosynthesis, 127, 129–30; of biomass, 152–6, 191; of growth, 199–202, 282–4, 307
Light-saturation: of photosynthesis, 127, 129, 130–2, 139; onset of, (I_k), 129, 201; of growth rates, at different temperatures, 201
Limnoplankton, 3
Lipid, 9, 10, 29, 217
Lipid accumulation, 217; effect of, on cell density, 58
Loss processes, 225–76; definition of, 225–8; effect of, on photosynthetic models, 154; effect of, on growth dynamics, 228–30, 234–5, 240–44, 259–76; seasonal variations in, 203, 274–6, 314
Lotic habitats, 3
Lund, J. W. G., work of, 170–2, 174
'Lund Tubes', experimentation in, 88, 89–90, 97, 115–17, 121–2, 171, 187–8, 202–12, 214, 218, 226, 227, 240–5, 248, 259–61, 264–6, 270–1, 274–6, 287–8, 312–14, 324–5, 327
'Luxury' uptake of nutrients, 36, 160, 166
Lycopodium spores, as a tracer, 89–90, 238

Macronutrients of phytoplankton, list of, 157
Macroplankton, 4
Magnesium, 157, 175
Manganese, 37, 157, 175, 177
Mastigophora, planktonic, 6
Mean-crowding, Index of, 95, 109, 111
Megaloplankton, 4
Megaplankton, 4
Meroplanktonic organisms, 2
Mesoplankton, 4
Mesotrophic lakes: phytoplankton cycles in, 114–15; dominant species in, 117; nutrients in, 190
Metabolic rate, in 'algae', 36, 196
Metabolism, in 'algae', 29, 36, 85, 125
Metalimnia, 46, 91; formation of, 91; current velocities in, 91; density gradients in, 100; (for significance in phytoplankton selection, *see also* Mixed depth')
Michaelis-Menten kinetics: applied to growth rates, 121, 259, 161, 164, 178–80; applied to nutrient-uptake rates, 163, 164, 169
Microbial activity, 5, 181, 238
Microhabitats, 181
Micronutrients of phytoplankton: list of, 157; importance of, 158, 175
Microplankton, 4
Microstratification, 92; *see also* Stratification
Mitochondria, 28
Mitosis, 169, 193
Mixed depth: in relation to sinking rates, 90, 234–5, 237, 244, 245; in relation to sinking losses, 234–45, 276, 306; and control of vertical station, 99–105; and regulation of the underwater light climate, 139, 147–9, 245; in proportion to euphotic depth, 147–9, 151–6, 183, 288, 295, 300, 306; optical depth of, 155; as a factor in periodicity, 294, 295, 304; as a correlative of phytoplankton dynamics, 205–10, 276, 306; in proportion to Secchi-disk depth, 205–10, 211, 216, 245, 307
Mixing processes, 41–7, 303; differences between those of lakes and open oceans, 107–9; effects on suspension and settling rates, 48–50, 74–7, 90–115, 234–45; effects on photosynthetic behaviour, 131, 133–4, 147; effects on underwater light climates, 147–9, 284, 293; effects on phytoplankton growth, 178, 181–3, 211–13, 284, 293
Molecular diffusion, 18
Molluscs, planktonic, 7
Molybdenum, 37, 157, 175, 177
Monovalent: divalent cation ratio, 175
Morphology of phytoplankton, 2, 12, 18–28
Mucilage, 19, 27, 56; effect on algal density of, 59
Müller, J., work of, 1
Myxophyta, 5; *see also* Cyanobacteria
μ-algae, 3; *see also* Nanoplankton

NADP (Nicotinamide adenine dinucleotide phosphate), 124, 125
Nannoplankton, 3; *see also* Nanoplankton
Nanoplankton, 3; size categories, 3; as a food source for zooplankton, 256–7; growth of, in the presence of zooplankton, 265, 271; growth of, and other limiting factors, 302, 306; seasonal abundance in, 271, 302, 304, 313
Natural eutrophication, 185–322
Nekton, 1, 4
Net plankton, 3
Niches: concept of, 120; numbers of, existing simultaneously, 120, 278; diversification of, 121, 181–2, 278
Nitrate: in freshwater, 164–5, 182; in seawater, 165
Nitrite, 164
Nitrogen: algal content of, 36, 157, 160; relative to cell volume, 37; minimum requirements of, 36; sources of, 36, 164–6; uptake of, by algae, 166; fixation of, in Cyanobacteria, 167–8; selective interactions of, with phosphorus, 180, 299; as a factor in eutrophication, 186, 190–1; relationship to chlorophyll in lakes, 186, 190–1
Nomenclature of 'algae' and bacteria, 5
Norfolk Broads, 118, 148
Nostocales, 9; nitrogen fixation in, 167–8; seasonal abundance of, 298, 319
Nucleic acids, 56
Nucleus, 28
Nutrient availability to phytoplankton, 37, 183, 185, 205–7; effect on rates of increase of, 208–11; seasonal variations in, 181–2; as a factor regulating phytoplankton periodicity, 214–17, 271, 283, 320
Nutrient deficiencies, 113; as a factor in community composition, 180, 322; as a cause of death, 245
Nutrient demand, relative to supply, 211, 222–3
Nutrient depletion of water, 40, 82, 181–2, 286
Nutrient gradients, 178–80
Nutrient limitation, 36, 99, 158, 283, 298, 299; explanation of, 159–61; of growth, 178; as a factor in niche diversification,

Nutrient limitation (*cont.*)
121; interactions with water movement, 181–3; interactions among limiting nutrients, 158, 178–80; effects on phytoplankton growth, 205, 207
Nutrient loadings on lakes, 185–91, 322; effect of high values on phytoplankton, 206
Nutrient re-cycling, 37, 185–6, 211, 273
Nutrient requirements of phytoplankton, 157–62
Nutrient resource ratios, 158, 178; effect of, on interspecific competition, 178–80
Nutrient sources, 36, 157–8, 185
Nutrient uptake, 18, 36, 40, 82, 157, 181; endothermy of, 158; importance of gradients to, 158; kinetics of, 159, 161–4

Oligotrophic lakes: characteristics of, 184, 190; phytoplankton cycles in, 113–14, 285, 299, 319–20; biomass in, 323; dominant phytoplankton species in, 116, 289, 299; zooplankton in, 271–3
Optical depth: quantification of, 149; effect on production in lakes of, 151–2; in photosynthetic regulation, 154–5; of mixed layers, 155; as a correlative of phytoplankton increase, 205–11, 284
Orientation of phytoplankton, in flow, 74
Orthophosphate, 162; *see also* Phosphorus
'*Oscillatoria* lakes', 289–93
Osmoregulation, 28
Osmotic pressure, 63
Overwintering of *Microcystis*, 220, 297
Oxygen: algal content of, 157; photosynthetic evolution of, 123, 125, 126–30, 150–2; respiratory consumption of, 130, 134–5; lack of, in eutrophic hypolimnia, 165, 182, 186, 325

'Paradox of the Plankton', 120
Parasites of phytoplankton, 6, 250–3; effects of, 225, 251
Patchiness, 84, 105–12; horizontal extent and longevity of, 105, 107–12; critical patch size, 107–8; in the vertical, 95; as a consequence of advection, 108–10
Pathogens of phytoplankton, 225, 227, 249–53
Pearsall, W. H., work of, 158
Pelagic existence, adaptations for, 8
Pelagic zone, 1
Pennate diatoms, 11, 29, 282, 285, 310; (*see also* Diatoms)
Perennation of phytoplankton, 217–21
pH: in relation to carbon dioxide system, 136, 173–4; in regulating photosynthesis, 136; in regulating phytoplankton composition, 214, 300

Phaeophyta, 9
Phagotrophy, 2, 223
Phosphate, 162, 182, 185; (*see also* Phosphorus)
Phosphoglyceraldehyde, 125
Phosphoglyceric acid, 125
Phosphorus: algal content of, 36, 157, 160; relative to cell volume, 37; minimum requirements of, 36; as a factor limiting algal growth, 158, 178; sources of, 36, 162; determination of, in natural waters, 162–3; uptake of, by algae, 163; selective interactions, with silicon, 178–9; selective interactions, with nitrogen, 180; influence of deficiencies of on phytoplankton composition, 180, 211, 198–299; as a factor in eutrophication, 186–90; relationship to chlorophyll in lakes, 186–90; loadings of, on lakes, 187–8
Phosphorylation, 124
Photoautotrophy, 13, 37, 123–6
Photobacteria, 5, 9
Photochemical systems (PSI and II), 124
Photoinhibition, 128, 129, 132–4; rapidity of, 133; occurrences of, 139
Photooxidative death, 132, 139
Photoperiod: effect of, on phytoplankton growth, 200, 203, 211
Photorespiration, 125, 133
Photosynthesis, 'algal', 13, 85, 123–85; biochemistry of, 123–6; electron transport in, 124; determination of, 125–6, 133, 137–8; integration of measurements, 139–52; rates of, in closed bottles, 126–37; 139; in relation to depth, 127; in relation to light intensity, 127; in relation to light intensity, 127–34; light-limitation of, 128, 129–30; in relation to temperature, 130, 134–5; temperature limitation of, 129, 134–5; in relation to carbon dioxide supply, 129, 135–7; onset of light saturation of, 130–2; per unit of chlorophyll, 129; per unit of biomass, 130; rates of, in natural light fields, 139–52; in relation to buoyancy regulation, 63–4
Photosynthetic apparatus, 123–6, 130; damage to, 139
Photosynthetic capacities, 284; in relation to temperature, 130
Photosynthetic efficiency, 129, 131, 147, 284; enhancement of, 130, 285
Photosynthetic integrals, 139, 149–52
Photosynthetic intermediates, 124–5, 248; extracellular release of, 132–3, 137
Photosynthetic models for lakes: development of, 139–49; application of, 149–56

Photosynthetic production, in lakes, 149–50
Photosynthetic productivity, 150–2
Photosynthetic regulation, of natural phytoplankton, 153–6, 248
Photosynthetically-active radiation (*PhAR*), 13, 29, 129, 142, 143, 144, 146, 172; (*see also* Light)
Phycobilins, 27, 124, 145
Phycocyanin, 29, 124, 291
Phycoerythrin, 124, 291
Phycomycetes, 230
Phycoviruses, 250; *see also* Viruses
Physiological responses of phytoplankton, to small-scale environmental variability, 85, 202
Phytoplankton: definition of, 2; general features of, 8; representation of 'algal' groups in, 5, 9–12; classification of 'algal' groups in, 9–12; shape of, 18, 26-8, 67; size of, 18–26, 67, 69; weights of, 29–32; suspension of, 40, 48–82; resistance to grazing in, 72–3; 257; 262–6; spatial distributions of, 83–112; cycles of abundance in, 112–16; cycles of species composition in, 116–20; 279–321; diversity of, 121–2, 314; photosynthesis in 123, 156; photosynthetic pigments of, 29, 124; excretion in, 125; self-shading in, 143–7, 191; elemental composition of, 36–7, 157, 160; nutrient requirements of, 157–77; links with water chemistry, 158, 173, 175; responses of, to nutrient loading, 185; growth in, 192–202; growth rates of, 194–5; perennation in, 214–21; life-history strategies of, 221–4; in winter and spring, 281–93; under ice, 281–3; in summer, 293–302; in autumn, 302–7
Pigments in phytoplankton, 9–12, 29, 37–9, 124; adsorption spectra of, 124; variations in 130, 291
Plankton, definitions of, 1–4; taxonomic representation in, 4–6, 7, 9–12
Plasmalemma, 28, 157
Plastids, 5, 28, 124; numbers of, in 'algal' cells, 9; variation in numbers of plastids, 130; proportion of diatom cell occupied by, 38; shrinkage of, 131, 139
Polar lakes, 150, 213, 283
Polder lakes, 213, 290
Polyphosphates: in 'algae', 36, 56, 79; from detergents, 185
Polysaccharides, 29; formation of, 125
Population dynamics, 29, 40, 193, 202, 227–8; effects on washout on, 228, 230–44; effects of sinking on, 234–5; effects of parasitism on, 251; effects of grazing on, 253, 256, 259–73; net changes in, 274–5
Potamoplankton, 3
Potassium, 37, 157, 175
Prairie lakes, photoplankton of, 118–19
Productive lakes *see* Eutrophic lakes
Productivity, of phytoplankton, 29, 149–52, 284, 294
Prokaryotes, 5, 9, 28–9, 124; cell division in, 193
Proteins, 29–56
Protozoan parasites of 'algae', 253
Protozoan grazers of 'algae', 254, 255
Pycnocline, 46; *see also* Stratification
Pyrenoids, 11
Pyrrhophyta, 9; *see also* Dinoflagellates

Q_{10} values: for photosynthesis, 135; for respiration, 137; for assimilation and growth 201–2, 282

r- and *K*-selection, 221–4, 298, 300, 307, 315–16
Radiation income, to lakes, 44, 181; *see also* Light
Raphidophyta, 10
Redfield ratio, 36, 180
Redox reaction: in phytoplankton, 123; in lakes, 181
Residence time, 185, 189
Resource-based competition theory, 178–80, 283, 286
Resource 'spectra', 183
Respiration, 13, 85; rates of, 137; temperature adaptation of, 137; effect of size and shape on, 201
Respirational loss, 40, 126; as a fraction of photosynthetic rate, 137, 154–5; in natural communities, 149, 152–6; in fluctuating light regimes, 201, 284
Resting stages, 40, 199, 217–24, 325
Retention time, 185, 189
Resuspension: of pre-sedimented materials, 44, 235, 304; of 'algal' propagules, 217, 219
Reversion, in Periodicity: definition of, 317–18; instances of, 212, 313, 318–21
Reynolds numbers: of flow, 42, 90; of particles, 51, 64, 74
Rhizopoda, planktonic, 6, 253, 254
Rhodophyta, 9, 29, 124
Richardson's number, 45, 47, 90, 91, 94, 234
Rotatoria, planktonic, *see also* Rotifers
Rotifers, planktonic, 6, 254; feeding apparatus in, 254; food preferences of, 254–5, 262, 263; filtering rates of, 257–9; dependence of upon food supply, 271

Salinity, of freshwater, 175, 214
Samplers, 87–8

Sampling, design of programmes for, 107, 202
Schizophyta, 5; *see also* Cyanobacteria
Seasonal periodicity, 112–20, 279–321; general considerations of, 85, 112–13; biomass fluctuations, 113–16, 192, 203; species-compositional fluctuations, 116–17, 280–1; classification of species sequences, 117–20; classification, related to lake trophy, 298–9; patterns of, in lakes, 279–308; driving mechanisms for, 281; directionality of, 315–21; in relation to size and shape of 'algae', 289, 293
'Seasonal succession', 203, 315
Seaweeds, 8
Secchi disk: description of, 146–7; interpretation of measurements with, 147; correlation of measurements with phytoplankton distribution, 101–2; measurements relative to mixed depth, 205–10, 211, 216, 245, 307
Sediment accumulation in lakes, 184, 277
Sedimentation: of algae, 202, 225, 227, 238–45, 266; of diatoms, 88–93; removal of nutrients by, 162, 165, 171, 181–3; of propagules, 217–21, 324–5
Sediment traps, 217, 238–9, 242, 247, 248
Seiches, 41, 46, 84, 303
Selective ion-accumulation in cells, 54, 58–9
Self-shading, by phytoplankton, 143–7, 191
Seston, 1
Settling, of phytoplankton: in static columns, 48, 51–72; through mixed columns, 48–50, 74–7, 92; in whole lakes, 88–91, 233–45, 282
Sewage, 185
Sewage ponds, phytoplankton in, 115
Shade 'algae', 153, 291
Shape, of phytoplankton cells, 18, 20, 26–8, 67; effect on form resistance of, 52–9, 64–77; effect on sinking rate of, 51–2, 64–81; effect on growth rates of, 196–8, 201
Shear stress, 46
Shift, in Periodicity: definition of, 317–18; instances of, 212, 313, 318–21
Shock periods, 317
Siderochromes, 177
'Sieve' effect, 141
Silica, 28; content of, in planktonic diatoms, 33–5; effect of, on diatom densities, 56, 78–9; as a factor in the ecology of diatoms, 158, 159; equivalence to silicon of, 168; *see also* Silicon
Silicic acid, 168, 170, 178; equivalence to silicon of, 168
Silicification, 10, 11, 28, 32, 33–5, 168–73
Silicon: sources of, 157, 160; contents of, in diatoms, 160; determination of, 168; assimilation of, 169; uptake and depletion of, 171–3; -limitation of diatom growth, 171–3, 287, 294; in relation to light availability, 172; -uptake, as an index of diatom growth, 239, 242–4; selective interactions of, with phosphorus, 178–9, 288
Sinking behaviour, of phytoplankton, 48–82, 233–45; in static columns, 48, 50–72; in mixed columns, 45–50, 74–7, 233–45
Sinking loss: theoretical determination of, 234–5; field determination of, 237–45
Sinking rates, of phytoplankton, 40, 48–51, 66–72; adaptive mechanisms for reducing, 52–77, 139; effect of shape and size on, 52–77; effect of physiological condition on, 52, 54, 55, 74–81, 237–244; changes in, during population growth, 53, 237, 244; effect of cell density on, 54, 56–69
Sizes of phytoplankton, 18–26, 67, 69; effect on sinking rate of, 52–6, 64–72; effect on ingestibility to zooplankton of 72–4; effect on growth rates of, 196, 197, 201–2
Size-efficiency hypothesis, 271
Sodium, 37, 157, 175
Solar radiation, 140; *see also* Light
Spatial distribution of phytoplankton, 83–112, 192
Species diversity indices, 121–2, 314
Spectral analysis techniques, 84
SRP (soluble reactive phosphorus), 163, 205–10, 211, 304; *see also* Phosphorus
Starch, 9, 11, 29, 217
Stenothermy, 199
Stokes' Equation, 51, 52, 53, empirical tests of, 51, 74, 234
Stoneworts, 8
Storage products of 'algae', 9–12, 56, 58
Stratification: of phytoplankton, 96, 98–105, 290–1; of water masses, 41, 44–6, 83, 89, 91; stability and longevity of, 91, 115, 181; autumnal breakdown of, 303–4; perturbation of, 212, 223, 315–20; distribution of nutrients in relation to, 165, 166, 181–3; sinking responses of phytoplankton to, 244–5; effect on photosynthesis of, 132, 139, 147, 154–6; effect on growth of, 181–3, 205–11; as a factor in phytoplankton periodicity, 214–17, 223, 294, 317–20
Structural viscosity, 80–1
Succession, 315; characteristics of, 317–18; instances of, 318–21
Succession rate: determination of, 310–13

Sulphide, 175, 182
Sulphur, 37, 157, as an electron donor, 175
Surface areas, of phytoplankton, 19–28, 197
Surface area/Volume ratio, 19–28; conservatism of, 25, 197; effects of mucilage on, 26–8; role in assimilation and growth of, 196–7, 201, 203; role in respiration of, 201; role in interspecific growth differences of, 211; in relation to periodicity, 289, 293, 304
Surface electrical charges, 80–1
Surface incident radiation, 139, 140; *see also* Light
Surface-scum formation, 61, 64; *see also* Water Blooms
Survival of adverse conditions, 217–24
Survival strategies, 221–4, 278–9
Suspension of planktonic organisms, 2, 40–82, 88–112
Synchronization of cell division, 202
Swimming of flagellates, 77, 97, 233, 301

Taxonomy of phytoplankton 'algae', 8–12
Temperature, effects of: on density of water, 45; on phytoplankton periodicity, 122, 284, 295; on photosynthesis, 129, 134–5; on respiration, 135; on growth rate, 178, 197–202, 205–10; on light-saturated growth, 201; on net increase in lakes, 205–11
Temperature, seasonal variation in, 203; effects of, 284
Temperature gradients, 84–92, 98–100, 102, 104, 105, 181–2; under ice, 281
Temperature limitation: of photosynthesis, 129, 134–5; of phytoplankton growth, 183, 197–202, 282, 304
Temperature regulation: of phytoplankton periodicity, 214–17
Temporal distribution of phytoplankton, 85, 112–20, 192, 277–328; in response to water movements, 88–95, 99, 104; in response to seasonal environmental change, 100–2; 112–20, 277, 279–321; in response to long-term changes, 277–321; in relation to nutrient supply, 158, 180
Thermal bars, 283
Thermal stratification *see* Stratification
Thermocline, 46; *see also* Metalimnia, Stratification
Thylakoids, 124
Tilman, D., work of, 178–80
Toxicity: of metals to 'algae', 177; of substances produced by 'algae', 245, 248–9
'Trace elements', 37, 158, 175–7
Track autoradiography, 246
Transmissometry, 87, 104
Transport, of particles, 43
Trichomes, of Cyanobacteria, 100, 102, 167
Tripton, 2
Tropical lakes: phytoplankton in, 116, 213, 308–9; nitrogen deficiences in, 165
Turbidity, 104, 128; contribution of 'algae' to, 142–6; autumnal increase in, 304
Turbidometry, 87
Turbulence, 15, 41–2, 45, 47, 48–50, 181, 284; effects of, on phytoplankton settling, 48–50, 74–7, 90–3, 234; effects of, on photosynthesis, 138; *see also* Mixing processes

Ultraplankton, 3
Ultraviolet light, 132
Unicellular algae, 13, 14, 49
Unproductive lakes *see* Oligotrophic lakes
Upwelling of currents, 84, 89, 109, 171

Vanadium, 157, 175
Vacuoles, intracellular, 28, 38, 54; *see also* Gas vacuoles
Veliger larvae, 7
Vertical distributions of phytoplankton: in relation to density gradients, 87–105; heavier than water, 88–93; lighter than water, 93–5; isopycnic with water, 96–105; graphical representation of, 87
Vertical extinction coefficients of light penetration, 139, 141–7, 284; minimum, 142; attributable to 'algal' presence, 144–6; in relation to expressions of euphotic zone, 147, 149
Vertical migrations of plankton, 96–100, 103–5
Viruses, in plankton, 6, 7, 245, 246, 250
Viscosity of water, 51, 52, 53
Vitamin B_{12}, 177
Volumes of phytoplankton, 20–1, 22–8
Volvocales, 12, 27, 98; seasonal abundance of, 117, 295, 208, 300, 319; cell division in, 193; growth rates in, 195

Washout, 202, 225; determination of, 228; dynamic effects of, 229–33, 284, 288; and *Oscillatoria* spp., 283
Water-blooms, 61, 64, 94–5; photosynthetic behaviour in, 128–9, 139; collapse of, 165, 246; as a trigger to sporulation, 219; increase in autumnal incidence of, 307
Water movements, 41–7, 83; effects on horizontal distribution of phytoplankton, 106–12; effects on vertical distribution of, 87–105; effects on photosynthesis on, 137–9; effects on nutrient availability, 181–3

Wave action, 41, 44, 235
Wind: action of, 84, 91, 105, 181, 223; role of, in seasonal destratification, 303; drift by, 93; -driven diffusivities, 108; -stress, 41, 46, 85
Wind speed, 41–2, 91, 95; fluctuations in, 107; and patchiness, 95, 107–11
Windrows, 42, 106
Winkler titrations, 126

Xanthophyceae, 11, 194
Xanthophylls, 29, 38, 124

Zeta potential, 80–1
Zinc, 157, 175, 177
Zooplankton: definition of, 2; representation in, 6, 7; grazing by, 27, 181, 202, 205, 225, 228, 246, 253–73; food preferences among, 254–7, 263; filter-feeding by, 255–74; population dynamics of, 266–70; adequacy of food supply to, 268–71; predation by fish on, 271–2; interaction with phytoplankton, 266–73
Zoospores, fungal, 250–2